ACCLAIM FOR JOHN NOBLE WILFORD'S

T H E

Mapmakers

"Wilford's tour d'horizon of the history of maps is itself a map of human genius." —*Minneapolis Star Tribune*

"A classic history of cartography. . . . An excellent book." —*The Boston Globe*

"If you've ever lain in the middle of the living room floor, mesmerized over a map, dreaming of the places, wondering about the names and the people who lived there . . . then you know the call of this book. . . . It's a dreamer's dream of a book, bound to bring hours of pleasure." —*The Phoenix* (Minneapolis)

"*The Mapmakers* is that rare variety of science writing that manages to respect technical matters by explaining them in evocative language." —*St. Petersburg Times*

"Absorbing reading, even for the scientifically and mathematically challenged." —*Fort Worth Star-Telegram*

"Wilford manages to tell the entire history of mapmaking in a very readable way, full of stories of the adventures and misadventures that got maps made. . . . His simple, relaxed style helps us enjoy the stories of how they came about." —*Post & Courier* (Charleston, SC)

"A winning chronicle of mapmakers over time and space." —*Scientific American*

"Wilford leaps between the fantastical and the highly detailed with ease. . . . A highly readable, definitive take." —*The Hartford Courant*

JOHN NOBLE WILFORD

THE
Mapmakers

John Noble Wilford is a science correspondent for *The New York Times*. He has won two Pulitzer Prizes—one in 1984 for his reporting on space and science, and the other in 1987 as a member of the *Times* team reporting on the aftermath of the *Challenger* accident. He was the McGraw Lecturer at Princeton University in 1985, and Professor of Science Journalism at the University of Tennessee in 1989–90. In 1998, he was elected to the American Academy of Arts and Sciences. Wilford has worked for *The Wall Street Journal, Time,* and, since 1965, the *Times.* He is the author of numerous books.

We Reach the Moon

Scientists at Work (editor)

Spaceliner (with William Stockton)

The Riddle of the Dinosaur

The New York Times Guide to the Return of Halley's Comet
(with Richard Flaste, Holcomb Noble, and Walter Sullivan)

Mars Beckons

The Mysterious History of Columbus

THE *Mapmakers*

THE Mapmakers

REVISED EDITION

by

John Noble Wilford

VINTAGE BOOKS
A DIVISION OF RANDOM HOUSE, INC.
NEW YORK

SECOND VINTAGE BOOKS EDITION, DECEMBER 2001

Copyright © 1981, 2000 by John Noble Wilford

All rights reserved under International and Pan-American Copyright Conventions. Published in the United States by Vintage Books, a division of Random House, Inc., New York, and simultaneously in Canada by Random House of Canada Limited, Toronto. Originally published in hardcover in the United States by Alfred A. Knopf, a division of Random House, Inc., New York, in 1981 and revised in 2000.

Vintage books and colophon are registered trademarks of Random House, Inc.

The Library of Congress has cataloged the Knopf edition as follows:
Wilford, John Noble.
The mapmakers / by John Noble Wilford.—Rev. ed.
p. cm
Includes bibliographical references and index.
ISBN 0-375-40929-7 (alk. paper).
1. Cartography—History.   I. Title.
GA105.3.W49 2000
912'.09—dc21
99-049957

Vintage ISBN: 0-375-70850-2

Book design by Ralph L. Fowler

www.vintagebooks.com

Printed in the United States of America
10 9 8 7 6 5 4 3 2 1

for Nancy

Note on Units of Measure

With a few exceptions dictated by the context, this book uses the metric system for units of length and mass. The unit of length is the meter, which is about 39 inches, or slightly more than one yard. Some other lengths: the centimeter (one-hundredth of a meter, or 0.39 inch) and the kilometer (1,000 meters, or 0.62 mile). The unit of mass is one gram. There are about 28 grams in an ounce, and a kilogram, 1,000 grams, is about 2⅕ pounds.

Contents

Who has not spread out a map on the table and felt its promise of places to go and things to see and do? Ah, so that's Zanzibar, a real place, as real as Dar es Salaam across this stretch of blue or Timbuktu up there in the emptiness of the Sahara. Let's see, we could cross at Dunkerque, here (so this is where it happened, the glorious retreat), and be in London, there, first thing the next morning. Now, here's the plan: we sail from Philipsburg, here, out across the Anegada Passage and put in at Road Town, over here on Tortola. Look at these names, will you—Oodnadatta, Ilbunga, Rumbalara, Bundooma, Rodinga, Alice Springs—you can almost see the lonely cattle stations, the dingoes and kangaroos, the dusty ringers drinking beer at some forlorn pub known as the Southern Cross. And look at this speck of land, Bouvet Island, several thousand kilometers from the tip of South Africa, where no one lives and few people have ever set foot; yet here it is on the map, inviting dreams, speculations, perhaps exploration.

Joseph Conrad understood this feeling. In *Heart of Darkness*, Conrad has Marlow saying: "Now when I was a little chap I had a passion for maps. I would look for hours at South America, or Africa, or Australia, and lose myself in all the glories of exploration. At that time there were many blank spaces on the earth, and when I saw one that looked particularly inviting on a map (but they all look that) I would put my finger on it and say, 'When I grow up I will go there.'"

Who does not have etched in the mind images of countries and of the world based on maps? Until recent times, indeed, the world was more familiar to us as a map than in reality. As he approached the end of his flight in Earth orbit in 1962, John H. Glenn remarked: "I can see the whole state of Florida just laid out like on a map." A number of astronauts, and then all of us who saw the photography from space, marveled at how much the Florida peninsula, the meandering Mississippi, the islands of Britain, the boot of Italy, or any of the geographical shapes resembled the maps everyone had grown up with. We had taken it for granted that maps were faithful reflections of reality; yet we were somehow amazed when reality turned out to be true to the maps.

This reaction is an unspoken tribute to the mapmakers, past and present. Speak of any beckoning new land, and there have always been people setting forth, the Lewises and Clarks and the Frémonts, to map and incorporate the new world into the mind of the old. Speak of the remaining unknowns, and there are the ships setting out to take a seismic or sonic measure of the ocean floor and thereby map it, or the spaceships embarking to map the Moon, Venus, Mars, and the satellites of Jupiter. Speak of anything spatial, and there was, is, or will be a mapmaker seeking to make it more understandable through a mosaic of points, symbols, lines, shadings, and coloring—that is, through a map.

But who are these people who make the maps that touch the little chap in us all? How did the art and science of mapmaking evolve? Who were the important pioneers in developing the map as one of the most useful forms of human communication? These are the questions addressed in this book. For this is the story of how Earth and then Earth's nearest neighbors came to be mapped.

It is a story with a multitude of diverse characters: scholars and scientists, soldiers and sea captains, explorers and adventurers, monks and clockmakers, schoolteachers and spies, aviators and technicians, rich men and poor, those who stand out in history and those who are generally forgotten. They were chosen for inclusion in this book because of achievements that promoted (or in a few cases, hindered) the development of certain important types of mapping or that illustrated the expanding reach and growing precision of cartography.

It is a very human story of heroics and everyday routine, of personal and national rivalries, of influential mistakes and brilliant insights, of technological innovation and a passion to explore and understand Earth and the universe.

Parts of the story have, of course, been told before, and without such earlier scholarship this book would not have been possible. But the story of the mapmakers in its full historical sweep seems to have been neglected. Antiquarians publish handsome portfolios of old maps, but treat them more as objects of admiration and preservation than as products of a fascinating history. The many atlases provide ample evidence of cartographic achievement, but little of how it was done. Historians write volumes about the explorations of new worlds, but usually regard the cartographic aspects as incidental to the political, economic, or adventure themes. The few historians of geography and cartography seem to be more interested in the maps themselves than in the people who went into the field, often at great risk, to get the information communicated through maps; moreover, most of these historians, in their writings, stop short of the tremendous advances in maps and mapping of the twentieth century. And, unfortunately, professional mappers themselves, the surveyors and cartographers, have seldom seen fit to write of their work except in the bloodless language of the specialized journals.

I was inspired to attempt a fuller account of maps and the mapmakers—of cartography and its allied science, geodesy—after working with a mapping party in the Grand Canyon. To see how much hard work and new technology went into mapmaking was a revelation. To consider how much more difficult the task must have been in earlier times aroused my curiosity about the pioneering mapmakers. I shall always be grateful to Bradford Washburn, director of the Museum of Science in Boston and leader of the Grand Canyon mapping party, for introducing me to the Canyon and to cartography.

As I learned during subsequent research, this is an appropriate time in history to tell the story of the mapmakers. For only now, at the end of the twentieth century, can it be said that Earth has been mapped, though not always well or everywhere completely. The first two parts of the book are devoted to the broad trends and signal achievements of cartography and geodesy prior to the twentieth century. The final two parts include some earlier history but are focused on the twentieth century, a time of new technologies, when so much that was once thought unmappable has come to be mapped.

While I was on a visit to the Library of Congress in 1997, it struck me that the time had come to update and expand this book, first published in 1981. The old maps on the shelves and in the vaults of the library's Geography and Map Division, repository of the world's largest collection of maps and atlases (4,600,000 map sheets, 60,000 atlases, and 300 globes), pulled me again into cartography's rich past. I admired a twelve-volume set of maps printed and hand-painted by the Blaeu family in the golden age of Dutch cartography in the sixteenth and seventeenth centuries. Here, too, were a 1607 vellum map of Samuel de Champlain's explorations of North America and Pierre L'Enfant's original 1791 plan for the new capital city of Washington, with Thomas Jefferson's handwritten changes.

The reading room snapped me back into the present. A couple of young men were hunched before computer screens, scrutinizing a map their fingers had just commanded into existence. The map could show them the trail to a trout stream in the mountains, the streets of their neighborhood, or the incidence of cancer by county in any particular state. The map could be ephemeral, vanishing at the next click on the keyboard. It could be altered on the screen—details added or highlighted, scale and perspective changed—and then printed out. Someone sitting at the computer screen could generate a custom-made map in the time it must have taken early chartmakers to draw a single rhumb line as a bearing for mariners navigating a wide sea.

Ralph E. Ehrenberg, chief of the map division, said that he had seen many radical changes in cartography since he was a young man converting aerial photographs into maps, the cutting-edge technology in the middle of the twentieth century. Speaking in soft librarian tones, he mused: "Today, you can put in one of our floppy disks and, with each image that comes on the screen, you are creating a new map. To me that's revolutionary."

Yes, it was time to bring the story of cartography forward to the end of the twentieth century. So much has happened in mapmaking in the last two decades that mapmakers themselves, I discovered, are as astonished as anyone by the transformation.

Joel L. Morrison, a former president of the American Congress of Surveying and Mapping, told me that at first he had underestimated the implications of the new technologies. They are, he came to realize, drastically changing the nature of maps and the traditional role of professional mapmakers.

This edition includes completely new chapters on the new technologies as well as many other additions and amplifications based on recent scholarship. But the thrust of the story remains the same as in the original edition. It is that mapmakers through time have communicated a sense of where we live, where we have been and want to go, and where we are when we get there.

PART *One*

Prologue at Dana Butte

One hot afternoon in a June not long ago, our helicopter circled the stark pinnacle of Dana Butte. We found a landing place below the pinnacle and eased down in a tumult of dust, settling onto the sunbaked talus. We stepped out onto the hard red flakes of an immemorial sea floor and began unloading the gear. Two-way radio. A bright orange metal target. Quartz reflectors. Theodolite. Then we set out for the pinnacle.

At first the going was easy, one shallow ledge piled on another and another like an uneven stack of pancakes. Bending forward, under the burden of our gear, we felt the slope steepen, the footing become more precarious. Rocks gave way, sliding from their timeless resting places and falling out of sight into the chasm below. Upward, more slowly, more cautiously now. We stretched to cross from one rock to the next, so much sky below. A shuddering sight. Better not think about it. We anchored our feet and hefted the gear over the top. Then, chinning the ultimate ledge, hiking one leg over the top, and the other, we surmounted Dana Butte.

We had reached the high ground where we had a job to do, the establishment of a mapping station. For we had come to measure and to map the Grand Canyon of the Colorado.

We pulled ourselves upright and tested the footing. Careful now, the pinnacle may be solid, but it is no more than two, three meters wide. And how barren! No vegetation sprouts from its hard, rust-colored mantle. No nest of eagles, no Anasazi shards, nothing.

This place stood out in defiance of the winds and floods that had shaped it and everything around as far as the eye could see. It stood res-

Measuring angles from atop Dana Butte, Grand Canyon, 1972

olute and solitary, 1,700 meters above sea level but reaching barely halfway to the Canyon rim. Perfect for our purposes. We could see and be seen from all directions, from Hopi Point and Yaki, Yavapai Point and Cheops and Ra. Here we could make more measurements needed to give the map its basic frame of reference, its mathematical skeleton.

Standing speechless, entranced by the prospect, we became strangely conscious of everything and of nothing. The sound of silence: a soft whistling of wind, murmuring molecules, and nothing more. The prickly shiver of quiet excitement: in the Canyon there are still many places where, in the old prospector's phrase, the hand of man has never set foot—and, until our visit, Dana Butte most certainly was one of those places.

No one before us had ever seen the Earth from quite the same perspective. We were, for one brief, exalted moment, Hillary and Tenzing, Armstrong and Aldrin. We were at one with every explorer, everyone who, like Darwin, longed "to set foot where no man has trod before." In every direction we looked, we could find nothing of man or man's handiwork, except for the bulb-nosed helicopter down in the flats. We knew what Joseph Wood Krutch meant when he said of the Canyon: "It is still the world as though man had never been."

There was in the view from Dana, in the silence and the aloneness, an intimation of what must have been the primeval wonder—the wonder of earliest incomprehension, when early man stood and looked across the African plain or out over the sea. Out of the primeval wonder came myth and legend. It must have been a common experience. The Navajos, while living unknown to the children of Noah, conceived their own flood story to explain the world they saw from the Canyon rim. Theirs was the primeval wonder, which precedes speculation and reason and exploration.

Wonder eventually gave birth to cartography, the art and science of making maps. Through maps people sought to explain where they were and the way things were. Through maps, as time went by, they sought to represent the world they knew and the worlds they could only imagine. Through maps they distilled knowledge and inspired the acquisition of still more knowledge.

At Dana Butte, on untrodden ground that hot afternoon in June, we sensed the beginnings of wonder and of cartography. We came away knowing the ageless compulsion to reach out and, through mapping's ever-widening embrace of worlds, to reduce wonder to a scale more susceptible of human comprehension.

The Map Idea

The origin of the map is lost to history. No one knows when or where or for what purpose someone got the first idea to draw a sketch to communicate a sense of place, some sense of *here* in relation to *there*. It must have been many millennia ago, probably before written language. It certainly was long before the human mind could conceive of the worlds beyond shore and horizon, beyond Earth itself, that would be embraced through mapping.

All the evidence suggests that the map evolved independently among many peoples in many separate parts of Earth. Before Europeans reached the Pacific, the Marshall Islanders were making stick charts. Sticks were lashed together with fibers to depict prevailing winds and wave patterns; shells or coral were inserted at the appropriate places to represent islands. The smaller stick charts were carried by the islanders in their outrigger canoes, while the larger ones were kept on land for instruction. When a Tahitian communicated his knowledge of South Pacific geography to Captain Cook by drawing a map, it was clear that he and his people were quite familiar with the map idea. Pre-Columbian maps in Mexico indicated roads by lines of footprints. Cortés traveled through Central America guided by a calico map provided by a local cacique. It has also been discovered that centuries ago Eskimos carved accurate coastal maps in ivory, the Incas built elaborate relief maps of stone and clay, and early Europeans drew sketch maps on their cave walls.

One of antiquity's great civilizations, China, fully grasped the map idea thousands of years ago. Most of our knowledge of this comes not from any extant maps, all of which are of more recent vintage, but indirectly through references in Chinese documents. A town-building prospectus submitted to the emperor in 1020 B.C. mentions an illustrative map, now lost. The Zhou dynasty in the immediately succeeding centuries had a standing order that each principality be mapped, and a royal geographer always accompanied the emperor on his tours of the realm. Maps, it seems, had already become instruments of bureaucracy and political power.

The oldest extant Chinese map came to the attention of scholars in the 1990s. Examining artifacts excavated two decades earlier in Hebei province, scientists discovered a large bronze plate engraved with the Zhao Yu Tu ("map of the area of the mausoleum"). The map, dated to the fourth century B.C., marked locations of buildings in the five mausoleums of Emperor Wang Cuo, his empress, and his concubines. Historians of cartography were impressed by the map as evidence of rather advanced practices in mapmaking, such as the use of symbols and numerals to show distances. Du Naisong, of the Palace Museum in Beijing's Forbidden City, said: "It is not only the oldest map ever found in China, but the oldest numeral-bearing map in the world."

Archeological excavations in Hunan province have uncovered silk maps in the tombs of a prime minister, his wife, and one of his sons, who had died in the second century B.C. According to A. Gutkind Bulling, an American authority on Chinese archeology, the maps were prepared by cartographers to the king of Ch'ang-sha, whose kingdom included modern Hunan and adjacent regions. "Their great significance," Bulling wrote of the maps after their discovery in 1973–74, "lies in the fact that they are in part surprisingly accurate and detailed and show that the art of cartography was well advanced at this time."

One of the maps, measuring 96 by 96 centimeters, depicts at a scale of between 1:170,000 and 1:190,000 (the ratio of distance on the map to actual distance) much of the topography of what is modern Hunan and regions as far south as the South China Sea. In some parts of the map the accuracy suggests to Bulling that on-the-spot surveys were employed. Another hallmark of the map's sophistication, compared to most ancient maps, is its use of standardized symbols and legends, which indicate the existence of a cartographic tradition based on considerable thought and practice. The names of all provinces are placed in squares, and cities and villages in circles. The names of tributary rivers are written near their confluence with larger rivers. Irregular double wavy lines depict mountain ranges. Roads are drawn in rather thin lines. Thicker lines delineate the more than thirty rivers on the map, the thickness not only reflecting the importance of a river but also its direction of flow.

The second map, describing the defenses of the kingdom, is embellished with some color—blue rivers, red lines enclosing towns, as well as the basic black markings. Each military encampment is represented by a rectangle outlined in red and black, the size of which indicates its importance and number of troops; the commanding officer's name is inscribed inside each rectangle. Other symbols, usually assorted triangles, represent walled fortresses, supply depots, and observation towers. According to Bulling, the map shows the organization of defense in a war that began twelve or thirteen years before the death of the man in the tomb. It also hints at the war's toll: next to the names of some towns are written phrases such as "35 families, all moved away" or "108 families, none back" or "now nobody."

If these two maps are any indication, Chinese cartography in ancient times was more advanced than that in contemporary cultures elsewhere in the world.

The earliest direct evidence of mapmaking comes from the Middle East, where archeologists have discovered several maps inscribed on clay tablets that are centuries older than the Chinese examples, though more primitive in execution. One of the earliest of these clay maps was found at Nuzi, in northern Iraq, and dated 2300 B.C., the age of Sargon of Akkad. The map shows settlements, streams, and hills or mountains, the latter indicated by a scalelike pattern. Written information on it gives the dimensions of cultivated plots of land. In addition, the Nuzi map specifies its orientation by naming three of the four cardinal points, each given by the appropriate wind: the west wind is at the bottom, the east at the top, the north on the left, but the south wind is missing. Ancient maps were oriented east (hence the expression "orientation") presumably because that was the direction of sunrise; medieval Christian maps resumed the practice because east was supposed to be the direction to Paradise.

Clay tablets from about the same period as the Nuzi map contain surveying notes for the purpose of taxing property. Cadastral (real-estate) maps seem to have been a staple of ancient cartography in Mesopotamia and Egypt. Plans of properties and towns in ancient Babylonia were inscribed on clay tablets, a number of which survive, the earliest dating from 2000 B.C. An interesting example is a clay map of

Ancient Chinese topographical map, second century B.C., *of what is modern Hunan and regions as far south as the South China Sea (the dark shape at the top of the map)*

Kassite Nippur, dated about 1300 B.C. It describes a number of privately owned fields, each recorded with the name of the owner, separated by streams and irrigation canals. In the center of the map is the king's field, designated as "field between the canals, the holding of the palace." Another surviving example is an Egyptian map of the Nubian gold mines between the Nile and the Red Sea. Drawn on a papyrus roll in about 1300 B.C., the map shows the gold-bearing basin east of Coptos marked in red; in addition, it shows the main road, the temple of Ammon, and some houses.

The city map seems also to have been a prevalent type of cartography in ancient Mesopotamia. In the late nineteenth century an expedition from the University Museum of the University of Pennsylvania discovered a clay map of the city of Nippur on the Euphrates in the sec-

Early Mesopotamian map of the world on a clay tablet

ond half of the second millennium B.C. Such a map was the answer to an archeologist's prayer, for it shows the outlines of the entire city, including its foremost temples, the central park, the "canal in the heart of the city," the river, moats, walls, and the city gates.

The Babylonians may have been the first to produce a map of the world—after a fashion, that is. The earliest extant world map is a Babylonian clay tablet from the sixth century B.C., on which Earth is shown as a flat circular disk surrounded by ocean and several mythical islands. This world is little more than the kingdom of Babylonia, schematically portrayed, with the city of Babylon shown as a long rectangle and with the Assyrians shown to the east and the Chaldeans to the southwest.

Whatever the mapmaker's ambitions, a paucity of knowledge and a chauvinistic conceit rendered the Babylonian world-mapping efforts meaningless, except perhaps as a philosophical or political statement. What the Babylonians knew, their property lines and city walls, they mapped with some attention to accuracy; what they did not know, the lands beyond their own, they chose either to ignore or to fabricate. This was to be true of world mapping for many centuries, even to some extent until the nineteenth and twentieth centuries.

The Greeks were soon trying their hand at mapmaking, with about as much sophistication at first as the Babylonians. Herodotus told a story of the use of one such world map in political decision-making. At the time of the Ionian Revolt, 499–494 B.C., Aristagoras of Miletus, the leader of the revolt, went to Greece to persuade Sparta, the dominant military power of the Hellenic world, to join the revolt and dispatch an army to fight against Persia. Aristagoras took with him a map of the world engraved on a bronze tablet, the work of Hecataeus. The tablet showed "the whole circuit of the earth, the seas and the rivers." According to Herodotus, Aristagoras appealed to the greed of Cleomenes, the Spartan king, by pointing out on the map the Persian lands that could be his and then delivering the first documented map-illustrated discourse on economic geography.

Aristagoras told Cleomenes that he could see for himself on the map that next to the Ionians were the Lydians, who had a wealthy country. Then came the Phrygians, farther east, who were richer in cattle and crops than all other known nations. And there, adjoining them, were the Cappadocians, and next to them the Cilicians, with their territory extending to the coast. He pointed to the island of Cyprus, where the people paid an annual tribute to the Persian king of five hundred talents, and to the land of the Armenians, who had cattle in abundance.

Farther east lay Cissia and the city of Susa, where the great king lived and kept his treasure. "Why," Aristagoras concluded, "if you take Susa, you need not hesitate to compete with God himself for riches."

Tempted, Cleomenes asked how long it would take to go from Ephesus to Susa following the route of the Persian royal road. At this point, Herodotus wrote, Aristagoras made his big mistake, for he told the truth. When Cleomenes heard that the journey would take at least three months, he refused to commit any Spartan troops to the enterprise and ordered Aristagoras to leave the city before sunset.

Since none of the early world maps by the Greeks has survived, an assessment is impossible, but the Herodotus story suggests that the concept of a map of the world drawn to scale was unfamiliar to people of the fifth century B.C. Herodotus scoffed at the maps he saw, though his own ideas of geography were also off the mark. However, his ideas about how to represent geographic knowledge in graphic terms indicated some progress toward scientific mapping. In his writings Herodotus discussed the arrangement of places along parallels and meridians; the preparation of itineraries, the forerunners of modern road maps; and the idea of breaking down the world map into sectional maps.

The map idea was taking hold in Greek culture. Aristophanes, in his comedy *The Clouds,* produced in 423 B.C., had a geometrical instrument brought on stage "to measure up land." One of the characters asked: "Do you mean the allotment land?" No, he was told, the whole world. His attention was then called to a world map that had also been brought on stage. Another character said: "Here you have the circuit of all the earth. D'ye see? Here is Athens."

Socrates is supposed to have employed a world map to deflate the ballooning ego of his pupil Alcibiades. As it was told by Aelian in *Various Histories,* written in the third century A.D., Alcibiades boasted of his wealth and landholdings so much that Socrates finally took him to a place in Athens where there was a plaque showing the circuit of the world. He asked Alcibiades to point out Attica, which the young man did, and then to show him his property in Attica, which the young man could not do. "These then make you boast," said Socrates, "though they are not even a part of the earth."

This story illustrates the limitations of a small world map and the potential value of a map for putting things in perspective—and people in their place, both literally and figuratively.

It must be apparent from the foregoing review that the map idea is both venerable and ubiquitous, for there is something fundamental about the map: it is a basic form of human communication.

In the first volume of the authoritative *The History of Cartography,* one of the editors, Brian Harley, wrote: "There has probably always been a mapping impulse in human consciousness, and the mapping experience—involving the cognitive mapping of space—undoubtedly existed long before the physical artifacts we now call maps."

Just how fundamental the map is came home to two American cartographers when they sought to develop a general theory of cartography. In their short and insightful book *The Nature of Maps,* Arthur H. Robinson and Barbara Bartz Petchenik of the University of Wisconsin described mapping as "the form of symbolization with special utility for encoding and transmitting human knowledge of the environment" and defined a map as "a graphic representation of the milieu." So far, simple enough. But as they contemplated the question of what there can be about the map that is so profoundly fundamental, the cartographers were astonished and frustrated. They discovered that no one seems ever to have given the question much thought, least of all cartographers and geographers. The fundamental nature of the map is simply taken for granted. Indeed, the term *map* is often used metaphorically to explain other types of knowing and communicating. In everyday conversation, the word *map* is used to convey the idea of clarification: someone maps out a plan or maps out his future or, if he seems to be a little dense, has someone ask him, Do I have to draw you a map?

Robinson and Petchenik explored some of the possible reasons for the map's being so fundamental as to be a common metaphor. They considered and generally rejected leads provided by philosophers, psychologists, and communications theorists. They noted that when an anthropologist uses the phrase "laying out a map of culture" he seems to "mean that it is possible to take isolated incidents, experiences, and so on, and arrange them intellectually so that there is some coherence, some total relation, instead of individual isolation." Clarification, coherence, plotting of information on a piece of paper—all are ingredients of the map idea. They concluded from the evidence they reviewed that mapping is a fundamental way of converting personal knowledge to transmittable knowledge, adding: "The basic significance of maps, then, seems to lie particularly in the fact that maps are surrogates of space."

Continuing this thought, the authors wrote: "There is fairly widespread philosophical agreement, which certainly accords with common sense, that the spatial aspects of all existence are fundamental. Before an awareness of time, there is an awareness of relations in space, and space seems to be that aspect of existence to which most other things can be analogized or with which they can be equated."

At the end of their book Robinson and Petchenik were still short of their goal of a theory of cartography, but at least they had arrived at an understanding of one important thing about maps. "The reason for the common use of mapping as a metaphor for knowing or communicating," Robinson and Petchenik wrote, "has finally become clear: the concept of spatial relatedness, which is of concern in mapping and which indeed is the reason for the very existence of cartography, is a quality without which it is difficult or impossible for the human mind to apprehend anything."

The uses of maps in human communication continually increase and diversify, reflecting the range of interests, knowledge, and aspirations—of what can be or should be "apprehended."

Once, the mapmakers' range was quite limited. The only maps that could be trusted were confined to immediate surroundings. Those that essayed to describe distant lands and seas were, until the modern era, exercises in conjecture based on inadequate surveying, wishful thinking, theological dogma, or sheer imagination. Such maps, where spaces that should have been left blank were adorned with imaginary continents and sea serpents or other fanciful creatures, inspired Jonathan Swift's well-known satirical verse:

> So Geographers, in Afric-maps,
> With savage-pictures fill their gaps;
> And o'er unhabitable downs
> Place elephants for want of towns.

Bear in mind that, even though elephants have vanished from today's maps, no map is as entirely trustworthy as Beryl Markham imagined. The aviator and evocative author once wrote: "A map says to you, 'Read me carefully, follow me closely, doubt me not. . . . I am the earth in the palm of your hand.'"

These words ascribe too much authority to maps. Maps are not reality; they are—have to be—selective representations of reality. All maps, of course, are smaller than the territory they depict. Only in a work of fiction by Jorge Luis Borges could there be "a map of the Empire that was of the same scale as the Empire." Lewis Carroll, in *Sylvie and Bruno Concluded,* expanded on this kind of absurdity. Considering the largest map that would be useful, Mein Herr conceived of the "grandest idea of all!" A map of the country was made on the scale of a mile to a mile. "'It has never been spread out, yet,' said Mein Herr. 'The farmers objected: they said it would cover the whole country and shut out the sunlight! So we now use the country itself, as its own map, and I assure you it does nearly as well.'"

Mapmakers must choose what to show and how to show it, and what not to show. They deconstruct the world or a part of it, then reassemble selected components. Stephen S. Hall has written of mapmaking as a process of subtraction. "Maps," he said, "boil down physical domains to their starkest, most essential, most comprehensible qualities, and in that sense they become the perfect graphic emblems of reductionism."

Although the sum of these reductive processes may in most cases be considered a fair and reliable map, Mark Monmonier, a Syracuse University geographer, cautioned that even a good map "tells a multitude of little white lies."

The most conscientious mapmaker perforce falls short of telling the whole truth, because of limited knowledge, restrictions imposed by the particular map format, and a strict devotion to the intended purpose of the work. Some things are left out: a map of waterways subtracts highways, a highway map does not show every creek, and a geological map excludes nearly everything built upon the land. Space is much reduced: cities become mere dots or squares. Colors are often unnatural, to accentuate different places and features. These artifices are attempts to call attention to the salient information the map is supposed to communicate, even if it sometimes confuses the unwary. Huck Finn and Tom Sawyer, up in a balloon, found this out when the color of the land did not change abruptly, as on their maps, when they floated from Illinois to Indiana.

One of the most memorable maps of the twentieth century was Saul Steinberg's drawing of the world as seen from Manhattan. New Jersey looms large, but beyond is the rest of the continental United States, foreshortened, with Asia sinking behind California. A playful comment on New York provincialism, the drawing merely exaggerated the distortion that to one degree or another exists, and always has existed, in most conceptual maps of world geography. Such maps tell almost as much about the societies that created them as about the world itself. In their maps, the ancient Babylonians and Chinese placed themselves at the center of everything, in part because they knew little else. Even today, world maps made by Europeans nearly always place Europe close to the center, and those by Americans have the Western Hemisphere at the center. This self-centeredness reflects the idea, expressed by J. M. Blaut, a University of Illinois–Chicago geographer, that in history and cartography "the world has an Inside and an Outside." Only after World War II did the Outside begin to insert itself firmly into the European and Anglo-American consciousness, and in maps.

Political perceptions still intrude in maps, sometimes deliberately, often unconsciously. In the Cold War, the Soviet Union was often projected on world maps in a way that made it appear much larger than it was in reality, stretching menacingly across Eurasia like a call to Western arms.

Not all maps are as blatantly political. Nor should a recognition of maps as subjective works lead to thinking of them "as conspiratorial devices of the powerful." Making this point, Jeremy Black, a British historian, said that understanding the map as a medium with the potential to deliver a message simply returns maps "to the social and political contexts in which they have meaning, not meaning without controversy, but meaning in controversy." Even seemingly objective, neutral maps, like other visual images, communicate influential impressions of reality, not reality itself.

As interpretations of place, which change over time, maps are valuable to historians. "The map," wrote Norman J. W. Thrower in *Maps and Man*, "is a sensitive indicator of the changing thought of man, and few of his works seem to be such an excellent mirror of culture and civilization."

As practical aids to human thought and action, maps are indispensable here and now, whatever their limitations. We would be lost without them, literally and figuratively. In the cockpit of her airplane over Africa, Beryl Markham continued in thought her paean to maps: "Were all the maps in this world destroyed and vanished under the direction of some malevolent hand, each man would be blind again, each city be made a stranger to the next, each landmark become a meaningless signpost pointing to nothing."

Yet she felt regret that a map can be "a cold thing . . . born of calipers and a draughtsman's board." A coastline there "speaks of no mariner, blundering full sail in wakeless seas, to bequeath, on sheepskin or a slab of wood, a priceless scribble to posterity." What of this mountain or valley or desert, what price in toil and life was paid to put each of them on the map? "Here is a river," she thought, "that some curious and courageous soul, like a pencil in the hand of God, first traced with bleeding feet." A map may be only paper and ink, Markham concluded, "but if you think a little, if you pause a moment, you will see that these two things have seldom joined to make a document so modest and yet so full with histories of hope or sagas of conquest."

Now, the range of the mapmakers is more extensive and intensive, as can be seen in the table of contents of any comprehensive atlas. In addition to the political maps, showing political boundaries and cities and highways, the *National Atlas of the United States of America* includes maps illustrating earthquake epicenters and gravity anomalies, topographic relief and land forms (flat plains, tablelands, low hills, and mountains shown in different colors), soils and vegetation, sea temperatures and salinity, precipitation and groundwater, farms and farm sizes by principal crops, metals and fuels, election results and population densities.

The application of the map idea, thus, seems to be as limitless as it is ubiquitous. As Robinson and Petchenik wrote: "Anything that can be spatially conceived can be mapped—and probably has been."

The Librarian
Who Measured Earth

B efore Earth could be mapped it had to be known, and the knowledge came slowly, at first indirectly, and only after the human mind could rise above myth and transcend narrow experience.

In our time, when men have looked upon Earth from afar, seeing it as a small, glistening sphere spinning in the black sea of space, it requires a long backward flight of the imagination to appreciate earlier perceptions of Earth. They were visions of wonder and myth, and often they were marvelously wrong.

It once went without saying that the Earth must be flat. Some believed it was a flat platter floating on water, or was square. Others thought it flat and supported by four elephants standing on the back of a giant turtle. Homer probably reflected the conventional wisdom of his time when he described Earth as a flat disk surrounded by an ever-flowing ocean. On the rim of Earth rested the high vault of the sky, with mighty Atlas tending pillars that gave additional support to the inverted hemisphere. The Sun rose daily from Oceanus and every night sank beneath its waves. The Moon and the stars made their nightly journeys from one edge of Oceanus to the other.

This conception had understandable origins. Anyone who has stood on a Kansas prairie and looked at the sky from horizon to far horizon shares what must have been the feelings of those long ago. When the shepherds and sailors of the Mediterranean studied the night sky, it must have seemed something like a great inverted bowl. As far as they could know, the bowl did indeed rest on a flat Earth, somewhere out there on the dark horizon. Was not the Earth flat in every direction they walked or sailed? And everywhere ancient travelers went, they soon reached broad expanses of water. Was not the flat Earth surrounded by the circumfluent ocean?

As time went on, however, a revolutionary idea took root among the Greeks. They began to assume that the universe was comprehensible— a bold, optimistic, thoroughly modern assumption. The Greek word for universe became *cosmos,* meaning "order." The people of Mesopotamia had had intimations of order as they observed the peregrinations of the heavens. Discovering a predictable pattern to the movements of the Sun, the Moon, and the stars, they learned to gauge the passage of time and to anticipate the onset of seasons. This was the beginning of astronomy, the first science, the pursuit of which by the Sumerians and then the Babylonians led to the creation of those tools of systematic thought, mathematics and writing. But the Greeks went a step further, and an arrogant step it was. For not only did the Greeks observe the order of the heavens, they suspected that their minds just might be equal to the task of explaining the order and thereby understanding the nature of the Earth.

Their approach represented a merger of the Babylonian methods of detailed observation, particularly in astronomy, with the Greek methods of logical analysis. This was a truly major breakthrough in thinking, the germ of the scientific method as we know it today. It led to the first reasoned conception of the Earth's being round, not flat, and then to the first known measurement of the circumference of the round Earth. This measurement was one of the most remarkable achievements of ancient Greece and of early science. It was fundamental to cartography.

The change in Greek thinking began to take shape around the sixth century B.C. At the time, the richest city of the Greek world was Miletus, a bustling, cosmopolitan seaport in Asia Minor on the Anatolian coast of what is now Turkey. The Ionian Greeks of Miletus thrived on trade and industry. People came there from all over the known world, trading ideas as well as goods and having a profound effect on the Milesians. They became an independent, inquisitive people and were eventually freed from a strict belief in the Olympian gods. "With the Milesians," wrote T. B. Farrington, a British classicist, "technology drove mythology from the field."

Out of this stimulating environment emerged Thales of Miletus, who by most accounts was the first man who sought to explain the Earth and the sky by reason.

Thales made a fortune in olive oil and then settled down to the scholarly life, studying Egyptian geometry and Babylonian astronomy. This enabled him to gain considerable fame in his day by predicting an eclipse of the Sun in 585 B.C. But when Thales contemplated the Earth his reasoning was ambitious but flawed. He sought some observable unifying explanation for all things. If there was order to the universe and it was comprehensible, then it should be possible to discover the fundamental principles by dismissing mythology and observing nature. Thales concluded that everything must originate as water. Perhaps he had in mind the life-giving properties of rain and of the great floods of the Nile. Though simplistic, the idea was at least based on rudimentary scientific reasoning. But Thales proved himself a transitional figure by clinging to the belief in a flat-disk Earth. For he proposed that the Earth must be floating on water, the fundamental substance, like a slab of wood.

Other Milesian philosophers advanced other concepts. Anaximander believed that Earth was a cylinder, on the top of which rested the disk-shaped habitable world. Anaximenes conceived of Earth as a rectangle buoyed up by compressed air.

Pythagoras came closer to the truth, although apparently for esthetic rather than scientific reasons. One of the earliest mathematical geniuses, Pythagoras was born on Samos, an island near Miletus, but migrated to Italy and founded a school of philosophy at Crotona. It was there, sometime before the end of the sixth century B.C., that he supposedly decided that the Earth is neither flat nor cylindrical nor rectangular, but spherical. He and his followers believed that the sphere was the perfect shape, observed that the Sun and Moon were spherical, and concluded that the Earth also enjoyed this perfection.

The idea of a spherical Earth probably arose independently in many cultures, but as far as we know it was such influential Greek philosophers as Plato and Aristotle who established the idea in Western thinking. "The sphericity of the earth," Aristotle wrote in *Meteorology*, in the latter half of the fourth century, "is proved by the evidence of our senses."

Like others before him, Aristotle had observed the apparent movement of the stars. But Aristotle noticed something else. As one traveled north or south, new stars appeared over the horizon ahead and others disappeared below the horizon behind. The sky appeared somewhat different at different latitudes on Earth. This suggested to Aristotle that the Earth is a sphere, and not a very large one at that. Otherwise, it would require much longer journeys, longer than between Egypt and Athens, to see such difference in the stars.

Two other phenomena attracted Aristotle's attention. Ships sailing out from port vanished hull first in whatever direction they traveled; if the Earth were flat, a ship would get smaller and smaller, become a dot, and then disappear. A lunar eclipse furnished another clue. The shadow that fell on the moon—the Earth's shadow—was always curved. "Consequently," Aristotle concluded, "if the eclipse is due to the interposition of the earth, the rounded line results from its spherical shape."

On the strength of Aristotle's compelling logic, the Greeks accepted as fact the Earth's sphericity, but they had no answer to the next question: What is the size of that sphere?

At first there were only estimates. Though great logicians, the Greeks sometimes shirked experimentation and systematic observation. Plato went no further than a broad guess. In *Phaedo,* he wrote of Socrates saying: "I believe that the earth is very large and that we who dwell between the Pillars of Hercules [Gibraltar] and the river Phasis [in the Caucasus] live in a small part of it about the sea, like ants or frogs about a pond, and that many other people live in many other such regions."

Aristotle, whose observations of the stars suggested to him that the size of Earth was not great, reckoned the circumference at about 64,000 kilometers. Archimedes later reduced the estimate to 48,000. But neither indicated how he reached his estimate.

The Chinese, unknown to their Hellenic contemporaries, must have pondered similar questions. According to various tales, the Chinese believed in a square Earth with a rim ocean. In the time of a fabled emperor, some four millennia ago, so the legend goes, two brothers measured Earth. They walked from north to south and then from east to west, and both times they got the same result—134,000 kilometers. If nothing else, this tale proves that the Greeks were not alone in their curiosity about the size of Earth, or in their ignorance about how to get the answer.

The first known scientific measurement of Earth did not come until, in the third century B.C., a librarian named Eratosthenes had an inspiration.

Eratosthenes lived between about 276 and 196 B.C., after the death of Alexander the Great and before the rise of Rome, in a time when Alexandria towered over all other cities in the Hellenic world. Founded in 332 by Alexander, the city became richer than either Miletus or Athens at its peak, more powerful than any other single city of the day, Antioch or Smyrna, Ephesus or Nicaea. It was a magnificent city by the Mediterranean and at the delta of the Nile, a city of stone palaces, broad avenues, spacious parks, thriving shops, and a teeming harbor known to seamen by the stately white marble beacon on the island of Pharos. To this center of commerce flocked Macedonians, Greeks, Egyptians, Jews, Persians, Syrians, and Negroes, giving it a vibrance and excitement all its own. As Herodas, a Greek poet of the third century, wrote: "Alexandria is the house of Aphrodite, and everything is to be found there—wealth, playgrounds, a large army, a serene sky, public displays, philosophers, precious metals, fine young men, a good royal house, an academy of science, exquisite wines, and beautiful women."

The rulers of Alexandria, and of all Egypt, were the Ptolemies. Of all the heirs to fragments of the conqueror's empire, they were the most astute and most successful. They were dictators who levied heavy taxes and held power through a large standing army. But, not unlike other rulers desiring prestige as well as power, the Ptolemies also had cultural ambitions. To this end they established a temple to the Muses, and in time it became the scientific and literary mecca of the Mediterranean and the greatest glory of Alexandria.

As it evolved, the Alexandrian Museum bore more resemblance to the modern research center, the so-called think tank, than to most museums. Astronomers, mathematicians, physicians, writers, historians, and philosophers from all over the Hellenic world were invited to the museum. Their salaries were paid from the royal treasury. They lived and worked and thought in buildings inside the royal quarter, the Bruchium. They were provided with a lecture hall, a dining room, laboratories, botanical and zoological gardens, an astronomical observatory, and quiet courtyards for contemplation. Although an Egyptian priest was nominally in charge, probably a royal bow to the native taxpayers, the scholars seemed to be independent of any religious influence on their thought. Nor was there any indication that the Ptolemies required "practical" results to issue from all these scholarly pursuits. Among the scholars thus attracted to the museum were Demetrius, Strato, Euclid, Archimedes, Apollonius, Callimachus, and Eratosthenes.

In about 240 B.C., Ptolemy III appointed Eratosthenes chief of the museum's library, probably the most coveted position of scholarship in the Hellenic world. The library was the repository of most of the world's recorded knowledge, the equivalent in papyrus rolls of some 100,000 books in the modern sense. Ptolemy III went to unscrupulous lengths to increase this collection. By royal order, every person arriving in Alexandria was searched for any rolls in his possession, and these rolls were confiscated and given to scribes for copying. The copies were eventually returned to the owners, but the originals remained in the royal library.

Eratosthenes must have been ideally suited to be a librarian. He was a man of many parts, one who had sampled and studied nearly all the learning of the day. He studied under a grammarian and a poet in his native Cyrene, west of Egypt. He moved on to Athens, where he immersed himself in philosophy, science, and mathematics. According to some accounts, Eratosthenes wrote a volume of verse, a history of comedy, and a chronology of major events in Mediterranean history.

If such versatility commended him to Ptolemy III, it brought upon him the unrelenting scorn of his peers, who were becoming increasingly specialized. They called him *beta* and *pentathlos*. The latter was a name given to athletes who distinguished themselves in five games, suggesting that Eratosthenes was a scholar of many talents—and perhaps, in the critics' opinion, master of none. To be beta was to be number two, after alpha, implying that he was second-rate in all the many fields of scholarship to which he applied himself. This damning characterization seemed to follow Eratosthenes throughout his long career, even after his inspired experiment for measuring the size of Earth.

But perhaps only a man like Eratosthenes, a polymath and intellectually venturesome, would have seized upon the story about the well, the story that supposedly gave him the idea for the experiment.

Among the travelers' tales circulating in Alexandria at the time was one about a well in Syene, up the Nile at the first cataract, where the Sun shone directly into its deep waters at high noon on the longest day of the year, June 21. Nothing so strange had ever been seen in Alexandria. To Eratosthenes the story meant that Syene (the Greek name for Aswan) must lie at the northern boundary of the tropics.

The assumption was based on observations by earlier astronomers. For centuries men had watched the Sun and the Moon and the stars and concluded that they must be fixed to the inside of a great celestial sphere that rotated about a motionless Earth. What could be more plausible? After all, anyone could see movement in the heavens, but who could say that he had seen or felt the Earth moving? Earth, the astronomers believed, was the stationary center of all things.

The ancient astronomers began to see that, from their perspective anyhow, the Sun not only swung around Earth each day, but that it also moved up and down in the sky through its 365-day cycle, being higher in some seasons and lower in others. They called this path of the Sun's apparent annual journey the ecliptic.

In plotting the ecliptic, the astronomers noticed that the Sun was nearly always at an angle to the celestial equator, the imaginary line with which they divided the celestial sphere. Further observations showed that the Sun seemed to migrate from about 24° south of the celestial equator to 24° north, then retreat back over the same course. This became known as the obliquity of the ecliptic.*

It means that only twice a year, on March 21 and September 23, does the Sun's path intersect the celestial equator. When this occurs, the hours of daylight and darkness are equal—the vernal equinox and the autumnal equinox. And twice a year, when the Sun reaches one of the extremes in its migration north or south, the Sun stands at its greatest angle to the celestial equator.

Each of the imaginary lines in the sky marking the extremes of the solar migration became known as a tropic, after the Greek word *tropos,* a "turn," for it seemed that at that line the sun stopped and reversed itself, or turned about. North of the celestial equator, this occurs on June 21, accounting for the longest day of the year for all northern people. South of the equator, this occurs on December 22, making for the shortest day in the north and the longest day for southern people—but then the Greeks of that time generally did not believe that people could exist down there.

By the time of Eratosthenes, men had translated what they saw in the sky to the Earth. They spoke of lands lying under the tropics and under the equator. This thinking marked the rough origin of geographical zones based on latitude and of the Earth's equator, on a plane with the celestial equator, as an imaginary circle dividing the Earth into

---

*The division of a circle into 360 equal parts, or degrees, originated with the Babylonians some 5,000 years ago. Babylonian astronomers arrived at this system after they had defined a year as approximately 360 days.

hemispheres. Such a conception gave cartography its first three reference lines for maps—the Tropic of Cancer, the equator, and the Tropic
of Capricorn. The northern tropic became known as the Tropic of Cancer because the Sun made its turn there at the time the Crab (Cancer),
one of the constellations in the zodiac, first appeared in the sky. Likewise, the southern tropic was named the Tropic of Capricorn for its
relationship to the first appearance of the year of the constellation
Capricorn, the horned goat.

Now, of course, we know that it is Earth, not the Sun, that is moving
and causing the changes in days and seasons. What Greeks of the time
of Eratosthenes did not know was that the Earth rotates from west to
east on its axis (once every 24 hours) and orbits the Sun (once every
365¼ days).* But Earth's axis is not perpendicular to the plane of Earth's
orbit around the sun. If it were, the Sun would shine vertically only on
Earth's equator.

Instead, Earth is tilted at an angle of about 23½°, very close to the
Greek estimate of 24° for the obliquity of the Sun's ecliptic. Except for a
slight wobble, Earth is tilted at the same angle all year, with its axis
always pointed in the direction of Polaris, the north star. Consequently,
when Earth is in one part of its orbit, the north pole is slanted toward
the Sun, and it is summer in the northern hemisphere and winter in the
south. At another point in the orbit, the north pole is slanted away from
the Sun, and it is winter in the northern hemisphere and summer to the
south.

But if the ancient Greeks were wrong about the cause of what they
observed, they were absolutely right about one effect: only in the tropics
is there ever a noon when sunlight can beam straight down into the
waters of a deep well.

With this in mind, Eratosthenes sensed that he had all he needed to
measure the circumference of Earth. He was aware that the Sun's rays
are, for all intents and purposes, parallel when they reach Earth. Therefore, if Earth is a sphere, the sunlight must hit different parts of Earth at
different angles, the curvature of Earth accounting for the difference.
This came home to him when he noticed that, on the day the Sun was
shining straight down the Syene well at noon and there were no shad-

---

*Heraclides of Pontus, in about 350 B.C., suggested that it would be easier to conceive of the
Earth rotating on its axis than that the entire heavens revolved around the Earth. Being
unable to prove it, however, Heraclides failed to overcome the popular notion of a stationary
Earth. Aristarchus of Samos (died 230 B.C.) espoused the heliocentric view of the universe,
but with little effect on the ancient mind.

ows in the town, he had seen walls casting shadows in Alexandria. If he could measure the angle of a shadow in Alexandria at noon on that day, Eratosthenes thought, he just might be able to measure the size of Earth without ever leaving the grounds of the library.

Eratosthenes did just that, using the Sun, the well, and a vertical column.

From what Eratosthenes could learn from travelers, Syene was supposed to be due south of Alexandria, which was particularly convenient. For it meant that the two places must lie on the same meridian and were therefore points on the same great circle of Earth. A meridian is half of any circle that crosses the equator at a right angle and encompasses the entire globe.

Moreover, the placement of Syene and Alexandria meant that if he determined the distance between the two he would know the exact length of an arc of the meridian—that is, the length of a part of the meridian and therefore a part of the circumference of the Earth. Eratosthenes was told that a camel caravan needed 50 days to make the journey and that camels usually traveled 100 stadia a day. The distance would thus be about 5,000 stadia. Royal surveyors, according to some versions of the story, may have paced off the route as an attempt to confirm the distance.

These two knowns—the location on the same meridian and the distance—left Eratosthenes with one unknown in his calculation. If the arc of the meridian between Syene and Alexandria measured 5,000 stadia, what fraction of the full circle did that represent? Was it a small fraction, or large? That was what Eratosthenes had to find out.

His next step was an ingenious exercise in elementary geometry.

On the next June 21, when he knew the Sun was shining vertically down the Syene well, Eratosthenes walked out to an obelisk used for telling time by the Sun. Called a *gnomon,* this vertical column rested on a horizontal base along a north-south line. The gnomon stood so exactly perpendicular to the ground that if you drew a line from the top of the gnomon down through its base the line would pass through the center of the Earth.

When the obelisk's shadow hit the meridian line, meaning that the Sun was as high as it would get over Alexandria, Eratosthenes bent down and carefully marked the edge of the shadow and then measured its length from the base of the gnomon. That was it. He had measured the Earth.

For Eratosthenes now knew the length of the shadow and the height of the gnomon, and this gave him two measured sides to a triangle. Knowing this, he could draw the third side of the triangle and thus find the angle between the top of the gnomon and the Sun's rays. It was a small angle—7°12′, very nearly one-fiftieth of a complete circle.

Since there were no shadows at Syene at that very moment, with the Sun directly overhead, Eratosthenes concluded that the distance between Syene and Alexandria was one-fiftieth of the distance around the Earth. Thus, he calculated, 50 × 5,000 stadia = 250,000 stadia, or 46,250 kilometers. The circumference of the Earth must, therefore, be about 46,000 kilometers.

This was about 16 percent too large; the circumference is now known to be slightly more than 40,000 kilometers. But Eratosthenes had come remarkably close, considering that he worked without the benefit of any modern measuring tools.

A certain amount of luck favored Eratosthenes, for, it turns out, he unknowingly made several errors. His one theoretical error, assuming the perfect sphericity of Earth, made little difference. More important, however, was the fact that Syene is not exactly on the Tropic of Cancer, but is about 60 kilometers to the north. Syene and Alexandria are not even on the same meridian, with Syene 3°3′ to the east. And, as could be expected, the camel caravans proved to be less than precise in their measurement of distances; the metropolis and the oasis are less than 5,000 stadia apart—that is, 725 kilometers apart, instead of 800. But the various errors must have roughly canceled each other out.

Eratosthenes, in his inspired experiment for measuring Earth, was certainly first-rate, alpha not beta. George Sarton, the historian of science, wrote: "There was among them a man of genius but as he was working in a new field they were too stupid to recognize him. As usual in such cases, they proved not his second-rateness but only their own."

By his measurement, the principle of which is still used today, Eratosthenes established two traditions of long standing in cartography. Those who have subsequently made great strides in measuring and mapping Earth, and even other worlds, have been like Eratosthenes, people of many talents. For them cartography has been only one avenue in their pursuit of knowledge. Eratosthenes also demonstrated to future mapmakers that they must first look to the heavens if they are to get their bearings on Earth. For many centuries progress in cartography went hand in hand with progress in astronomy.

What Eratosthenes did, moreover, earned him a reputation as the father of geodesy, the science of Earth measurement.

One of the oldest of sciences, after astronomy, geodesy has both theoretical and practical functions. Its more scientific mission is to determine as precisely as possible the size and shape of the Earth. From this, it is hoped, we may learn something about the Earth's evolution and of its unseen internal structure. On the more practical side, geodesists perform the painstaking measurements and calculations that provide the frame of reference for all good maps. This frame of reference ensures that every city, area, and ocean is drawn in its proper place on the map, one accurately located in relation to the other.

"Nothing can remain immense if it can be measured," Hannah Arendt wrote in 1958 in *The Human Condition,* enunciating what could be the geodesist's guiding inspiration. "Prior to the shrinkage of space and the abolition of distance through railroads, steamships, and airplanes, there is the infinitely greater and more effective shrinkage which comes about through the surveying capacity of the human mind, whose use of numbers, symbols, and models can condense and scale earthly physical distance down to the size of the human body's natural sense and understanding. Before we knew how to circle the Earth, how to circumscribe the sphere of human habitation in days and hours, we had brought the globe into our living rooms to be touched by our hands and swirled before our eyes."

It was Eratosthenes who first proved the "surveying capacity of the human mind" on an Earth-wide scale, condensing the size of Earth from unknown immensity to a measured dimension. He died at the age of eighty knowing with incredible accuracy both the size and shape of Earth—but without knowing much about the lands and seas that covered the Earth he had measured.

astronomy, and every day we speak of the rising and setting of the Sun rather than the turning of the Earth.

In his other important book, *Geography,* Ptolemy rejected the Eratosthenes measurement of Earth, and this led him to his second influential error. As scholars have often pointed out, Ptolemy's original contributions in science may have been few; he compiled and systematized knowledge, usually improving upon the ideas of others but sometimes, inevitably, fostering their mistakes as well. In one such case, Ptolemy based his maps and descriptions of the world on a smaller estimate of Earth's circumference that he derived from the Greek astronomer Poseidonius by way of Marinus of Tyre. As a result, Ptolemy's world was about three-fourths of the actual size. Moreover, he assumed that the "known" world in his day covered 180° longitude, from the Canaries or Fortunate Isles in the west to the easternmost tip of Asia. This compound error—the smallness of the Earth's circumference and the exaggerated eastern extension of Asia—would someday embolden men like Columbus to risk the unknown seas.

Even if he was not always right, it is unfair to remember Ptolemy only for his errors. His wide-ranging mind acted as a lens, and through it passed much of the knowledge of the ancient world. If the lens sometimes magnified distortions, it usually brought the knowledge of others into sharper focus and projected it into the future.

In his *Almagest,* for example, Ptolemy systematized astronomy and improved upon the trigonometry invented by Hipparchus. He made it easier to deal with the mathematics of the circle, an important step in geodesy and cartography, among other fields. In the course of explaining how to form a table of chords, which are straight lines connecting the extremities of an arc of a circle, he abandoned the practice of expressing divisions of the circle in awkward fractions and introduced a new system of subdividing a degree. These subdivisions, in the Latin translation, became known as *partes minutae primae* and *partes minutae secundae.* Hence the "minutes" (′) and "seconds" (″) by which a degree is now subdivided—a simple but invaluable concept.

Ptolemy was far ahead of his time in his views of how scientific research should be conducted. In an elaboration on advice by Hipparchus, Ptolemy said that for the best explanation of any phenomenon one should adopt the simplest hypothesis that it is possible to establish, provided that it does not contradict in any important respect the results of observation. Ptolemy also recommended that in investigations based on observations requiring the greatest precision the results should be

First Principles by Ptolemy

In the second century after Christ, when, as Gibbon wrote, "the Empire of Rome comprehended the fairest part of the Earth, and the most civilized portion of mankind," Alexandria had declined in power and glory, but not in its reputation as the center of learning. It was second in population to Rome but first in science, and the foremost scholar at the Alexandria library was Claudius Ptolemy.

Almost nothing is known of Ptolemy the man, except that, despite his name, he was apparently not related to the Egyptian royal family. Judging by records of his observations of eclipses and other celestial phenomena, his most active years of scholarship extended at least from 127 to 151. His interests ran from biography and music to mathematics and optics, but Ptolemy's fame and influence down through the centuries came from two monumental books, one on astronomy and the other on geography. Each book perpetuated a serious error.

Ptolemy is best remembered for his rejection of Aristarchus's theory that the Earth revolves around the Sun. Ptolemy's geocentric, or Earth-centered, scheme of things was adapted from Aristotle and was made a premise of Ptolemy's principal book, which is generally known by its Arab title, *Almagest,* "the Greatest." This was a compendium of Ptolemy's scientific theories and a synthesis of the ancient world's knowledge of astronomy—and the prevailing belief then, and for 1,400 years thereafter, was in a geocentric universe. Even now, Ptolemy's influence persists. Modern navies still find it more convenient to navigate by Ptolemaic

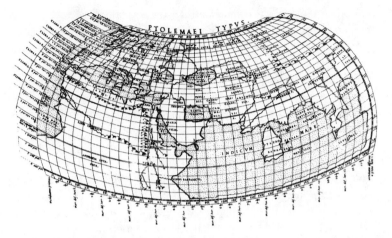

The world of Ptolemy as rendered in the sixteenth century

checked and rechecked many times over a considerable period. Sage counsel, a fundamental tenet of modern science and cartography, but Ptolemy would go unheeded in this regard for more than a millennium.

In his *Geography*, Ptolemy brought together an equally impressive body of knowledge and sound advice. One part of the book is a collection of maps, a world map and twenty-six regional maps. It is not certain whether Ptolemy himself drew the maps, but whoever did, they were based on Ptolemy's coordinates and drawn to his projection. Another part of the book, a long textual introduction, includes an exhaustive gazetteer of places known to Ptolemy and the ancients and a discussion of how to make the practice of cartography more systematic, more scientific. This was the first known recounting of the rules of scientific mapmaking, and it elevated Ptolemy considerably above the other outstanding geographer of the Roman period, Strabo, who wrote at the time of Christ. "More than any one of the ancients," concluded Lloyd A. Brown in *The Story of Maps*, "Claudius Ptolemy succeeded in establishing the elements and form of scientific cartography."

At the outset Ptolemy defined what he meant by geography, a word he considered synonymous with cartography. "Geography," he wrote, "is a representation in pictures of the whole known world together with the phenomena which are contained therein."

Such a definition stands the test of time, as William Warntz and Peter Wolff have noted in *Breakthroughs in Geography*. Ptolemy wrote of a pictorial representation: geography, or cartography, still deals with spatial characteristics of things, Warntz and Wolff observed, not "in the way of pure mathematics but in pictorial fashion." Ptolemy restricted the concern of cartography to the known world: "exploration and discovery, in other words, precede geography," according to Warntz and Wolff, though twentieth-century technology has considerably expanded the known and mappable and made it possible for mapping to precede human travel to some remote places on Earth and to worlds beyond Earth.

Finally, Ptolemy's definition envisaged cartography dealing not only with "those features likely to be mentioned in a general description of the Earth, such as the larger towns and the great cities, the mountain ranges and principal rivers," as he went on to write, but also with "the phenomena which are contained therein." Ptolemy was more foresighted than he could have imagined. His definition was broad enough to include the so-called thematic mapping of modern times, in which such phenomena as climate, population density, income distribution, birth rates, disease incidence, and variations in Earth's gravity are related to spatial location.

The following are some of the ingredients of scientific mapmaking elucidated by Ptolemy long before they were or could become the common practice of cartography.

The task of the cartographer, Ptolemy said, is "to survey the whole in its just proportions"—that is, draw maps to scale. Since a map is a controlled abstraction of a larger space, an inch or centimeter on the map represents thousands or hundreds of thousands of inches or centimeters of actual space.

The "just proportions" of maps today are usually scales expressed as the representative fraction. A distance on the map is expressed as a fraction or ratio of the distance in reality. This may be shown either as 1:1,000,000 or $\frac{1}{1,000,000}$. This means that 1 inch or 1 centimeter on the map represents 1,000,000 inches or centimeters on the Earth's or some other planet's surface. Sometimes, particularly on road maps, scale is explained with a small ruler printed near the bottom of the map. This is a graphic scale. On large maps of small regions the scale is often expressed in inches to miles or centimeters to kilometers.

There is considerable confusion regarding the terms "large-scale map" and "small-scale map." It is a common misconception, for instance, that a small-scale map shows more detail of a comparatively small area of ground. But the reverse is the case. Larger scale means larger detail: 1:5,000 is large scale; 1:5,000,000 is small scale.

Ptolemy lamented the casual approach to scale taken by his predecessors and contemporaries. When they attempted to draw a map of the known world, he observed, they tended to sacrifice proportion in order to get everything on the map. Thus, better-known areas of Europe and the Middle East were stretched in size to accommodate the many place names, while Asia and Africa were contracted rather than have too much of the map relatively blank. As a practical solution, Ptolemy recommended that a general map of the world be supplemented with regional maps. The maps of populous, well-known regions could be drawn to a larger scale to avoid crowding. This is the principle on which all modern atlases are based.

The mapping of the world, Brown stated, "depends upon a simple geometric proposition, namely, that the intersection of two lines is a point." These lines used in mapping are known as parallels (east-west) and meridians (north-south), or latitude and longitude. Such a network of lines is called a graticule or a grid or, more commonly, a coordinate system. To locate or "fix" a place on Earth in relation to any other place it is necessary to know its latitude and longitude; the place should be drawn on the map at the point where the two lines of known latitude and longitude intersect. Maps drawn with coordinates were fundamental to the scientific cartography of Ptolemy.

The Chinese at this time were also taking strides toward a systematic approach to mapmaking. The cartographer Chang Heng, a near-contemporary of Ptolemy, "cast a network of coordinates about heaven and earth, and reckoned on the basis of it." The source of Chang's inspiration for map grids is a mystery, although one story has its charm. It seems that a girl was employed to embroider one of Chang's maps on silk. When he saw the intersecting lines of warp and weft, Chang was struck with the idea of a netlike system of coordinates for maps.

Chang was followed by Pei Xiu, in the third century A.D., whose scholarship and advocacy of cartographic standards make him worthy of comparison to Ptolemy. As a minister in the imperial palace, this Chinese Ptolemy took advantage of his access to the many maps of his predecessors and, studying them, saw the need for a set of mapmaking rules. His "Six Essentials of Mapmaking" gave directions for preparing maps to proper scale, with rectangular grids and careful measurements. "If one draws a map without having graduated divisions," Pei said, "there is no means of distinguishing between what is near and what is far."

The terms Pei Xiu applied in describing his map coordinates suggest that there just might be some truth to the story from Chang's time. Pei used the words *ching* and *wei* for coordinates, which happen to be as well the words for the warp and the weft in weaving.

But the role of the Greeks in defining fundamental standards in cartography is better documented, and their achievements were the ones that later inspired and informed Renaissance mapmaking, which set in motion the practice of mapping in its modern, global manifestations.

Astronomers before Eratosthenes, as we have seen, had given cartography its first three east-west lines of reference, based on observations of the Sun's annual movement in relation to the surface of Earth. These lines—the equator and the tropics of Cancer and Capricorn—formed zones known to the ancients as *klimata*. To these lines Eratosthenes added several others when he prepared a world map. He drew east-west lines through familiar places that he supposed were on the same latitude: one line ran from the Pillars of Hercules through Rhodes, another crossed Alexandria. His nine north-south lines were likewise arbitrary.

Hipparchus, the astronomer and mathematician, then came along with suggestions for a more logical way of partitioning maps. Draw all klimata so that they would be truly parallel with and at equal intervals from the equator to the poles. And draw the north-south lines at right angles to the east-west parallels, equally spaced along the equator; they would be great circles passing through the two poles.

By the time of Ptolemy, the names "latitude" and "longitude" were being applied to these lines, and he explained why in a chapter on Marinus of Tyre, a geographer who seems to have lived just before Ptolemy.

"The greater distance we call longitude," Ptolemy wrote, "which is accepted by all, for the extent of our habitable Earth from east to west all concede is much greater than its extent from the north pole to the south."

Ptolemy, like Hipparchus three centuries before, saw the utility of a system of map coordinates and said that they should be established by careful astronomical observation. His advice to those who would map the world was to "contemplate the extent of the entire Earth, as well as its shape, and its position under the heavens, in order that one may rightly state what are the peculiarities and proportions of the part with which one is dealing, and under what parallel of the celestial sphere it is located." One could hardly ask for a better definition of geodesy or a better explanation of why many early mapmakers were astronomers and why progress in cartography often kept step with advances in astronomy.

For his maps Ptolemy adopted a system of parallels in which each line was chosen so that the length of the longest day of the year differs from one parallel to the next by a quarter of an hour. This put the first parallel 4°15′ north of the equator; he neglected the southern hemisphere because so little was known of its geography, though on his maps Ptolemy later added one parallel 8°15′ south of the equator. The tenth parallel, at 36°, followed tradition by being drawn through Rhodes. The 21st parallel, at 63°, running through Iceland or the Shetland Islands, was considered by Ptolemy and most ancients to be the northernmost zone of habitation. Ptolemy's system for meridians was to space them "the third part of an equinoctial hour" apart—that is, every 5°. Since to Ptolemy the known world spanned 12 hours, or 180°, his map was partitioned by 36 longitudinal lines.

On small-scale maps nowadays, parallels are usually delineated at intervals of 10° or 20°, though on the margins there are often marks to help the reader determine latitudes in between. On large-scale maps, each degree or even some fraction of a degree is marked by a parallel line. (Some such parallels take on a larger meaning in human affairs. Recall the crossing of the 38th parallel in the Korean War or the defiant cry of "Fifty-four forty or fight" in the American-British boundary dispute in the Pacific Northwest.) The standard spacing for meridians on today's small-scale maps is every 15°, which is also the basis for the width of the time zones.

Ptolemy conceded that reliable information necessary for pinpointing places by latitude and longitude was scant. He had devised a scientific framework for maps, but usually could not say exactly where places fitted in that framework. Of his predecessors, he said, only Hipparchus had recorded the elevations of the North Star at different places as a means of determining their latitudes. Ptolemy mentioned two devices then available for celestial angle-measurements in determining latitude. One was a gnomon, a type of sundial much like the one used by Eratosthenes. The other was a brass astrolabe with a graduated circle that, if properly used, "would show us each time how many divisions from the zenith the center of the Sun stood on the meridian line." But these instruments were too rarely used for north-south measurements, and east-west distances in Ptolemy's time and for centuries thereafter were unreliable. They were based almost entirely on travelers' accounts, which could rarely be trusted. Travelers either kept no records or, as Ptolemy warned, through "their love of boasting they magnify distances."

Centuries would go by before maps would be drawn according to Ptolemy's instructions for locating places precisely by their latitude and longitude. Like many of his excellent ideas, it was impossible to execute without considerable improvements in travel, celestial angle–measuring, and timekeeping. Even as late as 1740, it was estimated that not more than 116 places on Earth had been correctly located by astronomical observation.

P tolemy also addressed the problem of drawing the round Earth on a flat sheet of paper, a problem which every cartographer has coped with the best he could, knowing that there is no perfect solution.

A globe is, of course, the only possible medium for showing all geographical relationships in true perspective. "When the Earth is delineated on a sphere, it has a shape like its own, nor is there any need of altering [it] at all," Ptolemy wrote. But globes, as Ptolemy noted, have a serious drawback: they are limited in scale. A globe which would show a continent on the same scale as most standard maps (such as, say, the familiar National Geographic Society series) would have to be two, three, perhaps four meters in diameter. A globe on the scale of the U.S. Geological Survey quadrangle maps would stand higher than most Manhattan skyscrapers. For this reason, useful globes lack portability: one of any scale at which great detail could be shown would never fit in a ship captain's cabin, an airplane pilot's cockpit, or an astronaut's capsule.

Hence cartographers have had to devise methods of showing all or part of the spherical Earth on a flat surface. They do it by what are called map projections, by which they project spherical coordinates to a plane surface in a systematic manner. The grids (or graticules) of parallels and meridians are squeezed or stretched or otherwise distorted. For a sphere cannot be flattened without some distortion. Each projection, therefore, is a compromise between correct shape and correct size or between what is preferable in the lower latitudes or closer to the poles.

As long as mapmakers confined themselves to relatively small portions of Earth's surface, such as the lands around the Mediterranean Sea, they could ignore the problem and draw parallels and meridians as straight lines at right angles to each other. The resulting distortions were no greater than the other inaccuracies of ancient cartography. Ptolemy, however, wished to draw a map of the entire known world. He had to reckon with the fact that meridians of longitude are segments of great circles. He had to figure out some way—"a certain adjustment," he said—to retain a semblance of spherical properties on his flat map and to do it with a minimum of distortion.

Thus, Ptolemy wrote:

> Wherefore we shall do well to keep straight lines for our meridians, but to insert our parallels as the arcs of circles, having one and the same center, which we suppose to be the north pole, and from which we draw the straight lines of our meridian, keeping above all else similarity to a sphere in the form and appearance of our plane surface.
>
> The meridians must not bend to the parallels, and they must be drawn from the same common pole. Since it is impossible for all of the parallels to keep the proportion that there is in a sphere, it will be quite sufficient to observe this proportion in the parallel circle running through Thule and the equinoctial, in order that the sides of our map which represent latitude may be proportionate to the true and natural sides of the Earth.
>
> The parallel passing through Rhodes must be inserted because on this parallel very many proofs of distances have been registered and inserted in right relation to the circumference of the greatest circle. . . . By thus doing we shall insure that the longitude of our Earth, which is the better known, will be in right proportion to the latitude.

The result was a projection in which the meridians were straight lines equidistant at the equator and converging at the North Pole and the parallels were arcs of circles. In other words, faced with a choice of keeping angles correct or attempting to preserve areal relationships, Ptolemy opted for the latter. Ptolemy's projection comes closest to being what is now known as an equal-area projection: what are equal areas on the Earth's surface appear as equal areas on the map. A satisfactory method of preparing maps in which angles are preserved was not developed until the sixteenth century. (See Chapter 6 for a fuller discussion of projections.)

Within this projected framework of coordinates Ptolemy laid out the world as it was known to him through the writings of his predecessors, the reports of Roman expeditions, and the accounts of mariners and other travelers.

Following Marinus, Ptolemy drew the prime meridian through the Fortunate Islands, the assumed westernmost part of the known world; the islands should have been placed another 7° to the west. As already noted, he assumed that the known world extended 180° eastward, to the ends of the Eurasian landmass; but 180° east of the Canaries is the middle of the Pacific Ocean. Moreover, since Ptolemy considerably underestimated the Earth's circumference, his parallels and meridians were too close together and a given distance would cover too many degrees on his map. Ptolemy seems to have originated the practice of orienting maps so that the north is at the top and the east to the right. His reason was that the better-known areas of the world were all in the northern latitudes, and so on a flat map they would be easier to study if they were in the upper part.

Ptolemy's world consisted of three continents—Europe, Asia, and Africa. The British islands of Albion and Hibernia are set off to the west, somewhat misplaced and quite misshapen. Scandinavia appears as Scandia, an island of moderate size. Paris (Lutetia) is placed opposite the mouth of the Loire. The Vistula is the eastern boundary of Germany and beyond is Sarmatia. The northern coast of European and Asian Russia was beyond the realm of knowledge and therefore not shown at all, although Ptolemy does give some of the first definite information on the location of the Volga and also shows a knowledge of Asian mountain ranges and the Caspian Sea. Far to the east, Ptolemy describes the Sinae—the Chinese. Ptolemy somehow assumed that the land of the Sinae ran far to the south and then to the west until it joined the eastern coast of Africa, thus completely enclosing the Indian Ocean. Ptolemy's unknown land to the south of the Indian Ocean became the *terra incognita* which bemused and befuddled cartographers and explorers long after Ptolemy's word on matters geographic ceased to be gospel. As for Africa, Ptolemy knew almost nothing of the western coast or of the lands below the equator. Agisymba, inhabited by Ethiopians, he took to be the southern limit of the habitable world—16°25′ S. But Ptolemy came remarkably close to the truth when he described the Nile as being formed by two rivers flowing from two lakes a little south of the equator, a fact of geography that was not proven until the nineteenth century.

If Ptolemy's errors were many, often due to a lack of reliable information, the maps based on his scholarship were the culminating achievement of ancient cartography and a crucial legacy, errors and all, to cartographers and explorers of a much later time. It would be a *much* later time because, more than a century after Ptolemy, Alexandria became a battleground of revolt and the museum buildings, now over 500 years old, were destroyed. The temple housing the library survived, and a few scholars sought to carry on the museum's traditions of research.

Finally, in 391, Christian mobs sacked the library, burned the priceless contents, and converted the shell into a church. It was a symbolic victory of faith over reason. The words of Claudius Ptolemy, among others, were lost to the Western world for more than a millennium; and so too his ideas and ideals for a scientific cartography.

The Topography of
Myth and Dogma

Western scholars who followed the eclipsed Ptolemy were cloistered souls who rejected scientific inquiry as possibly pagan and certainly irrelevant. They could not bother with the latitude of the next city when Paradise was out there waiting to be mapped in all its glory. If they speculated about alien worlds, it was usually out of fear, not curiosity. For centuries people satisfied themselves with outrageous tales, told as fact, of lands where there existed fierce griffons and men without heads, dog-headed simians and birds that glowed in the dark.

Paradise, monsters, closed-minded ignorance. The West had slipped into its thousand-year slough of intellectual stagnation, the Middle Ages. From the fourth century through the fourteenth, no serious attempt was made in Europe to practice Ptolemy's recommended approach to scientific cartography. Maps produced in Europe in the Middle Ages were more ecclesiastic than cartographic, more symbolic than realistic. They reflected Christian doctrine more than observed fact. The result was, with one important exception toward the end of the period, a millennium without a significant advance in the mapping of the world.

These Christian mapmakers, as the American historian Alfred W. Crosby observed, were certainly not interested in showing that the Nile flows into the Mediterranean at precisely so many leagues south and west of Antioch. "Their map was a nonquantificational, nongeometrical attempt to supply information about what was near and what was far—and what was important and what unimportant," he wrote. "It is more like an expressionist portrait than an identification photo. It was for sinners, not navigators."

In the absence of scientific inquiry, someone like Gaius Julius Solinus was bound to come along and win an enduring, uncritical audience. A Roman grammarian of the third century, Solinus was an engaging spinner of tall tales, and quite a plagiarist. He borrowed shamelessly from Pliny the Elder, the first-century Roman scholar whose *Natural History* was the period's definitive encyclopedia of natural wonders. He borrowed so shamelessly that he has been dubbed "Pliny's Ape."

Almost invariably Solinus copied Pliny's wildest mistakes, anything that had to do with strange animals and monstrous races, the more preposterous the better. Adding some factual geography, probably also borrowed, and a few tales of his own fabrication, Solinus produced one of the most popular books of the Middle Ages. The book, revised in the sixth century under the title *Polyhistor,* left its misinformed mark on nearly every medieval map. Traces of Solinus even carried over into the eighteenth century.

In the East, according to Solinus, there lived horse-footed men with ears so long the flaps covered their entire bodies, making clothing unnecessary. There were one-eyed hunters and savages who quaffed mead from cups made of their parents' skulls. Farther out in Asia could be found gold and precious stones, but also griffons, vicious fowl that could tear an intruder to pieces. Some men in India were said to have only one leg each, though the foot was so large that it is often shown in medieval maps doubling as a parasol.

In Germany, Solinus continued, the Hercynian birds had feathers that gave off light in the dark. An animal resembling the mule had such a long upper lip that it could feed only by walking backward. Closer to home, in Italy, Solinus let his (Pliny's) imagination run almost as free. He told of pythons that grew fat on the udders of cows, of lynxes whose urine congealed into "the hardness of a precious stone, having magnetic powers and the color of amber."

For Africa the tales grew taller still. Solinus described hyenas whose shadows robbed dogs of their bark. He told of a beast in Libya, the cockatrice, that crept along the ground like a crocodile by its forequarters, while its hindquarters were suspended aloft by two lateral fins. The ants along the Niger River were as big as mastiffs, and the river itself boiled from the heat of the region, whose temperatures were hotter than any fire.

The belief in African monsters continued into the eighteenth century, their existence being explained in the following notice accompanying the de Mornes map of 1761: "It is true that the centre of the continent is filled with burning sands, savage beasts, and almost uninhabitable deserts. The scarcity of water forces the different animals to come together to the same place to drink. Finding themselves together at a time when they are in heat, they have intercourse with one another, without paying regard to the differences between species. Thus are produced those monsters which are to be found there in greater numbers than in any other part of the world."

Illustrations from Solinus appeared more and more on maps of the Middle Ages, and there they remained, since for many centuries explorers could not or would not go out to seek a truer picture of reality and no one in church or intellectual circles cared to dispute the veracity of Solinus. Of Solinus and his *Polyhistor,* Charles Raymond Beazley said in his comprehensive *The Dawn of Modern Geography:* "No one ever influenced [medieval geography] more profoundly or more mischievously."

The mischief of Solinus was compounded by the scholars of "Christian topography," one of the first of whom was a sixth-century monk named Cosmas. To Cosmas there was a simple explanation for everything, from the shape and extent of Earth to the nature of the universe, and it could be found in the Holy Scriptures.

Cosmas had been a successful trader who traveled over much of the known world, sailing the Red Sea and the Indian Ocean, visiting Ethiopia and Ceylon (Sri Lanka) and India. His journey to India earned him the surname Indicopleustes, meaning "India traveler"; his name Cosmas may also have been one assumed in later life, after he converted to Christianity and wrote extensively on religious cosmography. Cosmas Indicopleustes retired to a monastery in the Sinai and wrote a strange book entitled *Christian Topography.* "I open my stammering and unready lips," Cosmas wrote, "trusting in my Lord that He would vouchsafe me of His spirit of wisdom."

Whereas by observation and deduction the Greek sages had come to think of the Earth as spherical, Cosmas looked only to the Scriptures and found his guidance in the words of the Apostle Paul. In Hebrews 9, Paul described the tabernacle as "the worldly sanctuary," by which Cosmas decided Paul meant that the tabernacle of the "blessed Moses" was "a pattern of this visible world." The candlestick represented the lights of heaven, the table was Earth, and the shewbread the fruit of the Earth.

Invoking Scriptural authority, Cosmas went on to describe the world as a flat parallelogram twice as long as it was wide. In the center of his flat Earth Cosmas placed Jerusalem, for it was written in Ezekiel 5:5, "I have set [Jerusalem] in the midst of the nations and the countries that are round about her." Beyond the ocean was another world, where man lived before the flood, a world now uninhabited and inaccessible. Far to the north of the inhabited world stood a great mountain round which the Sun and the Moon revolved, causing day and night. The sky consisted of four walls meeting in the dome of heaven, or the ceiling of the tabernacle.

While few medieval thinkers subscribed to his "tabernacle concept," Cosmas probably reflected a Christian belief about the shape of Earth. Did not the Scriptures speak of the "four corners" of the Earth?

Cosmas had nothing but scorn for proponents of round-Earth ideas: "With supercilious air, as if they surpassed in wisdom the rest of mankind, they attribute to the heavens a spherical figure and a circular motion and by geometrical method and calculations applied to the heavenly bodies, as well as by the abuse of words and by worldly craft, endeavor to grasp the position and figure of the world by means of the solar and lunar eclipses, leading others into error, while they are in error themselves, in maintaining that such phenomena could not represent themselves if the figure was other than spherical."

So much for Pythagoras, Aristotle, and Eratosthenes. So much for the "worldly craft" of geometry, the foundation of geodesy and cartography as begun by the Greeks.

With equal fervor, Cosmas sought to deny the possibility of the Antipodes, of any land inhabited by human beings beneath our feet. The Antipodes was the Mars of that day, firing imaginative speculation and learned controversy. Like many of the ancients, Cosmas believed the tropics so hot as to be impassable, probably hotter than the boiling waters of Solinus's Niger River. In that case, if there were any people in the southern hemisphere, they "could not be of the race of Adam." Besides, the Apostles were commanded to go unto all the world and preach the Gospel to every creature. They did not go to the Antipodes, Cosmas pointed out, and therefore the Antipodes did not exist.

Calling the whole idea an "old wives' tale," Cosmas presented a further argument against the existence of the Antipodes, an argument that at the time must have seemed compelling. He wrote: "For, if men, on opposite sides, placed the soles of their feet each against each, whether they chose to stand on Earth or water, on air or any kind of body, how could both be standing upright? The one would assuredly be found in the natural upright position and the other, contrary to nature, head downwards. Such notions are opposed to reason and alien to our nature and condition."

Many other Christian thinkers of the age did not feel as strongly as Cosmas about whether Earth was round or flat. They concerned themselves more with matters spiritual and were, in fact, discouraged from considering the physical nature of the Earth. As Saint Francis warned: "There are many brothers who strive to acquire knowledge. . . . Those brothers whom curiosity drives to science will find, on the Day of Judgement, that their hands are empty."

The world map conceived by Cosmas seemed to bear the unmistakable imprint of such an unscientific age.* It showed a rectangular world surrounded by the ocean and surmounted by the heavens, a faithful image of the tabernacle, but entirely devoid of practical information. Although only later copies of the Cosmas map survive, they represent the earliest-known maps of Christian origin.

---

*Not that the modern and presumably more enlightened age has not had cartographic fantasies and its own tales of monsters.

The major tenet of a bizarre little sect called Koreshanity, which flourished in the United States late in the nineteenth century, was that we actually live on the inside surface of a hollow Earth. The sect's founder, Cyrus Reed Teed of Utica, New York, drew inside-out maps of his inverted cosmos.

Another hollow-Earth theory had an ardent proponent in John Cleves Symmes, Jr., of Ohio, a hero of the War of 1812. So convinced was he of the theory that on April 10, 1818, he wrote a letter to all congressmen and to the world's leading scientists. It read:

To All The World:
    I declare the earth is hollow and habitable within; containing a number of solid concentric spheres; one within the other and that it is open at the poles twelve or sixteen degrees. I pledge my life in support of this truth and am ready to explore the hollow, if the World will support and aid me in the undertaking.
    I ask 100 brave companions well equipped to start from Siberia, in the fall season, with reindeer and sleighs on the ice of the frozen sea. I engage that we will find a warm and rich land, stocked with thrifty vegetables and animals, if not men, on reaching one degree northward of latitude 82; we will return the succeeding spring.
                                                                    Jno. C. Symmes

The hole at the pole became known as Symmes's hole. Through it, according to an elaboration of the theory, the Ten Lost Tribes of Israel had probably gone underground. Down inside the inner spheres lived curious men and monsters. But Symmes never got the necessary support for his expedition.

No true Christian in the time of Cosmas or for centuries thereafter doubted the existence of an earthly Paradise, the Eden of Adam and Eve. Cosmas believed it lay in the world beyond the ocean, distant and unreachable. Others, in myth if not reality, went to heroic lengths to find Paradise, as in the case of Saint Brendan.

Brendan was an Irish monk of the sixth century who seems to have been an inveterate founder of monasteries and a dreamer about great voyages of discovery. Once an angel appeared before him in a dream and promised him success if he should venture forth to see the earthly Paradise for himself. Brendan is said to have sailed westward from Ireland with a crew of sixty men. In their five years at sea, they encountered many wondrous things—a palace where the devil himself appeared but did them no harm, an island of birds who were fallen angels, an island of smoke and fire (the gates of hell? or a volcano?), a glittering temple of crystal rising out of the sea (an iceberg?), a fire-eating dragon. Eventually they reached a beautiful island where they met a holy man. To Brendan this was the "Promised Land of the Saints."

The holy man directed Brendan to the cave of a dead giant. The holy man brought the giant back to life, baptized him, and, with Brendan, heard his story of having known of Christianity in his previous life. Then the giant, whose name was Maclovius, asked to be returned to his eternal rest, and his request was granted by the holy man of the island, which Brendan said was Paradise.

The legend of Saint Brendan's voyage, told as allegory and as fact, spread through Christendom. Saint Brendan's Island appeared on maps for the next 1,200 years, located first one place and then another over the length and breadth of the Atlantic: in the vicinity of the Canaries or the Madeiras, to the west of the Cape Verde group, or in the latitude of northern Newfoundland.

To Isidore of Seville, in the early seventh century, belongs the distinction of having fixed the presumed location of Paradise with great detail and authority. Isidore was an encyclopedist, theologian, archbishop, and the foremost expert on Paradise in the Middle Ages.

In his *Etymologies,* a multivolume compilation of all knowledge accessible to him, Isidore followed the pattern of Cosmas in using the Scriptures to support the framework of his world map. Isidore divided the Earth into three parts—Asia, Europe, and Africa, which to him represented the divisions of the Earth allocated to Noah's three sons, Shem, Japheth, and Ham, respectively. Isidore made some effort to blend geographic fact into the doctrinal framework. Asia, he said, is "bounded in the east by the sunrise, on the south by the Ocean, in the west by the Mediterranean, in the north by Lake Maeotis [Sea of Azov] and the river Tanais [Don]. It contains many provinces and districts whose names and geographical situations I will briefly describe, beginning from Paradise. . . . Paradise is a place lying in the eastern parts."

This seemed plausible. The east was the legendary land of spices, incense, and great wealth. The east was the source of the morning light. But, perhaps more decisively, it was linked with Paradise in the book of Genesis: "And the Lord God planted a garden eastward in Eden; and there he put the man whom he had formed."

Isidore went on to profess a detailed knowledge of Paradise, much of which he inferred from Biblical sources. Paradise, he said, was favored with "continual spring temperature." It was "planted with every kind of wood and fruit-bearing tree, having also the tree of life."

"From the middle of the Garden," Isidore continued, "a spring gushes forth to water the whole grove and, dividing up, it provides the sources of four rivers. Approach to this place was barred to man after his sin, for now it is hedged about on all sides by a sword-like flame, that is to say it is surrounded by a wall of fire that reaches almost to the sky."

Thus, in the small map illustrating his book, Isidore placed Paradise at the farthest reaches of the Asian mainland. It was partitioned from mankind by a fiery wall. Succeeding mapmakers either followed Isidore's example or located Paradise on an island off easternmost Asia. Few medieval maps in the Western world failed to include Paradise.

So great was Isidore's authority and the medieval desire to believe in such places that, as late as the fourteenth century, Sir John Mandeville wrote an account of the terrestrial Paradise "that is toward the East at the beginning of the earth." In his *Travels and Voyages,* one of the most popular and far-fetched of medieval books, Mandeville told of many fabulous sights he had seen in his years of travel and some tales he had heard, which he passed on as truth. Some of his stories ring of Solinus revisited. And though he reflected the changing medieval thinking about the shape of the Earth (". . . our Lorde made the earth all rounde in the middest of ye fyrmament"), Mandeville wrote of Paradise in a vein little different from Isidore's.

"Of Paradise," Mandeville said, "can I not speake properly for I haue not bene there, but that I haue heard I shall tell you. Men say that Paradise terrestre is the highest lande in all the worlde, and it is so high that it toucheth nere to the cyrcle of the Mone, for it is so high yet Noes floude might not come thereto which covered all the earth about."

Instead of being surrounded by Isidore's wall of flame, Mandeville's Paradise was "enclosed all about with a wall, and that wall is all covered with mosse." He told of the four rivers flowing out of Paradise, identifying them as the Ganges, Nile, Tigris, and Euphrates. "Many great lordes," Mandeville wrote, "have essayed many times to go by those rivers to Paradise, but they might not spede in theyr way, for some dyed for werynesse of rowinge, some waxt blynde and some defe for noise of the waters, so no man may passe there but through speciall grace of God."

As surely as Paradise lay somewhere to the east, there existed in medieval topography a place far to the north that inspired fear and dread—and more mythical entries on the map. This was the land of Gog and Magog, the walled home of the hordes who, it was prophesied, would someday break loose and, according to the Bible, "come up against my people of Israel, as a cloud to cover the land." The specter of these murderous armies from the north haunts the literature of Judaism, Islam, and Christianity. It is the specter of the Apocalypse and of Armageddon.

Ezekiel had prophesied the terrifying invasion. Gog, the prophet wrote, shall come forth in the latter day "from thy place out of the north parts" leading a mighty army. There will be "a great shaking in the land of Israel . . . and the mountains shall be thrown down, and the steep places shall fall, and every wall shall fall to the ground." The Lord would turn back Gog, destroying his army on the open field. In this way, the Lord said, "will I magnify . . . myself . . . in the eyes of many nations."

As the story evolved, Gog and Magog assumed an ominous reality. These forces of evil lived in the bleak lands bordering the northern sea. They were the Antichrist, the Huns, almost any marauding nomads. Fortunately, according to tales told by the Syrians, Alexander the Great had sealed off Gog and Magog with a wall of iron and brass. Variations of the story made their way into the Koran. "Build us a rampart between us and [Gog and Magog]," the Koran says. "And he said, Bring me blocks of iron; and when he had filled the space between the mountains, he caused them to blow upon it with bellows, and heated it fiery hot, and poured molten brass upon it. And Gog and Magog were not able to scale it, neither were they able to dig through it."

Even knowing of Alexander's ramparts, Christendom did not rest assured, for it was said that Gog and Magog would eventually break free and "make the earth tremble." In the book of Revelation, moreover, it was predicted that Satan, after his thousand-year imprisonment in the bottomless pit, would free the numberless horde to wreak death and destruction on the world.

This dread event was so much on the medieval mind in the thirteenth century that even Roger Bacon, one of the more enlightened men of the day, recommended a diligent study of geography so as to anticipate the most likely time and direction of the invasion. Mapmakers did their part. They generally placed the land of the imprisoned invaders along the northern shore in Russia. Since they knew nothing of the area, they saw no reason not to embellish the void with fanciful drawings of Gog and Magog.

Ibn Khaldun, a late-fourteenth-century Arab geographer, produced a map showing the "Wall of Gog and Magog" in such a way as to suggest that the Great Wall of China could have been the origin of the whole legend.

Sometime in the twelfth century rumors reached Europe that stirred new hope among the fearful. It was said that somewhere out in Asia reigned a Christian monarch of unimagined wealth and power. He was named Presbyter John, better known as Prester John. Hopefully, if only liaison could be established, this mighty Christian would come to the aid of western Christendom in its struggles with the Mongols and Saracens. Perhaps, if the terrible day should ever come, Prester John would join forces with western Christians against the unleashed Gog and Magog.

A bishop of Syria brought the first recorded news in 1145. At a meeting with the Pope, Bishop Hugh of Jabala said that Prester John was "a direct descendant of the Magi" and enjoyed such glory and prosperity that he used a scepter of solid emerald. This king, the bishop said, lived "in the extreme Orient, beyond Persia and Armenia" and he had emerged victorious "after a most bloodthirsty" battle with the Persians.

But the way Bishop Hugh elaborated upon the story suggests that he had an ulterior motive. Word of Prester John may have already reached Europe, and if so the bishop wanted to dispel any notion that the fabled ruler could be relied on in the Syrian Christians' fight against the Saracens. The main purpose of his journey was to appeal for military aid from the West. He could not allow the Europeans to be lulled into complacency by thoughts of Prester John's power to rescue eastern Christianity. Sounding a cautionary note during his conversation with the

Pope, the bishop told how Prester John had "moved his army to the aid of the Church in Jerusalem, but when he had come to the river Tigris he had not been able to take his troops across it in any vessel . . . [and] had been forced to return home."

The legend grew, nonetheless, encouraged by a letter Prester John was supposed to have addressed to the emperor of Byzantium. In the letter, copies of which circulated throughout Europe, Prester John claimed to "exceed in riches, virtue, and power all creatures who dwell under heaven. Seventy-two kings pay tribute to me. I am a devout Christian and everywhere protect the Christians of our empire, nourishing them with alms. We have made a vow to visit the sepulchre of our Lord with a great army, as befits the glory of our Majesty, to wage war against and chastise the enemies of the cross of Christ, and to exalt his sacred name."

Prester John's empire dominated "the Three Indias" and reached "through the desert toward the place of the rising of the Sun, and continue[d] through the valley of deserted Babylon close by the Tower of Babel." It was a wondrous place, as Prester John boasted in the letter: "Honey flows in our land, and milk everywhere abounds. In one of our territories no poison can do harm and no noisy frog croaks, no scorpions are there, and no serpents creep through the grass. . . . In one of the heathen provinces flows a river called the Physon, which, emerging from Paradise, winds and wanders through the entire province; and in it are found emeralds, sapphires, carbuncles, topazes, chrysolites, onyxes, sardonyxes, and many other precious stones. . . . If you can count the stars of the sky and the sands of the sea, you will be able to judge thereby the vastness of our realm and our power."

This was heady stuff for the medieval mind. Popes dispatched missionaries, kings sent ambassadors. They had to find this Christian ruler and this fabulous wealth. Describing the reaction, George H. T. Kimble wrote in *Geography in the Middle Ages:* "Such was the trust put in these stories that it is no exaggeration to say that the quest for the realms of Prester John soon became, as it was to remain for more than a century, a major factor in promoting the Asiatic enterprise."

On Easter Sunday, 1245, John of Plano Carpini, a Franciscan monk, set out from Lyons for Asia. He was a man in his sixties, accompanied by another monk, riding donkeys and armed only with a letter from the Pope. Friar John traveled across Poland, Russia, and Central Asia into the land of the Mongols, covering nearly 4,800 kilometers in 106 days and arriving in time to witness a lavish coronation. This and many other

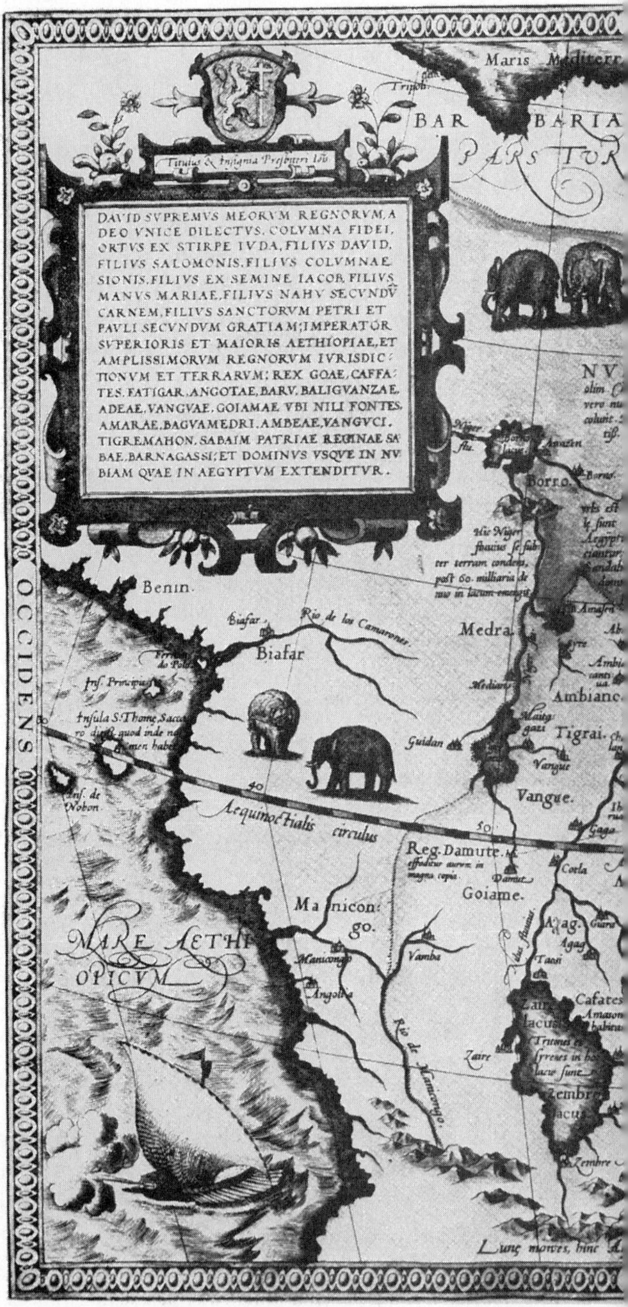

The empire of Prester John by Ortelius, 1573

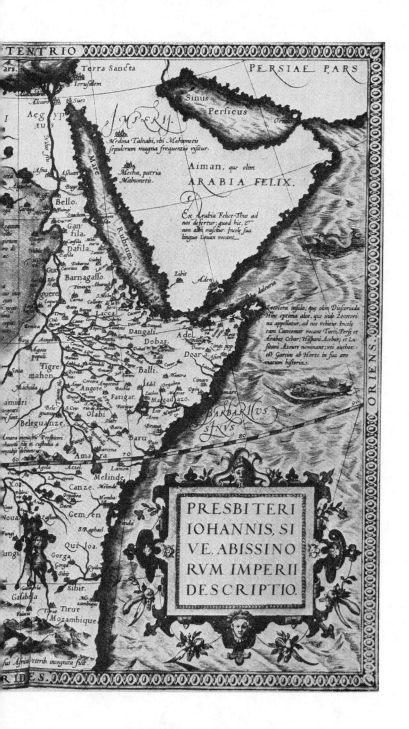

TENTRIO

ars.

Terra Sancta

PERSIAE PARS

Ierusalem

Sinus

I

Persicus

Ormuz

Aegyptus

Sues

Alcair

IMPERII

Nilus flu.

Afra

Assuam

*Medina Talnabi, vbi Mahumetis
sepulcrum magna frequentia visitur.*

Mecha, patria
Mahumeti.

Baga

Aiman, que olim

ARABIA FELIX.

Bello.

Ganfila.

*Ex Arabia Felice Thus ad
nos defertur; quod hic, &
non alibi nascitur. Incole sua
lingua Louan vocant.*

Dafila.

Mare Rubrum.

Zibit

Aden

Gue

Barnagasso

Iaberbut

Tigre
mahon.

Lacca

Dangali.

Dobas.

Adel

Doar.

*Socotora insula; que olim Diescoriada
Hinc optima aloe, que inde Zocotori-
na appellatur, ad nos vehitur. Incole
eam Cuccomar vocant Turis, Persi et
Arabes Cebar; Hispani Acebar; et Lu-
sitani Azeure nominant; vti author
est Garcia ab Horto in sua aro-
matum historia.*

amidri

Balli.

Adel

Beleguanze.

Olabi.

**BARBARICVS
SINVS**

Amara.

Xoa.

Canze.

Melinde.

Gemsen.

Noua

Gorga.

Qui Ioa.

Sibir.

Galabela

Tirut

Mozambique

sus Africa veteribus incognita fuit.

RIDES.

ORIENS.

**PRESBITERI
IOHANNIS, SI
VE, ABISSINO
RVM IMPERII
DESCRIPTIO.**

Mongol customs, along with a full account of Genghis Khan's exploits, were recorded by Friar John in one of the more trustworthy travel books of the age. But, after the fashion of Solinus and Mandeville, he could not resist retelling a few monster tales. Nowhere did Friar John run across any traces of Prester John, but he would not abandon the myth.

Other envoys followed, sometimes sifting fact from fancy. William of Rubruck reported being told that none of the Solinus-like monsters and monstrous races had ever been seen. He was also one of the first travelers to deflate the Prester John myth. Such a king might exist, Friar William said, but the stories of him were gross exaggerations. Another friar, Odoric of Pordenone, returned from "the country of Prester John" with much the same report. "But as regards him," he wrote, "not one-hundredth part is true of what is told of him as if it were undeniable."

At about the same time, the Polos of Venice were making their journeys through Asia to China. Marco Polo returned with the story that there had indeed been a Prester John, but that he had been slain in "the greatest battle that ever was seen." The great ruler had been overwhelmed by the superior forces of Genghis Khan. From that time forward, Marco Polo wrote, Prester John's "kingdom passed into the hands of Chinghis Kaan till the whole was conquered."

The increasing traffic back and forth to Asia spelled the beginning of the end of Europe's medieval isolation. More and more people, whether in search of Prester John or the wealth of Cathay, were going forth to see the Earth as it really was. Their knowledge of geography widened. Strangely, however, little of this new knowledge found its way onto the maps of the period. Mapmakers were slow to learn the essential relationship between exploration and cartography.

Nearly a century and a half later, in 1459, Fra Mauro produced an impressive world map incorporating geographic details from Marco Polo. Historians still argue over the credibility of Polo as a writer and authority on Asian geography. Did he actually visit China, and how much of his account is to be trusted as fact? In a recent study of Marco Polo, the British historian John Larner concluded that the weight of evidence and reasonable inference supported the view that the Venetian merchant apparently did live in China for twenty-four years in the late thirteenth century, was familiar with the court of the Great Khan, and was on the whole a reliable guide. "Never before or since," Larner wrote, "has one man given such an immense body of new geographical knowledge to the West."

Written in the engaging chivalric-epic style of the day, Marco Polo's book gained a wide audience in several languages, rare for a medieval work, but it was a long time before mapmakers paid heed. Larner found that the first known maps to be influenced by the book were drawn around 1380, although most maps were still in a highly schematic circular form. In his world map of 1436, the Venetian cartographer Andrea Bianco showed "the Empire of Cathay" and many islands of Asia as revealed in Polo's book. Bianco is thought to have helped in the preparation of Fra Mauro's *mappamundi*, commissioned by the king of Portugal. The book's influence on Fra Mauro is revealed in the part of the map showing China, the Yangtze and the Huang He rivers, and names of towns derived almost exclusively from Marco Polo, whose own words are often used in the annotations.

Fra Mauro's map was a transitional piece of work, limited and flawed but earnest and ambitious. The historian David Woodward called it "the culmination of the age of medieval cartography." Larner said the monk "should be hailed as a great hero of the Italian Renaissance."

Although his devotion to reality elevated him considerably above earlier medieval mapmakers, Fra Mauro still felt compelled to make a place on his map for Prester John. Since the Polos and other Western travelers had not encountered his realm in Asia, perhaps it had been a mistake to look there in the first place; perhaps the elusive Christian potentate was to be found in Africa. So ran the thinking. And so in the section of his richly ornamented map that shows Ethiopia, Fra Mauro inscribed: "Here Prester John makes his principal residence."

In other respects, Fra Mauro showed greater fidelity to what was actually known, and represented this knowledge with a commendable sense of proportion. For years the monk-cartographer labored in this vein in the quiet of the monastery of San Michele, an island close to Venice. On occasion he received visitors bringing back geographic information from their travels or those simply showing up out of curiosity about the map-in-progress. Once, it is said, a Venetian senator paid a call at the workshop, and had trouble finding Venice on the new map. When the monk pointed to a tiny dot, the sight of such an insignificant representation for his magnificent and powerful city appalled the senator. "Why so small?" he demanded to know. "Venice should be bigger and the rest of the world should be smaller." Honest maps have a way of putting us in our place.

O ut of myth and narrow doctrine, travelers' tales, and some vestiges of classical knowledge the cartographers fashioned what passed for maps in the Middle Ages. Fewer than 600 of these maps and sketches, produced between the years 300 and 1300, are extant. But they suffice to illustrate the style of medieval cartography, the state of geographic knowledge, and the purpose served by maps in those pre-exploration, pre-scientific, pre-jet-travel times. The style was symbolic, ornamental, and often beautiful; the geographic content, impoverished and usually misleading; the purposes, a representation of the mind more than of the Earth.

"In the aggregate," Kimble wrote, "it is probably true to say that the great majority of these *mappaemundi** are to be regarded as works of art and not of information. Their authors were creating something very different from the modern cartographic sheet that stands on its own merits as an essentially utilitarian document, scientific in construction. They would have branded any man a fool who might have supposed that he could determine the distance from London to Jerusalem by putting a ruler across a map."

The most common shape of the medieval map was the circular, disk, or wheel form, to preserve the "circle of the earth," as in Isaiah 40:22. This became known as the T-O map.

In most such maps, the whole of the Earth was surrounded by a circumfluent ocean—the O. Within the O lay the three known continents, separated by interior bodies of water drawn, often with diagrammatic neatness, in the shape of a T. Since nearly all medieval maps had east at the top, Asia filled the upper half. It was separated from Africa by the Nile and from Europe by the Don (Tanais). The two rivers formed the top of the T, and the Mediterranean formed the upright of the T. With such a configuration, the Christian cartographer could easily give Jerusalem prominence at the center of the map.

The T-O concept probably originated with the Ionic philosophers in pre-Christian times, but the circular-disk conception also had scriptural support: "It is He that sitteth upon the circle of the earth." Isidore of Seville apparently introduced T-O to medieval cartography, and some of the Arab cartographers of the period also produced T-O maps. Over the

---

*The term *mappamundi* (plural *mappaemundi*) is from the Latin *mappa,* a tablecloth or napkin, and *mundus,* the world. The term came to apply to any map of the world.

T-O map by St. Isidore of Seville, ninth century

centuries the maps became more ornate, the simple T-O framework being embellished with color, scrollwork, and drawings of assorted geographic lore, secular and sacred. The largest and finest of the circular world maps to survive intact is the Hereford map, which modern scholars consider a summary of medieval ideas and a landmark in medieval cartography.

Like many of its kind, the Hereford map was drawn by a monk on vellum, possibly a bullock's skin, for the adornment of a cathedral. The map, 1.3 by 1.6 meters, was completed in about 1275 by Richard of Haldingham and is still housed at the Hereford Cathedral in England. In its topographic and iconographic detail, it is as much an expression of the Middle Ages as the Gothic cathedral itself.

Set apart at the top of the map is the figure of Christ, accompanied by angels, presiding at the Last Judgment. On the left an angel exhorts the blessed to "Arise, and come to everlasting joy," and on the right the lost are being led away with the command "Rise and depart to hell-fire prepared." The map, it is believed, originally served as an altarpiece.

An inscription around the border attributes the survey to Julius Caesar, which conformed to a traditional belief that Caesar had ordered the first world survey. "The measurement of the world was begun by Julius Caesar," the inscription reads. "All the East was measured by Nicodoxus, the North and West by Theodoxus, the southern parts by Policlitus."

A wide spectrum of medieval notions fills out the map. The Earth is a circular disk with Jerusalem, of course, at the center. The map is oriented to the east—and Paradise. The four cardinal points are marked by grotesque dwarfs within circles; the supplementary points, making up the rest of the twelve winds of ancient conception, are each marked by a circle containing an animal's head. The three known continents are shown in a modified T configuration, except that "Europe" and "Affrica" are mistakenly transposed. In the centuries since Eratosthenes, Syene had shifted across the map a considerable distance from the banks of the Nile. Not far from the site of the Canary Islands are drawn the "Six Fortunate Islands, which are the islands of Saint Brendan." Spain has a rather triangular shape, and Italy's boot is poorly defined. England and Wales are depicted as an elongated island to which Scotland is joined by a narrow isthmus—a reminder of how little medieval cartographers knew or seemed to care about the topography of their own countries, much less the rest of the world.

The Hereford cartographer must have had access to reports brought back by travelers, probably some of the Crusaders. He includes on the map a number of well-traveled routes through Europe to the Holy Land, proving to Gerald R. Crone, map curator of the Royal Geographical Society, that "the practice of cartography was not entirely dead" in the Middle Ages. The routes are dotted with drawings of castles, shrines, and fortifications. Paris, Rome, and Antioch are the cities given great prominence, their names being written in red. Two crossings of the Alps are sketched through the scalloped pattern representing mountains.

Solinus would have recognized his hand on the Hereford landscape. Mythical creatures and monstrous races embellish great stretches of the map, the mermaid and unicorn and dragon mixing with the headless men of Africa, four-eyed Ethiopians, the man with a parasol foot, and others from the Solinus repertoire. To these are added drawings of the Garden of Eden, the Tower of Babel, Noah's Ark, and the Apostles out preaching.

As with all medieval *mappaemundi,* the Hereford map must be judged as a work of art, not of information. No attempt was made to draw it to scale or to fix the latitude and longitude of important places. With few exceptions, no effort was made to incorporate much of significance learned by contemporary travelers and mariners; instead, it contains little more than unsubstantiated legends and geographic information that had been around more than five hundred years.

But before we condemn all medieval cartography—with supercilious air, as Cosmas would say—we should recognize two important types of maps that evolved during the Middle Ages. They were as practical as the *mappaemundi* were theoretical. They were the road map and the sailing chart.

I t has been said that the Greeks measured Earth by the stars and the Romans by milestones. As their empire expanded, the Romans realized that maps were essential not only for military operations but for commerce with their far-flung citizenry. This led them to produce the first known road maps. Augustus Caesar put his son-in-law Marcus Agrippa (63–12 B.C.) in charge of a mapping project that took surveyors some twenty years to complete, a project that may be the source of the Hereford cartographer's belief that Julius Caesar had ordered the first world survey. Agrippa's survey involved thousands of kilometers of highways, extending from Britain into the Middle East. Along the way the surveyors placed milestones, the benchmarks of Roman roads. When the survey was completed, Agrippa prepared a master map of the world as it was known to the Romans. The circular map was engraved in marble and set up near the Roman forum. It became a source of information and, as it illustrated the breadth of empire, an object of Roman pride. Maps have served this dual function in all subsequent empires.

Copies of Agrippa's map were modified and enlarged over the years, and what may be one of them has survived—the Peutinger Table.* This example of Roman and early medieval road maps came to light in the sixteenth century, and was named after its owner, Konrad Peutinger, a map collector and the town clerk of Augsburg. Scholars believe it to be an eleventh- or twelfth-century copy of a third-century map.

The Peutinger Table, encompassing the Roman Empire, is constructed to fit into a long, narrow case, presumably so that it could be carried easily by soldiers and travelers. A parchment seven meters long and only one-third of a meter wide, it depicts the known world from Britain to the Ganges; it originally consisted of twelve sections, but the first one, showing most of Britain and the Iberian Peninsula, is now lost. Major roads are drawn in straight lines, with no attempt to show their true course or scale. Distances between places are written in. Smaller

---

*While this translation of *Tabula Peutingeriana* is common usage now, "Peutinger Map" or "Peutinger Road Map" are more informative renditions of the Latin.

A section of the Peutinger Table, showing Rome and environs, believed to be an eleventh- or twelfth-century copy of a third-century map

cities are depicted by drawings of little houses, and the greater cities—Rome, Constantinople, Antioch, Alexandria—are marked by vignettes or medallions. Temples, lighthouses, spas, bathing facilities, forts, and imperial residences are also noted. Sketches of trees mark the forest regions of Germany and Syria. Some Christian geography, such as "the desert where the Israelites wandered forty years," has been added, probably by the medieval copyist.

Even though this was a map intended for practical use, the shapes and sizes of countries bear little resemblance to reality. The northernmost lands are squeezed flat to fit in the narrow format, and Gaul is quite unrecognizable. The Rhine flows west, and the Nile flows west to east. But the roads and mileages are thought to be accurate enough to have been useful to a medieval traveler.

Another road map of sorts was developed by medieval Christians. This was the pilgrim's guidebook, forerunner of the tourist guide, which can be thought of as a prose map.

The earliest extant pilgrim guide, the *Itinerary from Bordeaux to Jerusalem,* was written by an unknown traveler in 330. Though it contained no maps, its directions and notations of places and distances were probably useful for any traveler between Europe and the Holy Land. The locations of hostels and places for changing horses and donkeys were described in detail. From Arles to Milan, according to the itinerary, was 760 kilometers, with twenty-two overnight stops and an average of three "changes" a day.

Some of these guidebooks were undoubtedly illustrated with crude maps, and by the thirteenth century such itineraries were often rendered in map form. An English monk, Matthew Paris, not only produced *mappaemundi* but also drew strip maps of pilgrimage routes within England (the main route from Dover to St. Albans to Durham) and of the route from London to southern Italy.

In purpose, if not exactly in presentation, the automobile road map—the map most familiar to the modern traveler—is the lineal descendant of the Peutinger Table and of the Paris strip maps. For an automobile race on Thanksgiving Day, 1895, the Chicago *Sun-Herald* published a simple map of the 86-kilometer course from Jackson Park to Waukegan and back to Lincoln Park. Frank Duryea tore the map out of the newspaper and referred to it frequently as he drove to victory. This is generally considered the first American automobile road map. Over the next two decades, however, motorists used route books that resembled the Bordeaux-Jerusalem *Itinerary* more than the Matthew Paris strip maps. Some of the route books included rudimentary maps, but their instructions were mainly written, to wit: "At the fork with saloon in the angle, take the right road leaving the trolley"; or "Cross the trolley tracks in the center of town and turn left at the first red barn."

By the 1920s, the growth of the highway network made the route books too voluminous for easy handling, and the increasing use of road-numbering systems made it feasible to represent all the necessary travel instructions on maps. Gradually, more information was added, including mileage between towns, construction, insets of city maps, indexes to towns, speed limits, toll bridges and ferries (plus their rates), airports, and even hiking trails. All this could be contained on a single sheet of paper, another example of the unexcelled utility of maps when it comes to summarizing and displaying vast amounts of information in a limited space.

But it was the Romans who first saw the need for road maps and the Christian pilgrims who first put the idea into practice. The appearance of the post-Roman road maps of the thirteenth century, moreover, marked a turning away from the literary, didactic, and ecclesiastical tradition of medieval mapmaking and a turning toward what we now take for granted as the elemental function of cartography—the communication of geographic facts and relationships in useful graphic form.

At about the same time, and in much the same evolutionary way, the nautical chart came into being. Travelers at sea, like those on land, saw the utility of translating oral and written records of sea routes into handy graphic form. This, more than any other development in the late Middle Ages, energized the craft of cartography and prepared the way for its greater role in world affairs.

A round the turn of the fourteenth century, when seafaring flourished anew throughout the Mediterranean, men known as cosmographers set up shop at every port of importance, at Amalfi and Pisa, Genoa and Venice. They may have been the first to pursue mapmaking as a full-time commercial craft.

The cosmographers studied the written descriptions of ports, coasts, and customary routes. They met incoming ships, talked with the captains, and recorded itineraries of countless voyages. They probably examined some kinds of maps roughed out by the mariners, those sketches showing coastal outlines, promontories to steer by, harbor plans, islands, shallows, reefs, and anything else that caught the sailor's practiced eye. Out of all this the cosmographers prepared the first accurate coastal sailing charts, the so-called portolan charts.

The charts probably began as graphic extensions of the written descriptions found in pilot books, the Greek *peripli* and the Italian *portolani*—hence the name "portolan charts." These were sailing and harbor guides summarizing the accumulated knowledge of generations of sailors. A periplus for the Mediterranean and Black seas, prepared in the fourth or fifth century, suggests the style and content: "From Phoenicus to Hermaea, 90 stadia; anchor with the cape on your right; the place has water in a tower. . . . It is 20 stadia from Hermaea to Leuce Acte; nearby is a low island which is distant two stadia from the land. Boats carrying merchandise can anchor here, entering by the west wind, but near the shore below the promontory is a wide roadstead for vessels of all kinds. Here is a temple of Apollo, a famous oracle. Near the temple there is water."

Major coastal nations continue the practice of producing pilot books. The National Ocean Survey of the United States issues a series of *Coast Pilots,* which are supplements to nautical charts and contain information on harbors, port regulations, sailing directions for entering harbors, and radio service.

If some of these early pilot books were supplemented by charts, the evidence is lost to history, but it would stand to reason that they were. A sea chart was mentioned in connection with the Crusaders' attack on Tunis in 1270. Marco Polo reported that he derived his knowledge of the coast of Ceylon from mariners' charts.

The oldest surviving portolan chart, dating from about 1300, is almost surely a copy of still earlier maps. The Pisan Chart, as it is called, seems too thorough and too faithful to reality over a wide area to have been the pioneering work of one man. The Italian cartographer Petrus Vesconte depicted the Mediterranean coastlines and ports with such fidelity that few significant alterations would be made on maps for the next three centuries. In this and subsequent portolan charts distances were drawn to scale, but no attempt was made to show parallels or meridians or to take into account the sphericity of Earth. Thus, they only partially met Ptolemy's standards for scientific cartography.

Over the next few decades the dockside cosmographers must have worked hard improving their charts and enlarging their scope. From single sheets of parchment they advanced to producing atlases containing from four to twelve large charts. From charts of the Mediterranean alone they moved to charts including the Black Sea in its entirety (and with almost as much accuracy), to charts showing the coasts of Britain and Ireland, and to charts showing, for the first time in existing cartography, Scandinavia as a peninsula. The Laurentian Portolano of 1351, a collection of eight sheets, covered the world from Britain and the Canary Islands to as far east as the middle of the western coast of India. Its remarkable rendering of the African coastline, particularly at Sierra Leone and the Cameroons, suggests that here too the cosmographers had access to some firsthand knowledge.

But they allowed themselves a few medieval flourishes. Important towns were illustrated by small pictures, and states were marked by appropriate coats-of-arms: Castile, the quartered field with a castle on red and a red lion on silver; Aragon, a red standard on a gold field;

Venice, the gold lion of St. Mark on red. The Red Sea was almost invariably colored red, while other bodies of water were either colorless or wavy-lined blue or green. Since the cosmographers paid little attention to interior regions, they sometimes let Gog and Magog, Prester John, and other ecclesiastical touches creep in for want of anything else to fill the blank spaces.

In one obvious respect, the portolan charts represented a return to Ptolemy. Nearly all of the charts were oriented north, not east. This was the first unmistakable effect of the extensive application of the magnetic compass as a tool in making maps and in using maps.

No one knows who discovered the magnetic property of the lodestone, who discovered that the stone's attractive power could be imparted to steel or hardened iron, or who discovered that the magnet could be used in determining geographic directions. The Chinese may have been the first, as early as the eleventh century, to use a magnetic needle for indicating direction. At about the same time, in 1269, a French Crusader, Petrus Peregrinus de Maricourt, the self-styled Peter the Pilgrim, recorded a description of a compass that could be a rudimentary version of those instruments so familiar to every Boy Scout. He wrote of a magnetized needle thrust through a pivoted axis and placed in a box with a transparent cover, a cross index of brass, a divided circle, and an external alidade provided with a pair of sights.

For some inexplicable reason tradition credits an unknown navigator of Amalfi with the invention, early in the fourteenth century, of the mariner's compass, though it is impossible to assess the truth of the story. What may have happened is that someone at Amalfi, probably in 1300 or earlier, got the idea of attaching a pivoted magnetized needle over a compass card, which is a circle divided into the principal points of direction.

In those days, even though ancient astronomers had learned to divide the circle mathematically, directions were not marked by degrees, but in terms of winds. Every experienced seaman, however untutored, knew his winds. They meant more to him than any number from 0 to 360. Since the ancients recognized 12 primary winds, at first the medieval compass cards were circles divided into 12 directional points. Later, these were reduced to 8 (corresponding to N, NE, E, SE, S, SW, W,

Catalan map of Europe,
fourteenth century

NW), but for greater precision the card eventually had 32 points—8 primary winds, 8 "half-winds," and 16 "quarter-winds." And the way they were drawn, often with artistic flair, reminded sailors of a 32-petaled flower. Hence the compass cards became known as "wind roses." To this day the Portuguese call a compass card a *rosa dos ventos,* a wind rose, and any modern cartographer wishing to affect an old chart places in one corner an elaborate and full-blown wind rose.*

The emergence of the mariner's compass wrought an important change in cartography: it was now possible—not that it always happened—to base maps on direct observation by means of an instrument. From the deck of a ship a captain taking compass bearings could readily determine the location of any coastal feature, a harbor or cliff or island, in relation to magnetic north. If he took enough readings at short intervals, he could trace the line of the coast as it inclined toward or away from north. The cosmographer at the next port would surely welcome the results. Although not precision mapping, it surpassed all previous techniques of estimation. Otherwise it is doubtful that the portolan charts could have shown the shapes and orientations of many islands and peninsulas with such remarkable reality.

The compass gave another new look to the portolan charts. The most distinctive feature of the charts was the network of rhumb (that is, constant-course) lines that crisscrossed all major bodies of water. These lines radiated from a number of wind roses placed at intervals over the map. Lines from the primary points, the half-winds, and the quarter-winds were usually drawn in distinguishing colors—often as not gold, green, and red. They were what made the charts worth whatever price the cosmographers exacted. A mariner would head his ship on one or several of those lines until he reached the next landfall.†

As the demand for charts increased, the centers of chartmaking spread beyond Italy to Majorca and Barcelona, where the influx of Arabs and Jews introduced fresh ideas and new energy to the craft. They brought with them a knowledge of eastern and African geography, supplementing the knowledge of northern and western shores being gathered by the increasingly vigorous Catalan seamen. The mapmakers

---

*Variations of the wind rose serve a practical purpose on some modern maps. On nautical maps they aid in the correction of the compass. On aeronautical charts they serve as protractors in estimating distances.
†Rhumb lines apparently reminded Shakespeare of the lines radiating at the corner of the squinted eye. In *Twelfth Night,* Maria says to Malvolio: ". . . he does smile his face into more lines than is in the new map with the augmentation of the Indies."

of Majorca were the first to place Gengen (Goa), Melli (Mali), and Tembuch (Timbuktu) in roughly accurate positions.

Abraham Cresques, a Jew of Palma, was a foremost member of the Catalan school of cartographers. He described himself as a "master of *mappaemundi* and compasses"—that is, a cartographer and instrument-maker. Captains and royalty bought his maps. And so he felt qualified to take on an assignment from Charles V of France to produce a new world map. In 1375, Cresques completed the famous Catalan Atlas.

The atlas, with its six brilliantly colored maps, combined the detail and accuracy of the portolan chart and the scope of the medieval *mappamundi*. Cresques apparently employed the best contemporary information. The atlas reflected a much improved conception of the peninsular shape of India, included some towns and rivers in China (from Marco Polo's accounts, which were full of bearings and distances), and depicted both the Blue and the White Nile. The shores of northern Europe were drawn with greater definition. Much more than previous portolan charts, the Catalan Atlas moved inland to show towns and rivers—and almost no mythical creatures.

With the Catalan Atlas of 1375, Beazley wrote, "we reach the zenith of medieval map-work." After more than a thousand years, people were learning how to make maps to be used. They were coming to realize that uninformed speculation was no substitute for careful observation. Gradually and not always too reliably, they were learning to transmit information, not theory, through maps.

And if maps could reproduce in recognizable fashion the familiar shores and guide men on their more ordinary journeys of commerce, perhaps they could serve a larger and more ambitious purpose. Perhaps they could be implements of exploration, inspiring and guiding and recording the extraordinary journeys to worlds unimagined even by Solinus. In their "spirit of critical realism," Gerald Crone said, "the Catalan cartographers of the fourteenth century threw off the bonds of tradition and anticipated the achievements of the Renaissance."

1492

Toward the end of the Middle Ages Europe was awakening with new energy and new ideas—and some old ideas revived. When printing with movable type was introduced in Germany in the mid-fifteenth century, many of the classical writings were rediscovered, translated, and printed, including Ptolemy's *Geography*.

Europe had the Arabs to thank for preserving Ptolemy for posterity. For centuries Muslim scholars kept the flame of learning from flickering out, and during this period they looked upon European Christians as little more than barbarians. In the tenth century, the Muslim geographer Masudi wrote that Europeans were dull of mind and the "farther they are to the north the more stupid, gross, and brutish they are." In contrast, the Muslims studied and added to classical learning. Their scholars had, among other things, treasured Ptolemy's book of geography and used it in preparing their own world maps, travelogues, and itineraries for pilgrimages to Mecca.

Early in the fourteenth century, a Christian monk who frequented the secondhand bookshops of Constantinople came upon a copy of Ptolemy's text. His maps, though, had disappeared. So the monk commissioned local artists to reconstruct them from geographic coordinates recorded in the text. It would be another century, however, before the book and maps reached the heart of Europe, the gift of a Turkish diplomat visiting Italy.

Then maps based on Ptolemy proliferated and assumed an authority once accorded only to Holy Writ, and for all their shortcomings (the Mediterranean was too long, Italy sloped too easterly, Ceylon was bigger than India, and Eurasia was much too extensive) the Ptolemy maps offered a much more realistic and useful view of the world than the medieval *mappaemundi*—a case of a step backward being a step forward. Ptolemy's revival also proved influential by stimulating debate about the Earth's circumference and the relative dimensions of the continents and the oceans. The European mind, thus broadened by Ptolemy and stirred by the fantastic tales of Marco Polo, whose writings had grown in popularity, stretched beyond the shores and familiar seas. Europe was heading inexorably toward 1492. A reawakened Europe, Renaissance Europe, was entering the Age of Discovery.

So much has already been written about the Renaissance and the Age of Discovery that, for the purposes of this book, it should be necessary only to focus on two particular questions: (1) What influence did maps and the mapmakers have upon the Age of Discovery, especially on Christopher Columbus? (2) What influence, in turn, did the Age of Discovery have upon the mapmakers and their maps?

P rince Henry of Portugal exemplified Europe's new outlook at a time when it was still part medieval and part modern.
Henry was born in 1394, the third son of John I and his English wife, Philippa of Lancaster, the daughter of John of Gaunt. After a youthful excursion battling infidels in Morocco, Henry established himself in 1419 as governor of the Algarve, the southernmost province of Portugal. There, on the rocky, windswept promontory at Sagres on Cape St. Vincent— "Where endeth the land and beginneth the sea," as the medieval Portuguese poet Camões wrote—he founded a small court of his own, to which he attracted assorted navigators, astronomers, shipbuilders, instrument makers, and cartographers. They collected and studied the known portolan charts. They exchanged information with returning seamen. Under Henry's auspices, they designed more seaworthy sailing vessels, the two- or three-masted ships with lateen sails that became known as caravels.

As one of Henry's chroniclers, Gomes Eanes de Zurara, wrote: "His aspect, to those who beheld him for the first time, was severe; when anger carried him away his countenance became terrifying. He had force of mind and acute intelligence in a high degree. His desire to accomplish great deeds was beyond all comparison."

At Sagres, Henry secured his place in history, a king's third son who might have remained an obscure prince but who came to be immortalized in history as Prince Henry the Navigator. Although he himself made no voyages of discovery, Henry's fame rests on his ability to organize and encourage others to undertake such expeditions. Exploration has been called planned discovery, and in that sense, according to J. R. Hale, a historian of Renaissance exploration, Henry was "the first rational organizer of exploration as an expanding reconnaissance based on co-operation between pilots at sea and experts at headquarters." This, added Hale, was "the first example of a program of discovery being put into effect."

Portuguese seamen were sailing to the Atlantic island groups, the Azores, the Madeiras, and the Canaries, and were venturing as far down the west African coast as Cape Bojador, where the continent begins its westward bulge. These sea routes had been plotted on the portolan charts of the day. And so Henry decided to push beyond the known, launching one expedition after another to explore the coast of Africa south of Bojador.

His decision may have been influenced by a reading of Marco Polo. Those tales of gold and spices whetted many a European appetite. But, since trade over the land routes to the east had proved costly and increasingly risky, because of shifting political tides and particularly Islam's expanding and threatening strength, there was growing interest in finding an alternate route to the east by sea. Ptolemy's maps suggested a possible direct route from east Africa across the Indian Ocean. The problem was getting to east Africa, where Islam held sway. Ptolemy's maps seemingly closed the gates on another possibility, sailing down the Atlantic round the tip of Africa and out into the Indian Ocean. Ptolemy had insisted that Africa joined some great southern continent, Terra Incognita or Terra Australis, and whatever Ptolemy had said the Europeans tended to believe. It is still not certain whether Prince Henry consciously sought to reach India and Cathay by circumnavigating Africa, but all his efforts for the rest of his life were directed toward exploring ever southward down the west coast of Africa. Neither is it known whether Prince Henry or any of his contemporaries ever seriously considered the alternative, sailing west across the Atlantic to reach Asia.

Henry had another motive for exploring the African coast. Besides the quest for riches, he sought to spread Christianity. Henry was a grand master of the Order of Christ, the supreme order of the Pope, and he dipped into the order's funds to help finance expeditions as latter-day crusades. He had an idea, one of his medieval legacies, that he might establish contact somewhere in Africa with the fabulous Prester John.

In another sense, however, Henry and his captains proceeded in the face of medieval superstition. They dared test the belief that anyone who attempted to reach the equator would run into boiling waters and certain death. They were no doubt inspired by Henry's words: "You cannot find a peril so great that the hope of reward will not be greater." Henry's first objective was to explore beyond Cape Bojador to see what lay below Africa's bulge. After several unsuccessful attempts, the cape was rounded in 1434. Eventually, Henry's captains, venturing farther southward, cape by cape, river by river, always hugging the coast, were able to produce the first accurate maps of the west African coast. They then began trading for gold and slaves, which covered costs and brought home to Lisbon the economic potential of exploration.

By the time Henry died, in 1460, Portuguese caravels had charted the African coast as far south as Sierra Leone, past the site of Dakar, only 10° north of the equator. Though no great progress was made for a decade or more after Henry's death, Portuguese ships continued to explore and trade along the African coast. In 1473 Lopo Gonçalves safely crossed the equator. No boiling waters, no harm. A medieval bugaboo had been banished from modern thinking.

Then, in 1481, John II ascended the throne of Portugal. He was the great-nephew of Henry, a scholar and a vigorous man very much imbued with the idea of finding a sea route to Asia. John had probably examined maps like Fra Mauro's, which, contrary to Ptolemy, showed a navigable route around the southern tip of Africa. Such ideas spurred him on.

Advances in cartography were encouraged by John. He had geographers and mapmakers in his personal household. For some time Portuguese captains had known how to shoot the Sun and the North Star to determine latitude, and they returned from their African voyages with detailed sketches of the coast, showing reasonably accurate latitudes. Each time a new Portuguese expedition set forth it carried charts with a latitude grid, something other nations had yet to develop. In fact, other nations sought Portuguese pilots to accompany their ships.

One of John's captains, Bartholomew Dias, finally rounded the tip of Africa in 1488. Because of his stormy passage, he called the tip the Cape of Storms. But John read a greater significance into the achievement and, realizing that Ptolemy had been proven wrong and that a new sea route to Asia had been discovered, named it the Cape of Good Hope.

Then, in one of the most courageous voyages of all time, Vasco da Gama sailed from the Cape Verde Islands round the Cape of Good Hope and, in 1498, reached the western coast of India at Calicut. Da Gama had discovered a sea route to Asia—and it really was Asia.

Hale, the historian, compared da Gama and Columbus as discoverer-chartmakers and reached the following conclusion: "As a practical navigator, Columbus' uncanny ability to feel his way repeatedly to a given landfall puts him in a class of his own. On the theoretical side, however, Da Gama was much his superior as the charts and maps, which represent their discoveries, show. To follow Columbus' latitudes resulted in errors of more than ten degrees, while the Cantino map which followed those of Da Gama contains no error in the African coastline of as much as two degrees."

But da Gama had reached India after Columbus had reached the "Indies." One had discovered a new sea route to Asia, the other thought he had—but had really discovered something infinitely more significant, a "new world."

No maps used by Columbus are known to have survived the centuries. But a terrestrial globe constructed in 1492 by a German named Martin Behaim has survived, and it portrays Earth much as it must have been perceived by Columbus. There appears, however, to have been no connection between Behaim and Columbus, except that they very likely had access to most of the same sources of information in forming their similar world views.

It is well-nigh impossible to winnow fact from braggadocio in Behaim's life. He was from Nürnberg, where he was born about 1459—that much we know, as well as that he was the son of a patrician merchant at a time when Nürnberg was a flourishing city of commerce and art, of science and mapmaking. It was intended that he follow a commercial career, and toward that end he was sent as a young man to the Low Countries for study and apprenticeship. His travels led him eventually to Lisbon, probably in 1484, when it was alive with talk of maritime trade, African coastal exploration, and the search for a sea route to the East.

Behaim insinuated himself into the heart of Lisbon's commercial and intellectual life. Passing himself off as a pupil of Regiomontanus, the great Nürnberg mathematician, and as an expert in navigation, Behaim secured an appointment to John II's council of nautical advisers. Marriage into a prominent family opened even more doors to fame and possible fortune. It is believed that he was knighted at this time. It is known that he visited the Azores and the Canaries, and he may well have participated, as he claimed, in an expedition down the African coast. But to hear Behaim tell it, he was the navigational genius without whom Portugal might never have been the far-ranging maritime power that it was.

Behaim must have found a receptive audience upon his return to Nürnberg for a visit in 1490. For when his fellow townsmen heard his stories, the city council prevailed upon Behaim to construct a terrestrial globe on which recent Portuguese and other discoveries should be represented. Construction of the globe took slightly less than a year and cost the city something less than $75. The globe was commonly called Behaim's *Erdapfel*—literally, "Earth apple."

According to a written account by one of the city councilmen, the spherical shell, 50 centimeters in diameter, was made of a composition material fashioned over a mold. An artist named Glockenthon drew the map on strips of parchment, which were then fitted and pasted onto the sphere. The finished globe stood on a wooden tripod.

Two great circles are laid down on the globe—the equator, divided into 360°, and the ecliptic, studded with signs of the zodiac. The tropics and the Arctic and Antarctic circles also appear. The only meridian is drawn from pole to pole 80° west of Lisbon. The sea is colored a dark blue and the land a bright brown or buff with patches of green and silver, representing forests and regions supposed to be under ice and snow. Glockenthon sketched in 111 miniatures, drawings of kings seated on thrones, saints, missionaries, sailing ships, elephants and camels and fishes, and a merman and a mermaid, near the Cape Verde Islands. The globe was produced originally in six colors and contains more than 1,100 place names.

Behaim's *Erdapfel* was not the first globe—the ancient Greeks are credited with that—but it is the oldest terrestrial globe extant. By the time of Behaim's death back in Lisbon, in 1507, globe-making had become a widely practiced skill. Some of the globes were made of metal, on which the map was engraved. Others were of wood with a drawn or printed map glued on. Sixteenth-century cartographers soon established mathematical rules for the construction of globe gores, those sections of parchment or paper cut in such a precise way as to fit on a round surface. The globe-makers were responding to a new demand, created by the rush of geographic discovery, for more nearly accurate representations of the Earth's surface than could be set forth on a plane map.

But of primary interest here is that Behaim's globe represents a view of the world held by many of Columbus's contemporaries and, apparently to a large extent, by Columbus himself. Most certainly the globe lays to rest that tired old myth about everyone in Europe, except Columbus, believing the world to be flat. All educated people in western Europe assumed by 1492 that Earth was a sphere.

In designing his globe Behaim is believed to have relied considerably on a large world map by, or after, Henricus Martellus Germanus, a prominent cartographer of Florence. Martellus worked with Francesco Rosselli, the first known specialized map printer and dealer. Florence had become a center of map printing, and Rosselli's workshop was among the most advanced and prosperous in cartography; he was, as R. A. Skelton wrote, "a Rand McNally or Bartholomew of his day." Like most world maps of the time, the maps of Martellus, drawn about 1490, were amended and expanded versions of Ptolemy's world picture, with some results of contemporary discoveries grafted onto the classical and medieval views of the world.

Consequently, Behaim's globe retained many medieval touches, and this was nowhere more evident than in the inclusion of mythical islands out in the Atlantic. Behaim, like mapmakers before and since, could not abide open spaces. Beyond the Azores, which had been well charted by the Portuguese, he placed the island of Antilia—which meant "island opposite" Portugal, and was also called the Isle of the Seven Cities. This was such a standard feature of fifteenth-century maps that Columbus seems to have had every expectation of finding it—and possibly using it as a way station—en route to the Indies. An entry in Columbus's journal indicates that he made several vain attempts to sight the island about where Behaim had placed it, on the latitude 28° N.

Behaim's globe followed Ptolemy's concepts closely, more so than was justified in the light of discoveries in the latter fifteenth century that Behaim must have heard of in Lisbon. In one important and intriguing instance, however, Behaim, like Columbus, exaggerated Ptolemy's Asia, extending the reach of Asia farther east than it was generally believed to be in 1492. Behaim's Asia was stretched to make room for lands described by Marco Polo, particularly Cipangu, or Japan. Behaim, also like Columbus, believed that a very narrow ocean lay between the western shores of Europe and the eastern edge of Asia.

Thus, Behaim's "Earth apple" of 1492 shared, in several important aspects, prevalent geographical notions that were false and convinced

Christopher Columbus that he could safely cross the Atlantic to "the land of India and the great island of Cipangu and the realms of the Grand Khan."

W hen or where Columbus conceived the plan for his "Enterprise of the Indies" remains a mystery.

He was born in 1451 in Genoa, the son and grandson of humble but respected weavers. Like many young men of Genoa, he put out to sea, first in the Mediterranean and then in the Atlantic. On ships out of Lisbon he ventured north of Iceland in the winter of 1476–77 and also sailed down the African coast.

According to Samuel Eliot Morison, the admiral's authoritative and admiring biographer, there is no evidence that Columbus heard anything on his Iceland voyage of the earlier Norse expeditions to North America or of any English voyages out of Bristol. But the possibility that he did know of the Norse voyages has never been ruled out to everyone's satisfaction and remains a source of controversy. The problem for historians is not who reached America first: that Norsemen sailed to some part of America around the year 1000 has been accepted as fact since the early nineteenth century. Rather, the problem is how to assess the achievement of Columbus: did he sail into completely mysterious and uncharted waters, in courageous pursuit of an original vision, or did he somehow get the idea and perhaps some guidance from the Norse?

In Columbus's behalf it has been said that, while Leif Ericsson may have been first across the Atlantic, the Norse produced no maps that entered the mainstream of European thought, and an unmapped discovery is of little value to succeeding generations. However, this explanation has suffered what seem to be two sharp blows in recent years—when the Vinland Map came to light and when the John Day letter was discovered.

Few maps have provoked more scholarly excitement and public tumult than the Vinland Map. A sheet of patched and worm-eaten parchment, the map was acquired in 1957 by a Connecticut book dealer while on a buying trip to Barcelona. It showed Europe in clear detail, as well as the traditional distortions of Asia and Africa and the usual mythical Atlantic islands. In the upper left-hand corner, however, was a surprisingly accurate outline of Greenland and, west of there, a large island labeled "Vinlanda Insula." Above the island there was an inscription in Latin, which, translated, reads: "By God's will after a long voyage from the island of Greenland to the south toward the most distant remaining parts of the western ocean sea, sailing southward amidst the ice, the companions Bjani and Leif Ericsson discovered a new land, extremely fertile and even having vines, the which island they named Vinland."

For about eight years scholars at Yale University quietly examined the map and subjected it to such scientific tests as were then available. The Gothic cursive script was the style of mid-fifteenth-century monks. The bull's-head watermark was similar to one used by a mill operated in the Rhine Valley in the mid-fifteenth century. The geographical errors corresponded closely to those committed by Andrea Bianco in his *mappamundi* of about 1436. All the evidence pointed to about 1440 as the date the map was drawn.

If genuine, the Vinland Map was the only pre-Columbian map to represent the Norse discoveries in America or to make Greenland an island. It was incredible how well the Isolanda, Gronelanda, and Vinlanda on the map corresponded in outlines and relative positions to Iceland, Greenland, and Baffin Island on modern maps. No other map for another two centuries would show Greenland as an island and not a peninsula of Scandinavia.

In 1965, a few days before Columbus Day, Yale announced the existence of the map, calling it "the most exciting cartographic discovery of the century." But at an international conference the next year scholars found themselves divided on the question of the map's authenticity.

There the matter stood until 1974, when Yale made another announcement. The Vinland Map, the university said, was an elaborate and amazingly skillful forgery. Newly devised chemical tests had showed that the yellowish-brown ink used to draw the map contained large amounts of anatase. This was a form of titanium dioxide, which was not invented until the 1920s. The only thing mid-fifteenth-century about the map, the university said, was the parchment and many of the geographic concepts. But in 1996, with the publication of a revised edition of *The Vinland Map and the Tartar Relation,* the map's defenders mounted a counterattack with new evidence. They even offered an alternative explanation for the map's suspect ink chemistry and new radiocarbon dating, which appeared to show that the parchment could be from the mid-fifteenth century. The last has not been heard about the Vinland map.

Another source of confusion about what Columbus could have known has been the John Day letter, which was published in 1956. Louis

André Vigneras found the letter, written in Spanish and undated, in some Spanish archives. It was addressed by John Day, an English merchant, to a Spanish official, the "Almirante Mayor," who is believed to have been Christopher Columbus, for he sometimes went by that title. Historians concluded from clues in the letter that it must have been written in December 1497 or early 1498.

In the letter Day reported a successful voyage from Bristol by an unnamed explorer—presumably the John Cabot expedition of 1497— who reached Newfoundland and other coastal points of North America. But the most intriguing passage is the following: "It is considered certain that the cape of the said land [that found by Cabot in the 1497 voyage] was found and discovered in other times by the men of Bristol who found 'Brasil' as your Lordship knows."

In other times. What exactly did Day mean? The Spanish words *en otros tiempos* could also be translated "in times past" or "in past times." Did that imply a few years, or many years?

David Beers Quinn, a historian at the University of Liverpool, interprets the letter and related historical accounts in such a way as to suggest that English explorers may have reached America before 1492 and that, judging by the phrase "as your Lordship knows," Columbus may have known about their voyages. Two ships, Quinn pointed out, did leave Bristol in 1481 "to serch and fynde a certain Isle called the Isle of Brasile." There is no record that they failed. Could they have discovered some part of America? Other accounts spoke of caravels, possibly transatlantic, leaving from Bristol from about 1490 onward.

Morison dismisses the Day letter out of hand. He wrote: "There is no English or other evidence of the alleged pre-Cabotian discovery of America. The probability is that Day picked up the usual post-discovery yarns about 'of course, we know all about that, old So-and-So went there years ago,' etc. and that he thought it would please Columbus to report that Cabot's discovery was no part of the Indies but had been discovered before."

Quinn concedes that the letter is "suggestive rather than conclusive." But he concludes that there is some reason to doubt that Columbus sailed into the completely unknown. If the news of a prior English discovery "reached Christopher Columbus before he sailed to America in 1492," Quinn wrote, "then the English discovery could have been one of the more significant pieces of information which led Columbus across the Atlantic with the conviction that there was land to be found within the range of distance which he anticipated. The Isle of Brasil as found by the English can therefore join . . . the traditions of the Vinland voyages among the data which Columbus could have had at his departure."

According to his son Ferdinand, however, Columbus was influenced by a reading of the classics. From these he learned that Aristotle had written that it was possible to cross the ocean from Spain to the Indies in a comparatively few days. He read and pondered the prophecy in Seneca's *Medea:* "An age will come after many years when the Ocean will loose the chains of things, and a huge land lie revealed; when Tethys will disclose new worlds and Thule no more be the ultimate."

The Book of Ser Marco Polo and *Imago Mundi* by Pierre d'Ailly, a French theologian and cosmographer, were apparently Columbus's bedside reading. In 1410, d'Ailly had written: "The end of the habitable earth toward the Orient and the end of the habitable earth toward the Occident are near enough, and between them is a small sea." And: ". . . it is evident that this sea is navigable in a few days with a fair wind." Columbus underscored or made marginal notes wherever d'Ailly discussed the extent of Asia or the narrowness of the Atlantic. Like Ptolemy, though more so, d'Ailly underestimated the circumference of Earth and exaggerated the east-west length of Eurasia, which was exactly what Columbus wanted to read and wanted to believe.

The rediscovery of Ptolemy, by stimulating thought about the size and configuration of the world, had already awakened others to conceive of a new paradigm of global geography. It was the talk of Florence, a center of intellectual ferment where savants studied Ptolemy and Strabo, pored over new maps, and interrogated returning travelers from as far away as Sumatra and Java. Pondering this flow of new and rediscovered knowledge, the Florentines decided that the ancients as far back as Homer had been unduly restrictive in describing the world as an island encircled by a river called Ocean. They instead advanced a worldview that had never been clearly stated before, a theory inviting the test by action.

In a study of geographical thought in fifteenth-century Florence, Thomas Goldstein, a scholar of the Renaissance, described the new premise as the "intellectual conquest of the earth" and "nothing less than a complete revision of the basic concept of the earth." According to this view, the ocean need not be a barrier to travel but could be the medium for communication between peoples in faraway lands. Moreover, Goldstein wrote in 1965, the theory postulated "a second equally revolutionary premise—equally subversive of the entire medieval tradition of geography: that the navigable Ocean Sea included the Southern Hemisphere." If this proved to be true, it "meant the decisive abandonment of the traditional (ancient as well as medieval) concepts of the orbis terrarum—the idea that a severely foreshortened three-continent landmass (Europe, Asia, and Africa) was forever confined within its own limits by an all-encircling Ocean Sea, conceived of very much as we today look upon outer space, as another orbit, believed to be by its very nature inaccessible to the efforts of man."

This revolutionary concept spread from Florence and intrigued the seafarers in Lisbon, where it came to Columbus's attention. The principal agent of transmission was Paolo dal Pozzo Toscanelli, a physician, mathematician, and geographer in the circle of Florentine scholars. In 1474, by then an old man, Toscanelli wrote a letter to the king of Portugal, Afonso V, through a friend in Lisbon. He also sent along a chart, now lost. In the letter he explained the reasons for thinking that the shortest sea route to Asia lay to the west, straight across the ocean. Like Ptolemy, only more so, Toscanelli calculated the world down to a smaller size than it really is. Ships sailing westward from the Canary Islands, he contended, would reach the "nobel island of Cipangu" in 5,500 kilometers and Quinsay (the modern Hangchow) in 9,200 kilometers. Toscanelli suggested that the voyage could be broken by a stopover at "the island of Antilia which is known to you."

Later, after he was shown a copy of the letter, Columbus wrote to Toscanelli requesting more information. The year was probably about 1481. The Florentine replied with a letter praising Columbus's "great and noble ambition to pass over to where the spices grow" and with another chart illustrating his concept of the ocean's width. (The chart has been lost.) Another letter from Columbus to Toscanelli elicited no additional details. But Columbus felt encouraged, for he now had a letter and a map from an eminent scholar endorsing his own theory.

While mustering support from classical, medieval, and contemporary sources, Columbus made calculations of his own, and some nimble arithmetic it was. He gave his proposed venture the benefit of every doubt, which led him to make a gross miscalculation.

First, Columbus stretched the length of the Eurasian continent, from Cape St. Vincent to eastern Asia, from Ptolemy's already inflated 177° to 225°. To that Columbus added another 28° from the discoveries of Marco Polo, plus 30° for the reputed distance from China to the east coast of Japan. Columbus had now accounted for 283° of land. Since he proposed to sail to Asia from the Canaries, which were 9° west of Cape St. Vincent, Columbus confidently concluded that he had only 68° of ocean to cross before reaching Japan.

Second, Columbus scaled downward the estimated length of a degree of longitude, making it 25 percent smaller than Eratosthenes had calculated and 10 percent smaller than Ptolemy had taught. The value of a degree should have been easy enough to figure out—simply take the circumference of the Earth and divide by 360. But the true circumference, and thus the length of a degree, had been a subject of uncertainty and debate for centuries. Eratosthenes had been close to the truth with his estimate, but few who followed chose to accept his calculation, and particularly not Columbus. He favored an estimate by a medieval Muslim geographer, Alfragan, who found the degree to be 56⅔ Arabic miles—which would be 76.1 statute miles, or 121.8 kilometers. Columbus, however, assumed that Alfragan was using the much shorter Italian mile and concluded that the length of a degree was only about 52 miles, or 83 kilometers.

Third, Columbus estimated that if a degree of longitude on the equator measured about 83 kilometers, it would be only 74 kilometers on the latitude of 28°, which was where he proposed to cross the ocean.

Thus, as Columbus figured it, there lay only 4,300 kilometers of open water between the Canaries and Japan. This was 1,200 kilometers less than the Toscanelli estimates and more than 15,000 kilometers less than the actual air distance between Spain and Japan. Columbus had in his own mind reduced Earth to a size more congenial to his plan of action.

Having armed himself with scholarly endorsements and a favorable set of calculations, a view of the world which he is believed to have committed to a map, Columbus presented his first formal proposal in 1484 in an audience with John II of Portugal. John dismissed the plan because Columbus demanded a stiff price (noble rank for himself and his heirs and one-tenth of the wealth of Cathay) and because so much of the plan seemed to depend on the word of Marco Polo and on the doubtful existence of Japan. Learned men in Portugal, with some justification, considered Polo to be no more than an entertaining liar; nor had they been as moved by Toscanelli's map as had Columbus. Besides, John had his money on finding a route via the south of Africa.

Representations to the English and French monarchs made by Columbus's brother, Bartholomew, a Lisbon chartmaker, proved equally fruitless. It was eight years before Columbus finally persuaded Isabella of Spain to take the chance.

Those who had opposed Columbus had actually been more right than he. No one in the late fifteenth century could have sailed the distance between Spain and Japan. If it had not been for the totally unexpected discovery (assuming he knew nothing of Vinland) of an intervening continent, America, Columbus might have sailed on into oblivion. But, without knowing it, Columbus had made a quite accurate calculation of the distance from Spain to the West Indies, making more understandable his mistaken belief that the landfall he reached on October 12, 1492, was an island off the coast of the same Indies that Marco Polo had made known and that had been Columbus's destination and his vision.

That vision, J. R. Hale observes, set Columbus apart from his contemporaries. In his book *Renaissance Exploration,* Hale wrote:

> The pilot on an ordinary trading voyage in familiar waters needed only the map-type which had evolved from medieval needs: the portolan. But the portolan, being essentially a record of what was known, was no use to a voyager facing the unknown; he needed an indication of what he would find beyond portolan range and it is here that the humanist study of cartography—speculation in terms of modified Ptolemaic world maps—supplemented the medieval sea chart with Renaissance theory and made exploration (planned discovery) possible. The theories were often erroneous; sometimes a theory postulated a continent that was not there—Terra Australis; all were ignorant of a continent that was there—the Americas; but without a theory mariners like Columbus and Magellan would not have sailed, could not have found backers. This is the chief point, perhaps it is the only point, where an aspect of the intellectual Renaissance intervened to help a process that was leading naturally from medieval international trade to interoceanic commerce.

In *The Measure of Reality,* Alfred Crosby places this transformation in the context of a growing enthusiasm for measurement in Europe, a shift "from thinking of the world in terms of qualities to thinking of the world in terms of quantities." The change was reflected in the new mechanical clocks, music, and perspective painting, as well as more geometrically precise maps. "The choice of the Renaissance West," he observed, "was to perceive as much of reality as possible visually and all at once, a trait then and for centuries after the most distinctive of its culture."

Unlike their medieval predecessors, who seemed not to make the connection between discovery and mapmaking, often as not ignoring the fragmentary travelers' reports that were available, the Renaissance cartographers were quick to respond to the new discoveries by Columbus and his many successors.

The first Renaissance maps were, however, transitional. The cartographers continued to dot the Atlantic with imaginary islands; one of them, O'Brasil or Brazil, was not removed from some charts until the mid-nineteenth century. They were also reluctant to abandon Ptolemy altogether. Still believing in a more or less Ptolemaic size of Asia and of the globe, they bent and squeezed the American discoveries and left no room for an ocean the size of the Pacific, which Balboa would not see until 1513.

As a former mapmaker, Columbus drafted numerous charts of the islands and mainland he sighted during his four voyages—1492, 1493, 1498, and 1502. His principal tools were a compass, quadrant, and traverse table. A Spanish voyage to the New World in 1499 is known to have relied on a chart of the South American mainland that Columbus had sent home. But only one of the Columbus charts survives. Morison wrote: "As the new lands were more minutely explored, new charts were made and the Admiral's destroyed, all save one rough sketch of Northern Haiti. This has a sureness of touch that fully sustains his reputation as a cartographer, and gives us a keen sense of our loss in having none of the larger charts that he presented to the Sovereigns."

The first world map to incorporate the Columbian discoveries is believed to have been one produced in about 1500 by Juan de la Cosa, who sailed on Columbus's second voyage* and accompanied subsequent expeditions until his death at the hands of Indians in South America. La Cosa's map was drawn on oxhide in bright colors and followed the style of the portolans, having the spider-web rhumb lines. On the map La Cosa illustrated the first three Columbian voyages, showing in considerable detail all the West Indies islands charted by Columbus on the second voyage plus Puerto Rico, Jamaica, and Cuba. La Cosa's treatment of Cuba is particularly interesting. Persisting in the belief that he had reached Asia, Columbus insisted that Cuba was part of the mainland of Cathay and forced his crew to take an oath to that effect. La Cosa, nevertheless, depicted Cuba as an island. But La Cosa did join America and

---

*A Juan de la Cosa was master of the *Santa Maria* on Columbus's first voyage and also sailed on the second voyage. But scholars are not sure whether these were one and the same person. The map was prepared by the Juan de la Cosa known to have made the second voyage.

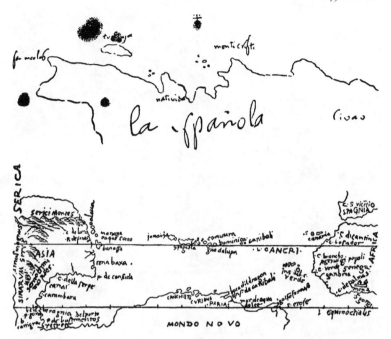

Two sketches: the northwest coast of Hispaniola by Christopher Columbus (top); the new and old worlds by Alessandro Zorsi (bottom)

Asia as one continent without an intervening ocean. The map also embodies the results of John Cabot's reported voyage in 1497 to North America, at Labrador, and the coastlines of South America as seen by Columbus and others. In all, La Cosa was able to piece together quite a bit of heretofore unmapped territory in a remarkably short time.

A second map was drawn in Lisbon in 1502 for Alberto Cantino, acting as an agent for the Duke of Ferrara, a prince fascinated by the new discoveries. The so-called Cantino map (the actual cartographer is unknown), a work of colorful beauty, reflects its Portuguese origins in depicting with reasonable accuracy the outlines of Africa, Madagascar, India, and Ceylon—geography that owed much to the Portuguese explorers. As for the New World, the map is less reliable than La Cosa's in its representations of the South American coastline and the West Indies. The cartographer, assuming the island chain to be an extension of the imaginary Antilia, labeled the islands *Las Antillas de Rey de Castella*—whence the names today for the Greater and Lesser Antilles.

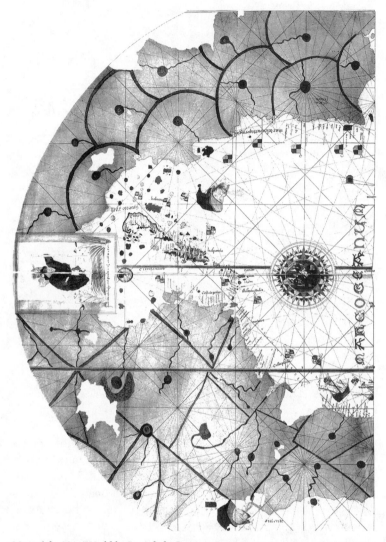

Map of the New World by Juan de la Cosa, 1500

Significantly, the cartographer of the Cantino map believed that the islands Columbus had discovered had nothing to do with Asia. Many others in Europe were coming around to the same belief, but not Columbus. The Alessandro Zorsi sketch maps, probably drawn about 1509 from information supplied by Bartholomew Columbus, show scrawled across the interior of Nicaragua the word "Asia." Though he did write of an "other world," when describing Venezuela,* Columbus apparently went to his grave, in 1506, resolute in the conviction that he had at least reached the fringes of his goal. The vision, which carried him so far, had blinded him to the magnitude of his real discovery. This probably cost Columbus one measure of fame, the honor of having the new lands bear his name. That honor went to Amerigo Vespucci, who hardly deserved it.

Born in Florence in 1454, Vespucci became prominent in Seville as a banker and ship's chandler, though he widely proclaimed his additional abilities at navigation; his ceaseless self-glorification brings to mind Behaim. He claimed to have made a major voyage in 1497, but no one believes the story. Somehow, perhaps with his purse, he did join a Spanish expedition to the New World in 1499 and returned even more full of himself. Through a royal appointment he made two Portuguese voyages under the command of Gonçalo Coelho to reconnoiter Brazil in 1501 and 1503.

After the voyages Vespucci wrote two letters to a friend in Florence, and they assured him of a certain immortality. The letters, published in 1504 and 1506, were widely read and translated into several languages. Vespucci was capable of vivid descriptions of scenery and native customs, including sexual practices, which no doubt improved readership. In one letter, he wrote typically of his navigational prowess: "I was more skillful than all the shipmasters of the whole world." In the other, significantly called *Mundus Novus,* he wrote that he surely had found "what we may rightly call a New World." As far as can be known, Vespucci was the first to make explicit references to the newly discovered lands as a New World.

A German cosmographer, Martin Waldseemüller, happened at the time to be preparing a new edition of Ptolemy. He was a professor in a small town, St.-Dié in Lorraine, but was well informed about the new discoveries and knew all about Vespucci. The Waldseemüller map of 1507, a woodcut print, therefore showed the Earth very much in accord with Ptolemy, except in the case of the New World. There he depicted a coastline from Newfoundland to Argentina. The northern continent is hopelessly narrow, but the southern continent is somewhat angular, giving a vague hint of its true shape. On the main map the two continents

---

*Columbus, seeing the freshness of waters in the Gulf of Paria, correctly concluded that a river (the Orinoco) flowed out of the "very great continent." Then he lapsed into a medieval cast of mind, declaring that the river flowed from the earthly Paradise.

are separated by a strait at Panama, though a secondary map by Wald-seemüller has them joined at the Isthmus of Panama. More important, this was presumably the first map to show the two continents clearly separated from Asia.

With the writings of Vespucci fresh in his mind, Waldseemüller wrote across the southern continent, in the region of Brazil, the word "America." In an introduction to the Ptolemy book, Waldseemüller explained himself: "Since another, fourth part [of the world] has been discovered by Americus Vesputius . . . I do not see why anyone should object to its being called after Americus the discoverer, a man of natural wisdom, Land of Americus or America, since both Europe and Asia have derived their names from women."

And America it became; such is the power of maps. This map has been called "the baptismal certificate of the New World."

Although Waldseemüller dropped the name from a 1513 edition of the map,* other cartographers had already given it such wide currency that there was no erasing America from the map. A 1515 globe applied the name to the northern continent, and by the mid-sixteenth century maps were showing a North and a South America. Vespucci, who died in 1512, would have liked nothing better.

The idea of circumnavigating the Earth apparently came to Ferdinand Magellan after he examined charts in Lisbon and listened to his geographer friend Guy Faleiron. The charts indicated the presence of a strait between the southern tip of South America and the shores of the supposed Terra Australis. Through that strait, he decided, a ship could pass from the Atlantic to the Pacific and thence to Asia. Had Magellan known how far south the strait and how wide the Pacific he might have changed his mind and stayed home.

Magellan, a native of Portugal, where he was born about 1480, had little trouble persuading the Spanish crown to sponsor the expedition. Since Columbus, monarchs had paid more attention to the suggestions of mapmakers and would-be explorers. And Magellan's plan accorded with Spain's continuing desire to find a western sea route to Asia. Thus Magellan embarked in 1519 on one of the most heroic voyages of all time.

His discovery of the strait was no mean feat, as it was not a direct

*Waldseemüller's 1513 edition was also notable as the first to separate modern maps from ancient ones. He consigned Ptolemy to where he belonged, the second century, and ceased trying to make all new knowledge conform to the general outlines of Ptolemy's world.

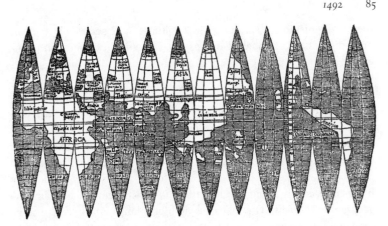

Wooden globe gores, Waldseemüller, 1507; first map known to have named the New World America

passage like Gibraltar but a labyrinth of islands and jutting capes. The strait now bears Magellan's name.

His crossing of the Pacific, which he named, was even more incredible. No one brought up to believe in the small-world concepts of Ptolemy could have imagined the enormous breadth of this new ocean. Magellan's course bypassed most of the island chains but took him to his discovery of the Philippines, where he was killed in a skirmish with natives. Under Sebastián del Cano, the expedition limped home by way of the Indian Ocean and the Cape of Good Hope, arriving in Seville in 1522—only one ship out of an original fleet of five, 35 out of 280 men surviving.

Magellan's voyage proved, once and for all, that the Earth was round and could be sailed around. The true scope and nature of the Earth began to take shape in the human mind, vaguely at first, then more surely, much as the solar system is being revealed to us today.

The effect on cartography was immediate. Magellan's discoveries were put on the map by Diego Ribero, a celebrated Portuguese cartographer working for Spain. One of his surviving maps, dated 1529, incorporated much of the information from the charts of the *Padron Real*, the official Spanish record of discoveries that was begun in 1508 and revised periodically as new explorations were reported. The map contains the whole circuit of the globe between the polar circles, and places the continents, in both latitude and longitude, more accurately than any preceding chart, although it retains the Ptolemaic exaggeration of the extent of Asia. "Nothing else about the chart is archaic," said J. H. Parry in *The Age of Reconnaissance.* "Its drawing of the Pacific, for example, incorporates all the discoveries of Magellan and of the Spanish explorers of the coasts of South and Central America. Ribero's chart represents a major landmark in the development of knowledge of the world."

Exploration and cartography were now joined. No longer, after Columbus and Magellan, could mapmaking afford to be a contemplative occupation, the pursuit of cloistered minds. Exploration, or planned discovery, brought it into the thick of contemporary affairs and geographic thought. The many newly found lands and seas had to be represented on maps, as swiftly, thoroughly, and accurately as possible, for a place is not truly discovered until it has been mapped so that it can be reached again.

Mercator Squares the Circle

No name in cartography is more widely known than that of Mercator. Generations of sailors who have braved long, lonely reaches of the sea have made safe landfalls at journey's end because of charts drawn according to Mercator's principles. Anyone who has examined the fine print in the legends of maps is familiar with the words "Mercator Projection." And many a schoolchild has been perplexed by a teacher's declaration that South America, which certainly looks no bigger on the map, is actually nine times as large as Greenland—all because a cartographer named Mercator decided four centuries ago to draw a map of the world in a new way.

Gerardus Mercator was the foremost cartographer of the sixteenth century, when the Age of Discovery flooded mapmakers with information and made increasing demands on their skill and ingenuity. Responding to the challenge, Mercator produced maps and globes that were as up-to-date as possible, faithfully depicting the new lands as they became revealed and eliminating nearly all of the medieval misconceptions. He also appreciated, more than his predecessors or most of his contemporaries, that maps were made not only to record discoveries but to be used in commerce with the new lands and for making other discoveries. For Mercator had future navigators in mind when he set out to develop the map projection that won him lasting fame.

Like Ptolemy long before him, Mercator attempted with his projection to achieve a commonsense answer to an insoluble problem: how to translate the sphere that is Earth with relative accuracy into the flatness of a map.

Mercator was born in 1512 in the small town of Rupelmonde, near Antwerp, the sixth child of German parents who had only recently moved to Flanders. He was christened Gerhard Krämer, but following the custom of Renaissance scholars he later adopted a Latinized version of his name, Gerardus Mercator.

After his parents died when he was young, a great-uncle in Rupelmonde assumed responsibility for Mercator and saw to it that he entered the nearby University of Louvain. There he came under the tutelage of Gemma Frisius, a renowned mathematician, from whom he learned geometry, astronomy, and geography. Shortly after his graduation, Mercator established a prosperous business making globes, sundials, astrolabes, and other scientific instruments. By the age of twenty-four he had also become an accomplished engraver of maps and charts.

One of his first was a six-sheet map of Palestine, which found a ready market among devout Christians. Three years of his own surveying went into his next map, of Flanders. In 1538 he published his first world map, which was one of the earliest to bear the names North and South America. The map also represented Asia and America as separate continents, an idea which increased in favor with cartographers long before the discovery of the Bering Strait proved it to be a fact. For this map Mercator divided the world into two hemispheres, east and west, and employed a sort of heart-shaped projection, the double-cordiform projection which was popular in the Renaissance. Places at the center of the map appeared in their true shape and scale, but toward the edges there was a falling-away effect, which distorted the shapes of lands to convey a feeling of the Earth's roundness.

Like many people of his time, Mercator was swept and tossed by religious tempests. Martin Luther had nailed his theses to the door of Wittenberg Castle Church in 1517, setting in motion the Reformation and Protestantism. Then came the Counter-Reformation, which struck with particularly cruel force in the Low Countries. Mercator the student is said to have expressed religious doubts. Mercator the cartographer was observed to spend considerable time out of Louvain on some strange business that he explained as surveying. This was surely a man to be viewed with suspicion. So in 1544 Mercator was arrested as a heretic. After he had spent several months in prison, some influential friends obtained his release, probably sparing Mercator the fate that befell others seized with him, who were beheaded, buried alive, or burned at the stake by the Inquisition.

The experience left its mark on Mercator. Whatever his religious views—the historical accounts differ on this point—he kept quiet about them and went about his work. But Louvain was no place for one so tainted by suspicions of heresy. In 1552 Mercator moved his family and business to more tolerant surroundings in Duisburg, a walled city in the wooded hills where the Ruhr joins the Rhine. There he enjoyed the support of Wilhelm, Duke of Cleves, and there he spent the rest of his life.

The burghers of Duisburg knew Mercator as a gentle humanist. A neighbor, Walter Ghim, described him as "a man of calm temperament and of exceptional candor and sincerity." Ghim observed Mercator in his workshop putting the finishing touches on a map of Lorraine and of Europe, doing his own delicate engraving on copperplate, a method of map publishing that had become the widely used replacement for the woodcut. Ghim was on hand also when Mercator began work on the famous world map of 1569. In a memoir, the neighbor Ghim wrote: "[Mercator] set out, for scholars, travelers, and seafarers to see with their own eyes, a most accurate description of the world in large format, projecting the globe on to a flat surface by a new and convenient device, which corresponded so closely to the squaring of the circle that nothing, as I have often heard from his own mouth, seemed to be lacking except formal proof."

Mercator's world map of 1569 bore the legend: "New and Improved Description of the Lands of the World, amended and intended for the Use of Navigators."

The geographic content of the map reflected the progress in exploration since the time of the Waldseemüller map of 1507. The west coasts of South and Central America were taking shape in the European mind. Baja California was firmly attached to the continent, though it and California would later be represented on some maps as an island. The great width of North America was beginning to be appreciated, though none of the interior; Mercator chose the North American interior as a conveniently blank space for a decorative cartouche. The outline of Asia, including Southeast Asia, was more properly delineated than on the earlier maps. And, in the Old World, Mercator had reduced the map length of the Mediterranean from Ptolemy's 62° to 52°—though this was still 10° too long.

Reliable as the map was in most respects, it contained one outlandish mistake. Mercator, along with many cartographers of the time, perpetuated the Greek concept of a great southern continent, Terra Australis. The supposed continent covered the entire polar region and extended almost as far north as the tips of South America and Africa. This idea would exert a driving influence on exploration over the next two centuries.

But the most significant cartographic feature of the map was its projection. Mercator's aim was to enable mariners to sail to their destinations by following a fixed rule. He wanted to produce a chart on which a navigator could draw a straight line between two points and immediately determine the constant course he must steer in sailing between those points. In actuality, because of Earth's roundness, such a line of constant compass bearing, called a rhumb line, is a long curve. It curves a little more than a great circle route, which is the shortest distance between two points on the surface of a globe, but which requires constant changing of the compass course. Mercator sought a way to convert the curving rhumb line to a more easily plotted straight line.*

It must have seemed an impossible idea. As Ghim said, Mercator likened the task to squaring the circle, and everyone knew that was an impossibility.

Mercator never explained when or exactly how he arrived at a solution and eventually his projection. In a note on the 1569 map, under the heading TO THE READERS OF THIS CHART, GREETING!, Mercator simply said that "we have had to employ a new proportion and a new arrangement of the meridians with reference to the parallels." It is generally believed that his methods were more empirical than mathematical. The assumption, according to William Warntz and Peter Wolff, is that Mercator came upon the idea while transferring features of a globe onto paper. In doing so, he discovered that the rhumb lines from the globe could be portrayed as straight lines perpendicular to the equator—not as lines converging, as they do, toward the poles. Thus, with the lines of latitude running parallel to the equator, as before, the meridians of longitude and the parallels of latitude would intersect each other at right angles.

---

*Navigation was a subject of much discourse in the century. Pedro Nunes, a brilliant Portuguese cosmographer, had described the loxodromic curve that comes from sailing the rhumb and had explained to mariners the method of great-circle sailing. As sixteenth-century mariners undertook longer and longer voyages, they recognized the inadequacy of charts still based more or less on portolan principles. Those straight lines radiating from wind roses might be sufficient for plotting courses on short journeys in a narrow range of latitudes, but not for extended ocean expeditions.

For this to fulfill his purpose, Mercator realized, it would be necessary for the spacing of the parallels of latitude to be made progressively larger away from the equator toward the poles. The spreading of the parallels would have to be in exactly the same proportion as the spreading of the meridians. This means that if at any place on the map a minute of longitude is spread to twice its value on the Earth, then the minute of latitude at that place is also spread to twice its value on the Earth. Such, in a manner of speaking, is the distortion formula for the Mercator projection.

It remained for an English scholar, Edward Wright, using the newly invented logarithms, to provide an analysis of the mathematics of the projection. This he did in a 1599 book, *Certaine Errors in Navigation*. The book included methods of computation and tables for the construction of the projection. While acknowledging that it was Mercator's concept, Wright said of the mathematical explanation: "The way how this should be done I learned neither from Mercator nor any man els."

The projection, like all projections, is not without its flaws. It is completely true to size at the equator, but at the 30th parallel, for example, it shows a scale increase of 15 percent and is far more distorted to the extreme north and south—which accounts for the "Greenland problem" on all maps drawn to this projection. (Recognizing this shortcoming, Mercator did not attempt to include the north polar region on the body of his 1569 map; instead, he showed it, in an inset, using a projection centered at the pole.) Consequently, there is no single scale for the entire map, there being more kilometers to a centimeter at the equator than at the higher latitudes. This made it impossible for navigators to calculate distances on the early Mercator-type maps, but eventually "meridional parts" tables were prepared by which navigators could convert chart measurements to nautical miles.

The flaws were the compromises Mercator had to accept to achieve his objective of translating the round Earth to a plane surface so that navigators would have a simple and reliable map on which to plot straight-line courses to distant landfalls. Often ignored is another important feature of this and other Mercator maps: the reintroduction of systematic latitude and longitude lines, which were conceived by the ancient Greeks but only sporadically and carelessly used during the intervening centuries.

Mariners, however, did not immediately grasp the significance of Mercator's map or, more particularly, his projection. Perhaps they could not believe the rhumb line had actually been straightened. Although the supply list for Martin Frobisher's first quest of the Northwest Passage, in 1576, shows that a copy of the 1569 Mercator map was purchased for one pound, six shillings and sixpence, most navigators ignored Mercator for nearly a century. Then they understood, as have mariners ever since.

The cartographer's task is to design maps that will show the least distortion or no distortion in those properties the map's intended user deems desirable. For this to be accomplished the cartographer must draw from a repertoire of projections. Wellman Chamberlin, a former cartographer of the National Geographic Society, wrote: "There is no over-all 'best' projection for all maps. Each map represents a particular problem. The projection that is ideal for mapping Chile, which extends more than 2,600 miles north and south and an average of less than 150 miles east and west, is not the ideal projection for mapping an area like the [former] Soviet Union, which extend[ed] 5,000 miles east and west near the top of the globe."

The Mercator projection belongs to that class of maps called conformal. Another term for it is orthomorphic, meaning "correct form." Such a map preserves the shapes of small parts of the mapped surface, though it cannot preserve the shape of an especially large country or continent. Angles around a point are shown correctly, but as we have seen, the scale varies from point to point.

Another major class of maps is called equal-area. Other names for it are equivalent and homolographic. While conformal maps are preferred by navigators, engineers, and military strategists, among others, equal-area maps are the choice of scientists, geographers, and others to whom a standard areal scale is more important than correct shape. For on equal-area maps a square centimeter of paper at one point represents the same number of square kilometers of ground as a square centimeter does in any other part of the map. The price paid for equal-area, however, is distortion of shapes and distances. One of the most popular of equal-area world maps is an elliptical projection devised by Karl B. Mollweide in 1805.

But conformal and equal-area represent only broad categories of maps. The needs of cartography have inspired a variety of more specialized projections.

A polar projection, for example, is one that is centered upon either the North or South Pole. Everything else on the map is falling away from the pole with radial symmetry. A polar projection is one of the most easily drawn maps, although it is of little value in portraying the populated parts of Earth. But Athelstan Spilhaus, practicing what he calls "geo-art," conceived a sort of variation of a polar projection to make an important geographic point: that all the world's oceans are really one, interrupted here and there by landmasses. He centered his map more or less at the South Pole. The rest of the world seemed to fall away from that point, with three petal-like seas—the Atlantic, Indian, and Pacific—blossoming out from the southern ocean. "The point is that no flat map can show the whole globe without distortion," Spilhaus said. "The trick is to choose the projection that will show you what you want to see."

Projections centered at some point on the equator are called equatorial or meridional. If centered elsewhere, they are called oblique or horizontal. (In a whimsical exercise of the art of map projections, a mathematician at the Bell Telephone Laboratories prepared a map so that the entire world seemed to be centered at Wall Street.) In either case, projections touching at a point are also called azimuthal—which is to say that all points on the map are in their true direction from the center of the map.

Some projections are based on contact along a line instead of at a point. Imagine that a sphere just fits into a cylinder. Let the line of contact between the cylinder and the sphere, the line of tangency, run along the equator. This is the case in the standard Mercator projection. But the line of tangency can just as well be any great circle, any 360° circle of the globe. Instead of using the equator as the contact line, the cartographer can twist the cylinder 90° and use a meridian, which then becomes the axis of the map. This provides the basis of the Transverse Mercator projection, which minimizes the higher-latitude distortions of the original Mercator maps and in fact has no distortion along the line of tangency. The National Ocean Survey employs this projection for its polar navigation charts. States with great north-south expanses (Maine, Indiana, Alabama, and Nevada, for instance) base their maps on the Transverse Mercator.

Another variation is known as the Oblique Mercator projection. Since the line of tangency can be any great circle, the axis of a map can be a line that crosses meridians and the equator at oblique angles. This method, though more difficult to execute, is used more and more to show great-circle routes, which the military and airlines follow as the shortest flight distances between two points continents apart.

OVERLEAF: *From a reproduction of Mercator's map of the world, 1569*

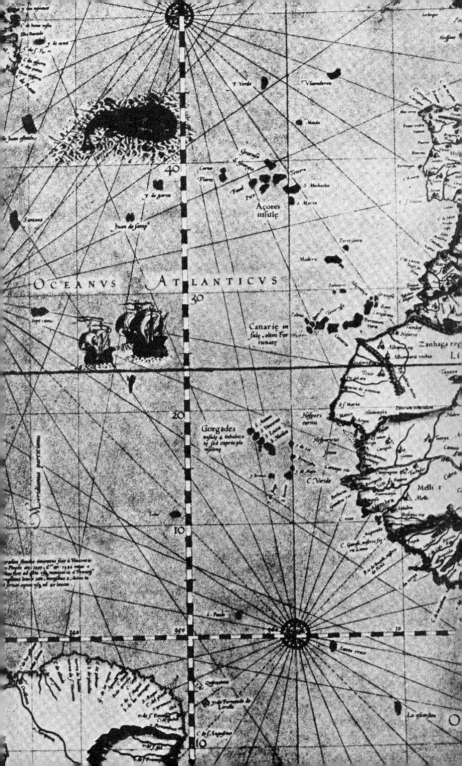

An important aspect of some projections is that they are perspective—that is, they have a point of view. The word "perspective" comes from the Latin "to see through." A perspective projection depends on where our eye is in relation to the globe. The simplest way to produce a conformal map, for example, is by stereographic projection onto a plane tangent to the globe. The point of view, or perspective center, in this case is a point on the sphere diametrically opposite the point of tangency. If the point of view is at the globe's center, however, it is a gnomonic projection. Its advantage is that all great-circle arcs are shown as straight lines on the map, making it indispensable to aviation and ocean shipping. But a point of perspective can also be at infinity. This gives the impression of the Earth or another planetary body as seen from a great distance away. A map drawn from this perspective, called orthographic, preserves neither area fidelity nor angles and, consequently, is of little scientific or navigational value. Such projections have usually been employed for maps of the Moon, particularly the pre-Apollo Moon, because it is the way we perceive the Moon from our point of view on Earth.

The preceding projections generally owe their conception to the cylinder and the plane. The Mercator and its variations show the Earth as if its spherical surface had been transferred onto a cylinder, which was then unrolled as a flat map. From whatever perspective, be it stereographic, gnomonic, or orthographic, the effect is that of projecting a curved image onto a plane surface.

Another major category of projections falls under the heading of conic, a method conceived in its simplest form by Ptolemy. With the resurrection of his ideas and maps during the Renaissance, Ptolemy's conic projection was employed for a number of maps of the new discoveries, most notably Waldseemüller's. Ptolemy's projection was most accurate in the middle latitudes and in showing areas extending east and west, which was what appealed to Ptolemy in the days before the known world became more longitudinal.

Johann Heinrich Lambert, an eighteenth-century Alsatian mathematician, provided a modification of the conic projection that made it one of the most valuable bases for maps. Lambert was the brilliant, self-educated son of a humble tailor. His scholarly interests ranged from philosophy to physics; the lambert (a unit of brightness) was named after him for his discovery of a method of measuring the intensity of light and the absorption of light.

When in 1772 he addressed himself to the mathematics of map-making, Lambert developed formulas for a half-dozen new projections. One was the Transverse Mercator. Another important one was called Lambert's Conformal Conic.

The simple conic projection derived from Ptolemy has one standard parallel, the line of tangency where, so to speak, the sweatband of the conic hat placed over the top of the globe came to rest. When the cone is spread out flat it looks like a fan. On the line of tangency the map should be errorless, as at the equator on the standard Mercator map. All the parallels on a conic map are concentric arcs spaced equally apart and true to scale. The meridians are straight lines converging toward the pole. Cartographers call the projection "a balance of errors," being neither perfect nor excessively wrong in shape, area, or scale.

Lambert's improvement was basically to make the conic map conform with two standard parallels. The cone placed over the top of the globe touches its first parallel, then cuts through the globe, and re-emerges at the second contact parallel. In this way, through complex mathematics devised by Lambert, the entire map is conformal, but scale is preserved along two parallels.

It was more than a century before cartographers fully appreciated the value of the Lambert projection. But now it is considered a preferred map for showing broad areas in the middle latitudes—the United States being a perfect example. Applying the Lambert projection to the United States, which extends from 49° to 25° N, the tip of Florida, the standard parallels usually chosen are 45° and 33°. On those two parallels, scale is exact. Between them, scale is no more than $\frac{1}{20}$ of 1 percent too small. On the Canadian border the scale is only $1\frac{1}{25}$ percent too large. At Key West the scale is its most distorted—2½ percent too large. Many states that extend mostly in an east-west direction (Connecticut, Pennsylvania, Tennessee, and Washington, for instance) are mapped with Lambert as the basis for their coordinates. Some aeronautical charts also are based on Lambert.

Another version of the two-standard-parallel conic projection is the one developed by H. C. Albers in 1805. Albers spread the parallels so that areas are exactly preserved instead of shapes—except at the two standard parallels, where both shape and scale are correct. Many maps of Russia, Central Europe, and the United States are based on Albers's Conic Equal-Area projection.

In cases where north-south accuracy is more important than east-

TYPE OF MAP PROJECTION *Lines of longitude (meridians)*	*Lines of latitude (parallels)*	*Graticule spacing*	*Linear scale*
MERCATOR *Conformal* Meridians are straight and parallel.	Latitude lines are straight and parallel.	Meridian spacing is equal, and the parallel spacing increases away from the equator. The graticule spacing retains the property of conformality. The graticule is symmetrical. Meridians and parallels intersect at right angles.	Linear scale is true along the equator only (line of tangency). Scale can be determined by using a degree of latitude, which equals 60 nautical miles, 69 statute miles, or 110–111 kilometers.
OBLIQUE MERCATOR *Conformal* Meridians are curved concave toward the line of tangency.	Parallels are curved concave toward the nearest pole.	Graticule spacing increases away from the line of tangency and retains the property of conformality.	Linear scale is true along the line of tangency only.
TRANSVERSE MERCATOR *Conformal* Meridians are curved concave toward a straight central meridian that is tangent to the globe. The straight central meridian intersects the equator and one meridian at a 90° angle.	Parallels are arcs concave toward the nearest pole; the equator is straight.	Parallels are equally spaced on the straight central meridian. Graticule spacing increases away from the tangent meridian. The graticule retains the property of conformality.	Linear scale is true along the line of tangency only.
LAMBERT CONIC CONFORMAL *Conformal* Meridians are straight lines converging at a pole.	Parallels are concentric circles concave toward the pole and centered at the pole.	Meridian spacing is true on the standard parallels and decreases toward the pole. Parallel spacing increases away from the standard parallels and decreases between them. The graticule spacing retains the property of conformality. The graticule is symmetrical.	Linear scale is true on standard parallels. Distance and directional measurements are good. Maximum scale error is 2½ percent on a map of the United States (48 states) with standard parallels at 33° N and 45° N.
ALBERS CONIC EQUAL-AREA *Equal-area* Meridians are straight lines converging on the polar axis.	Parallels are concentric circles concave toward the pole.	Meridian spacing is equal on the standard parallels and decreases toward the poles. Parallel spacing decreases away from the standard parallels and increases between them. Meridians and parallels intersect each other at approximately right angles. The graticule spacing preserves the property of equivalence of area. The graticule is symmetrical.	Linear scale is true on the standard parallels. Distance and direction measurements are good. Maximum scale error is 1¼ percent on a map of the United States (48 states) with standard parallels of 29½° N and 45½° N.

C Y L I N D E R S

C O N E S

Notes	Uses	Examples
Projection can be thought of as being mathematically based on a cylinder tangent at the equator. Any straight line is a constant-azimuth (rhumb) line. Areal enlargement is extreme away from the equator; the poles cannot be represented. Shape is true only within any small area. Reasonably accurate projection within a 15° band along the line of tangency.	An excellent projection for equatorial regions. Otherwise the Mercator is a special-purpose map best suited for navigation. It can be constructed secant at other latitudes to obtain exact linear scale. Such constructions are used for large-scale coastal charts.	
Projection is mathematically based on a cylinder tangent along any great circle between the equator and the pole. Shape is true only within any small area. Areal enlargement increases away from the line of tangency. Reasonably accurate projection within a 15° band along the line of tangency.	Useful for plotting linear configurations that are situated along a line oblique to the earth's graticule. Examples are: NASA Surveyor Satellite tracking charts, ERTS flight indexes, strip charts for navigation, and the National Geographic Society's maps "West Indies" and "Countries of the Caribbean."	 Line of tangency
Projection is mathematically based on a cylinder tangent to a meridian. Shape is true only within any small area. Areal enlargement increases away from the tangent meridian. Reasonably accurate projection within a 15° band along the line of tangency. Cannot be edge-joined for more than a few sheets.	Used where the north–south direction is greater than the east–west direction. Used as the base for the U.S. Geological Survey's 1:250,000-scale series and some of the 7½-minute and 15-minute quadrangles of the National Topographic Maps.	
Projection is mathematically based on a cone that is secant on two parallels. Areal distortion is minimal but increases away from the standard parallels. North or South Pole is represented by a point. Great circle lines are approximately straight. Retains its properties at various scales; sheets can be joined along their edges.	Used for large countries in the mid-latitudes having an east–west orientation. The United States (50 states) Base Map uses standard parallels at 37° N and 65° N. Some of the National Topographic Map Series 7½-minute and 15-minute quadrangles use standard parallels of 33° N and 45° N. Aeronautical charts for Alaska use standard parallels at 55° N and 65° N. The National Atlas of Canada uses standard parallels at 49° N and 77° N.	
Projection is mathematically based on a cone that is secant on two parallels. No areal deformation. North or South Pole is represented by an arc. Retains its properties at various scales; individual sheets can be joined along their edges.	Used for thematic maps. Used for large countries having an east–west orientation. The National Atlas of the United States, United States Base Map (48 states), and the Geologic map of the United States are based on the standard parallels of 29½° N and 45½° N.	

west, the American Polyconic projection is sometimes employed. It was conceived in 1820 by Ferdinand R. Hassler, the organizer and first super-intendent of the old Coast and Geodetic Survey. Faced with the task of mapping the young nation's coastal waters, then primarily along the Eastern Seaboard, Hassler devised a conic projection based not on one or two standard parallels, but on many—in fact, there is a cone for each parallel so that each is a line of tangency. A central straight-line meridian serves as the map's axis, but the other meridians are concave toward the central meridian, though equally spaced apart and converging toward the pole. The arcing parallels cut the central meridian at equal intervals but get farther apart at the east and west borders of the map.

The projection is neither conformal nor equal-area. It is a compro-mise that maximizes accuracy of scale, direction, and shape within small areas—900 kilometers east or west of the central meridian the scale error is 1 percent. The polyconic projection is still used for coastal surveys and for some of the Geological Survey's familiar topographic quadrangle maps, though it has been largely replaced by the Transverse Mercator.

Projections continued to cause confusion and contention through the twentieth century. Yet mapmakers kept try-ing new ways to portray the world with a truer perspective of actual geo-graphic relationships.

The University of Wisconsin cartographer Arthur Robinson became dissatisfied with existing projections and devised his own in 1963. For much of the century, the United States government and the influential National Geographic Society had used a projection invented by the American engineer Alphons van der Grinten in 1898. In that projection, as in Mercator's, the size of lands in the high latitudes was exaggerated in order to be more faithful to their true shapes. Greenland, Alaska, Canada, and the Soviet Union appeared larger than they actually are. Van der Grinten maps showed the Soviet Union to be 223 percent too large. Robinson, pondering a new map strategy, was perhaps responding to an American urge to cut the country's Cold War adversary down to size.

Robinson took a backwards approach in developing the projection, starting with artistic considerations. He visualized the best-looking shapes and sizes for a world map. "I worked with the variables until it got to the point where, if I changed one of them, it didn't get any better," he recalled. "Then I figured out the mathematical formula to produce that effect. Most mapmakers start with the mathematics."

To give the impression of the world's roundness, the Robinson projection is based on an ellipse, with the lines of longitude curving toward the poles. But the latitude lines are straight, evenly divided by longitude lines, as on the globe. In flattening out the globe, of course, the scale is true along only one great circle or along two lines of latitude, which are called the standard parallels. Everywhere else on the map the scale is larger or smaller than in reality. For his projection, Robinson chose 38° north and 38° south as the standard parallels. This provides the most realistic size and shape accuracies in the temperate zone, where most of the land and people are.

Maps based on the Robinson projection depict a flatter, more squat world, a sacrifice in shapes made necessary in order to be more accurate in terms of area sizes. The projection is thus superior to Van der Grinten's in several important ways. There is less exaggeration of size in the high latitudes. The area embracing the former Soviet Union—now Russia and the several republics spun off after the Soviet collapse—is shrunk to a size only 18 percent greater than reality; it no longer looks bigger than all of Africa. Canada's size is reduced from 258 percent too large to only 21 percent. Greenland remains a problem, its shape distorted by compression and its size reduced, though still far from its true measure. The forty-eight contiguous states of the United States, which were 68 percent too large on the Van der Grinten maps, are shrunk to a size 3 percent less than reality. One effect of the reductions in size of larger landmasses was to produce maps more faithfully reflecting the extent of the world's water surface.

After more than a half-century of using the Van der Grinten projection, the National Geographic Society in 1988 adopted the Robinson projection for its popular world maps. Then the Society again changed projections for its world atlas of 2000, choosing one developed in 1921 by Oswald Winkel of Germany as a truer representation of scale and shape. But this by no means ended innovation and controversy over projections.

In 1973, a German historian, Arno Peters, introduced what he described as a new projection and promoted it on the grounds that it corrected the European-centered bias of most projections. The Peters projection sought to represent more accurately the sizes of lands, particularly those in the equatorial regions. The result was considerable distortion of shapes. Robinson, one of many critics, complained that the Peters projection stretched out landmasses to "look like wet, ragged long winter underwear hung out to dry on the Arctic Circle."

Cartographers also noted that the projection was not really new, but borrowed from the Gall projection developed in 1855. The geographer Mark Monmonier said: "Peters said he had something new, which he didn't. He claimed the projection did not distort area, but it does."

Even so, the Peters projection had its fans. Peters argued that the end of European colonialism made a new cartography necessary. For the very reason that it seemed to correct Eurocentrism, his projection won favor with a wide international audience, and agencies of the United Nations and the World Council of Churches adopted it for their maps. To the British historian Jeremy Black the reception of the Peters projection "revealed the extent to which politics was more than simply a subtext in projections and, indeed, more generally, in maps."

If the Peters map promoted the Third World to prominence by demoting the United States and Western Europe to a kind of cartographic marginality, other projections at the time reflected another emergent geopolitical perception. These projections were centered on the North Pole. Maps drawn accordingly showed the world with all the continents and oceans radiating from an Arctic focus.

During World War II, the American cartographer Richard Edes Harrison adopted the North Pole–centered projection for his popular maps published in *Fortune* and *Life* magazines. These maps provided a fresh perspective on North America's geographic position in the global conflict. In particular, they presented the Arctic region in truer proportions, which illustrated with startling clarity the proximity of North America to Europe and the Soviet Union just across the roof of the world. Once, polar ice rendered the proximity inconsequential, but no longer could military strategists ignore that region. With advances in aviation, the shortest routes between the New World and the Old lay over the North Pole.

"Just as the Mercator projection had reflected the traditional (British) 'seaman's view' and (German) 'landsman's view,'" observed Alan K. Henrikson, an American historian, "so the Harrison map illustrated the emerging (American) 'airman's view.'"

While the United States and the Soviet Union were allies in the war, polar-projection maps may have comforted both by showing their close geographical relationship. By this perspective, supply lines to the beleaguered Soviet forces seemed short enough to engender optimism. In the subsequent Cold War, though, such maps were anything but reassuring: they charted the potential flight paths of long-range bombers and missiles. Yet in the brief time before the former allies became adversaries,

Harrison's version of a polar-focused world map was modified and adopted as the official emblem of the new United Nations.

In contrast to the Peters projection, polar projections continued the northern hemisphere bias in world maps. But Peters and his allies aimed their sharpest attacks at the venerable, durable Mercator projection, long the instrument of European discoveries and maritime power.

A year after Mercator introduced his projection, a friend and cartographic rival, Abraham Ortelius, published the first modern atlas.* He called it *Theatrum Orbis Terrarum* (Theater of the World).

Ortelius (1527–1598) was not so much a cartographer himself as an editor and publisher of maps. His workshop in Antwerp was one of the busiest and most prosperous in the Low Countries, which were then emerging as the center of the map and chart trade. He got the idea for the *Theatrum* while working on an assignment for an Antwerp merchant named Hooftman, a collector of old maps. Hooftman found the job of rolling up and unrolling his many large maps an increasing nuisance. And so he commissioned Ortelius, who made many trips about Europe collecting maps, to obtain as many single-sheet maps as he could find and to stitch and bind them in a single volume. Having in this way put together about thirty maps for Hooftman, Ortelius decided to do the same for a more general audience.

The first edition of the *Theatrum,* in 1570, incorporated seventy maps. These were the products of many cartographers, each of whom was given full credit by Ortelius; but Ortelius re-engraved each map to conform to a uniform format. The first printing was such an immediate success that Ortelius brought out a second edition in three months. Eventually there were forty editions in several languages. Scholars and cartographers wrote Ortelius letters of praise that often included suggestions for additions to future editions. Mercator complimented his friend on "the care and elegance with which you have embellished the labors of the authors, and the faithfulness with which you have preserved the production of each individual, which is essential in order to bring out the geographical truth, which is so corrupted by mapmakers."

---

*The Chinese were producing such map compilations as early as 1,800 years ago. Pei Xiu, the Chinese Ptolemy, prepared a collection of eighteen map sheets that may be the oldest known atlas. The atlas itself is lost, but its preface has survived. Only after the tenth century did atlas-making become a significant part of Chinese cartography.

Mercator also had plans for publishing a collection of maps in book form. The first two parts of the collection were issued in 1585 and 1590, with the final volume appearing in 1595, a year after Mercator's death.

The title for the collection represents the first use of the word "atlas" in connection with a geographical work. The full title read: "Atlas, or cosmographical meditations upon the creation of the universe, and the universe as created." In the introduction, Mercator included a genealogy of Atlas, the mythical figure who led the Titans in their war against Jupiter and was therefore condemned to support the heavens on his shoulders.

This was Mercator's final contribution to cartography. Henceforth, nearly all bound collections of maps were called atlases, just as nearly all nautical charts came to be drawn according to the Mercator projection.

Gerardus Mercator probably did more than any single man to complete the revolution in mapmaking that was begun early in the Age of Discovery. By converting maps from philosophical pictures and rude drawings to more useful tools, Mercator prepared the way for cartography to enter its modern and more scientific era.

PART *Two*

Yaki Point

The spirit of Matthes was with us every day of the expedition. François E. Matthes had been here before us and had mapped the Grand Canyon before aerial photogrammetry, helicopters, electro-optical distance measuring, or any other modern tools of mapping. He did it by plane table and transit, by foot and mule, by patience and fortitude.

At the lodge one evening, after an exhilarating day in and above the Canyon, Bradford Washburn, the leader of our project, talked about how Matthes had made the first really thorough topographic map of the Canyon.

As a topographer for the Geological Survey from 1896 to 1947, a year before his death, Matthes made his way into many of the West's most rugged regions. The Bighorn Mountains and Mount Rainier, Glacier National Park and Yosemite Valley. Yet, more than anything else, it was his work in the Grand Canyon, primarily in 1902–1903, that established his reputation. The beauty of Matthes's sketchings and the reliability of his surveys were remarkable in the early twentieth century. And nothing Washburn found, in remapping the Canyon in the 1970s, diminished that reputation or his reverence for Matthes.

In a memoir accompanying the 1906 edition of the map, Matthes described how he met the challenge of surveying the Canyon:

> The map was constructed wholly in the field, on a plane table—a drawing
> board mounted on a tripod. A few control points having been located by tri-

angulation and plotted on the map by their latitudes and longitudes, the table was set up at one of the triangular stations (that at Hopi Point marked on the map with a little triangle), and oriented with respect to north by sights taken to other triangulation points. Thereupon, each summit, each pointed spur, or other feature affording a good natural target was sighted, and the line of sight to each was drawn on the map along the rule attached to the telescope. The table having been moved to another station on the rim, the same features were again sighted, and to each a second line of sight was drawn. The intersection of each pair of lines accurately located the feature of the map.

Thus, through triangulation, the mapper's most widely practiced short cut, Matthes was able to fix the locations of many points and buttes and side canyons without ever setting foot on them. From some plane-table stations he sighted more than a thousand points. If he had had to go to every point he placed on the map, he could never have completed it in a reasonable time. Only the Inner Gorge, Matthes wrote, and some of the "tortuous side canyons not visible from the rims required special lines of survey."

Altitudes of places on the rim or down in the Canyon were computed in relation to the sea level on the California coast. Surveyors had run a line of levels from the coast along the route of the Atchison, Topeka & Santa Fe Railway right up to the rim of the Canyon.

This was the slow and tedious and arduous way by which much of the Earth first came to be surveyed and mapped—to be embraced.

We were reminded again of Matthes the next afternoon, out on Yaki Point. We thought of him and the countless other old-time surveyors when we found the monument. It was a small circular brass tablet implanted in the rock at the edge of the precipice. A triangulation mark. Like thousands of similar monuments established across the United States by the Geological Survey and the Geodetic Survey, it marked a "known point"—its latitude and longitude calculated and its elevation a known 7,260 feet.

"If we didn't have these 'knowns,'" Washburn said, acknowledging his debt to such predecessors as Matthes, "we would have to establish them ourselves by making star sightings to fix our position and by leveling from who knows how far away."

Washburn then placed a tripod holding the Wild T-3 theodolite directly over the mark. From the instrument he dropped a string with a plumb bob so that it rested exactly in the center indentation of the mark. He hunched over and squinted into the theodolite's 40-power telescope, focusing Point Sublime far across the Canyon to the northeast.

Washburn had set out to improve upon Matthes and the subsequent Geological Survey maps. His objective was to produce a more detailed map of 440 square kilometers in the heart of Grand Canyon National Park. Moreover, the "manuscripts" on which the finished map is based would be on a scale of 1 foot to 1 mile, or 30 centimeters to 1.6 kilometers. These should be indispensable to botanists, geologists, archeologists, and others doing research in the Canyon.

This was an exciting labor of love for Washburn, who was taking time from his duties as director of the Boston Museum of Science. Sixty-two years old at the time, he was an experienced explorer-cartographer who had produced the first detailed topographic map of Mount McKinley. Mapping a place like the Grand Canyon, he remarked, was "like mapping a mountain upside down."

The first step in the project, accomplished by a commercial aerial survey firm, had been to photograph the entire area from an aircraft at 5,000 meters. A mapping camera scanned the terrain lengthwise and crosswise, each strip of pictures covering a 5-kilometer-wide swath.

Now, Washburn and his small party were going in to "get control." That is, they had to establish the elevations of and the distances between known points in the Canyon. Everything else in the aerial photographs, and the resulting map, would be related to those 91 selected control points—the precise skeleton on which the detail from the aerial photograph would be hung.

The strategy for selecting the control points was simple: the points had to be fairly evenly distributed throughout the area to be mapped. Pick a few along the south rim, some more on the south side of the Colorado River down in the Canyon, others along the north side of the river, and still more along the north rim. The four lines of points were roughly parallel. At each point a piece of metal painted luminescent orange was placed as a target for observations. Each target would be clearly visible, it was hoped, from at least three other stations, for purposes of cross-checking the accuracy of measurements.

"In mapping, you have to start somewhere, and we chose Yaki Point," Washburn explained as he adjusted the theodolite.

He focused the cross hairs of the telescope on the orange marker over at Point Sublime. He then referred to an encased glass on which all the angles of the compass were etched. With such theodolites, it is possible to determine angles as fine as one-tenth of a second. In this way,

Washburn could determine precisely how far above or below any point was from another, as well as the horizontal angles.

"This is just good old trigonometry," Washburn said, entering a set of angles in his notebook. From the vertical angle between Yaki, with its known altitude, and another point Washburn could calculate the difference in altitude between the two points.

Shadows paraded down the canyon walls bearing their diurnal message. A swallow whooshed by, then another. A twilight chill drifted in with the wind. But there was time to point the theodolite for one more set of angles.

"This part of mapping is just dreadfully routine," Washburn was saying. "But it pays off in accuracy. Sometimes you make twenty sightings of a single angle to completely eliminate the chance of errors."

We thought of François Matthes. For all the modern technologies, the fieldwork in mapping could still be slow and trying. Yes, the short, dark-haired topographer of 1902 could have stepped right into the scene at Yaki Point, one more figure around the tripod for a moment frozen in time: one of us peering through the telescope, another kneeling with a notebook, still another with arm outstretched toward the distant terrain—a diorama celebrating the intrepid surveyor.

The Matter of a Degree

Jean Fernel was born in France in 1497, the year John Cabot reached Newfoundland and the year before Vasco da Gama reached India by rounding the Cape of Good Hope. He grew to manhood as Magellan was sailing on his voyage of circumnavigation, and he must have examined with interest the outlines of new lands as portrayed on the maps of Waldseemüller and, later, Mercator. Before Fernel died in 1558, Copernicus had exposed the venerable notion of an Earth-centered universe as being nothing more than human conceit, thus freeing humans to contemplate and eventually probe a larger and far more wondrous universe.

The modern age was dawning, and Fernel recognized what was happening and approved wholeheartedly. In 1530 he wrote:

> But what if our elders, and those who preceded them, had followed simply the same path as did those before them? . . . Nay, on the contrary, it seems good for philosophers to move to fresh ways and systems; good for them to allow neither the voice of the detractor, nor the weight of ancient culture, nor the fullness of authority, to deter those who would declare their own views. In that way each age produces its own crop of new authors and new arts. This age of ours sees art and science gloriously re-risen, after twelve centuries of swoon. Art and science now equal their ancient splendor, or surpass it. This age need not, in any respect, despise itself, and sigh for the knowledge of the Ancients. . . . Our age today is doing things of which

antiquity did not dream. . . . Ocean has been crossed by the prowess of our navigators, and new islands found. The far recesses of India lie revealed. The continent of the West, the so-called New World, unknown to our fore-fathers, has in great part become known. In all this, and in what pertains to astronomy, Plato, Aristotle and the old philosophers made progress, and Ptolemy added a great deal more. Yet, were one of them to return today, he would find geography changed beyond recognition. A new globe has been given to us by the navigators of our time.

Fernel himself embodied the new spirit of curiosity, the search for "fresh ways and systems." A brilliant professor of medicine in Paris, he was ahead of his time in the application of anatomical knowledge to physiological theory. He served as physician to the king of France and was celebrated by those who knew him as "the modern Galen." But in the history of cartography Fernel is best known for a carriage ride he once took from Paris to Amiens. The year was probably 1525.

It had occurred to Fernel, perhaps after reading the conflicting esti-mates of Ptolemy, D'Ailly, and Columbus, that no one really knew the length of a degree of latitude. What does one part of 360 mean in stan-dard linear measurement? Eratosthenes, as we have seen, had come close with his calculation that Earth's circumference was 46,000 kilo-meters, which should mean that a degree is 127.7 kilometers. But who could be sure of such an ancient estimate? Columbus had used esti-mates of about 83 kilometers. Where, Fernel wanted to know, lay the truth? Not being one to sit around and theorize, Fernel called for his carriage and set forth to find the answer by measuring an arc, or seg-ment, of the meridian.

At the starting point in Paris, Fernel made a measurement to estab-lish his north-south position, or latitude. He used an angle-measuring device, a crude quadrant, which was a version of the ancient astrolabe. The quadrant was a simple wood or brass instrument in the shape of a quarter circle, with a graduated scale of 90° marked along its arc. Astronomers had for centuries used variations of such an instrument, and now mapmakers and mariners were beginning to use it for finding latitude. Finding latitude involves measuring the height above the hori-zon of the Sun at apparent noon or of some prominent star—not in meters or kilometers, of course, but in angular degrees. Fernel aligned the sights of his quadrant on the noonday Sun. A plumb line extending from the point of suspension dropped across the arc of the quadrant, the point of the intersection on the scale indicating the angle of the Sun to the horizon.

The Sun, as the ancients had observed, seems to travel a set course through the sky, varying predictably from day to day during the year. Each noon the Sun appears a little higher or lower, according to the day and the place from which it is observed; tables of these variations were devised by ancient astronomers and refined somewhat by the time of Fernel. By referring to such tables, Fernel could convert his Sun angles into the latitude of his position north of the equator.

Next, Fernel measured with great care the circumference of one of his carriage wheels. The wheel became his odometer. Ever since the Romans, surveyors had used a wheel in one way or another to measure distances. As Fernel drove slowly north, he counted each revolution of the wheel—17,024 revolutions, we are told, between Paris and Amiens. At the end of the journey, Fernel took another angle of the Sun to find the latitude of Amiens. Paris and Amiens were almost exactly 1° of arc apart. The distance between the two, Fernel concluded, must be the length of a degree of latitude.

It could not have been the most accurate of measurements, considering the unevenness of terrain, the twists and bumps in the road, the quadrant's lack of precision, and the many opportunities for human error. But, by an astounding series of compensating errors, Fernel computed the length of a degree—and came very close to the truth. He then multiplied the length of his degree by 360 and arrived at a circumference of the Earth that was only one-tenth of 1 percent in excess of the true value.

As far as anyone knows, this was the first degree measurement in Europe. In geodesy, as in physiology, Fernel was ahead of his time.

A century would pass before scientists possessed the surveying tools and the mathematical sophistication to undertake more precise measurements of a degree. For yet another century degree measurements would engage some of the brightest minds of Europe, igniting learned debate, providing insight into the laws of nature, and prompting the dispatch of two of the earliest expeditions for the sole purpose of scientific discovery. This was a lively time for geodesy and a critical time for cartography. Until the matter of a degree was settled, and the more exact shape of the Earth made known, cartography could not move into its own modern age.

In the seventeenth and eighteenth centuries, surveying assumed a more integral role in geodesy and cartography. No longer were surveyors constricted by their primitive instruments to the performance of only small-extent work, the measurements of property lines, building sites, and roadways. No longer were mapmakers dependent almost solely on the meager fare that came their way from travelers' tales, explorers' crude sketches, and mariners' random compass bearings. The invention of more accurate measuring instruments and the evolution of new mathematical techniques elevated surveying to a skill of increasing range and precision.

The following are some of the surveying tools and techniques that were either developed or improved during the two centuries:

Plane Table. The origin of this simple tool is unknown, though the Romans are believed to have used some form of a plane table in laying out their famous roads. The plane table came into general use in Europe in the sixteenth and seventeenth centuries as the first mapping device for establishing and recording angles.

In England it was first known as a "plaine" or "playne" table to suggest its simplicity in both design and use. A plane table consists of a smoothly finished drawing board mounted on a tripod and a metal sight rule (alidade) for accurate aim on the object to be plotted. Later adaptations include telescopic sights and magnetic compasses. Corner clamps hold a sheet of paper on which the surveyor records his observations.

Triangulation. The plane table was employed in the first surveys by triangulation, one of the advances in mathematics that proved decisive in modernizing mapmaking. Triangulation is based on a simple geometric principle: if you know one side and two angles of a triangle, you can determine the properties of the rest of the triangle. Surveyors with plane tables practice a form of triangulation to fix distant places by intersecting rays. This is done by sighting on a distant object, a mountaintop or house or tower, from the two terminal points of a measured base line. The one known side of the triangle is the measured base line. Separate lines, or rays, drawn from the two sighting points give the two known angles. The intersection of the two rays at the remote object closes the triangle, making known the third angle (the three interior angles of a plane triangle equal 180°) and, with some calculation, the lengths of the other two sides.

It is thus possible to determine the distance to the remote objects without ever going there. And the object could be quite remote. One of the early applications of triangulation was the measurement of the distance from the Earth to the Moon. Astronomers measured the angle of the Moon above the horizon at two places on the Earth's surface—as far apart as possible and at the same moment of time. This gave them a triangle and a base line (the known distance between the two observation points). Since they had measured two angles of the triangle, they knew the third and thus had enough information to find the distance between the apex (the Moon) and the base line (the Earth) of the triangle.

With the same system a surveyor can accurately fix on his map the location of major features of a part of the Earth and can sketch in details in their true relation to those features and the base line. Though such plane-table surveying is still practiced, it has been superseded in many cases by photogrammetry.

Triangulation provided the means for even more ambitious surveys. For when you figure out the other sides of a triangle, you can use any one of them as the known line for the next triangle, and in that way extend a network of interconnecting triangles across a province, a country, or a continent. These triangles can, with more intricate calculation, give you the distance between the starting point and the other end of the network. Only once, when the initial base line is laid out, is it absolutely necessary to measure every meter of any line.

Angle-Measuring Instruments.   The quadrant, an instrument for determining angular elevations of heavenly bodies, continued to be used by surveyors to establish latitude, much as Fernel had done. But telescopic sights were added. Then smaller, more portable, and more accurate instruments were invented—the octant (with an arc one-eighth of the circumference of a circle) and the sextant (one-sixth of a circle).

For non-astronomical angle measurements, the theodolite became the surveyor's basic instrument. Leonard Digges of London is credited with its invention in 1555, but the instrument did not come into wide use until a telescope was added by Jonathan Sisson in about 1720. The theodolite is used for measuring both horizontal and vertical angles—the horizontal for determining distances by triangulation, the vertical for determining relative elevations also by triangulation.

The theodolite consists basically of a sighting tube in the form of a telescope and horizontal and vertical scales. These scales are so arranged that horizontal and vertical angles of the object sighted in the telescope can be determined. The instrument is mounted on a tripod and equipped with compass and bubble level. For more precise sighting the eyepiece of the telescope is fitted with crosshairs, which are made of human hair, metal wire, or, in the finest old instruments, spider silk.

Distance-Measuring Instruments. Though variations of the wheeled instruments known as odometers or perambulators were still commonly used, surveyors felt the need for much more accurate ways of measuring distances, particularly for their crucial base lines. This need produced the surveyor's chain. Invented in 1620 by Edmund Gunter, an English mathematician, the chain was 66 feet long, with 80 chains equaling one mile. Gunter's chain has now been replaced by steel tape, which in turn has been largely replaced by electronic distance-measuring instruments.

Elevation-Measuring Instruments. Three methods of determining the difference in elevation between two points came into use during this period—spirit leveling, trigonometric leveling, and barometric leveling.

The most common method is spirit leveling, or differential leveling, which dates back to about 1700. A leveling rod, made of wood with graduations marked in large numbers, is held vertically on a point of known or assumed elevation. A telescope device, equipped with a spirit level (a sealed glass tube filled with a liquid and a trapped air bubble, which stands at the center of the tube when the instrument is exactly level), is leveled and sighted on the rod. The line of sight to the rod establishes the elevation of the telescope in relation to the rod. With the telescope remaining where it is, the rod is then moved to a forward position for similar readings. The difference between the first and forward rod readings yields the elevation difference between the two points. In topographic surveys the elevation readings are usually related to mean sea level.

Trigonometric leveling, using a theodolite, is an indirect method of determining differences in elevation. This involves triangulation in which the angles are vertical instead of horizontal. At the turn of the eighteenth century, French surveyors employed this method in making the first reasonably accurate measurement of the height of a mountain. They measured the altitude of Mont Canigou, in the eastern Pyrenees within sight of the Mediterranean (sea level), with an overestimate of less than 1 percent of the present figure, which is 2,740 meters.

Barometric leveling became a widely used method in the eighteenth century, after Blaise Pascal had demonstrated that atmospheric pressure decreases with increasing elevation. The first barometer was developed in 1643 by Evangelista Torricelli. It is true that, since barometric pressures do not remain precisely constant, this method was never too accurate, but it proved useful on broad reconnaissance surveys, particularly in rugged terrain.

With these new surveying tools and a knowledge of triangulation, with the new telescope (Galileo in 1609) and the pendulum clock (Christian Huygens in 1657), began the first modern efforts to measure a degree of latitude.

Willebrord Snell, a professor of mathematics at the University of Leiden, was the first to practice triangulation in an extensive survey. In 1615, he ran a network of 33 interconnecting triangles 128 kilometers across the frozen meadows between the Dutch towns of Alkmaar and Bergen op Zoom. He measured the base line with an odometer. At the terminal points of the network, he made astronomical observations with a quadrant to determine the amplitude, or breadth, of this particular arc of the meridian; it was considerably more than one degree. Taking the amplitude of the arc and the triangles, Snell made his calculations.

They left much to be desired. His measurements of a degree and of the Earth's circumference fell short of the true value by more than 3 percent. But Snell had introduced a technique of measuring distances that would become standard in all cartographic and geodetic surveying.

Richard Norwood, a London mathematician, did much better using the newly invented surveying chain. On June 11, 1633, he observed the altitude of the Sun at noon over the Tower of London. Then he set out along the road from London to York, stretching the chain to its full length, moving forward and stretching the chain again, and repeating the tedious process the entire 288 kilometers to York. He made allowances for the turns in the road. On June 11, 1635, he took the Sun's angle at York, the difference of the two angles between the two cities yielding the difference in their latitudes. From the two latitudes and his chain measurements, Norwood determined the length of a degree to be 69.2 miles, or 110.72 kilometers. Amazingly, this was, as the proud surveyor himself believed, "within a scantling of the truth."

Even so, it became clear that the triangulation of Snell, not the plodding chain survey of Norwood, held greater promise for degree measurements. More ground could be covered more quickly. And the trigonometric calculations for triangulation had now been simplified as a result of another advance in mathematics. The invention of the logarithm tables, perhaps the most universally useful mathematical discovery of the seventeenth century, cut in half the time it took to calculate distances from a string of connecting triangles. The French seized upon triangulation and in turn developed it into an increasingly refined skill.

Soon after the Royal Academy of Sciences was established in 1666, it pondered the continuing uncertainty as to the size of the Earth and decided to settle the matter once and for all. Jean Picard, one of France's foremost astronomers and mathematicians, drew the assignment to make a scientific measurement of an arc of the meridian and from that to determine the Earth's true circumference.

Picard carried out his survey in 1669 and 1670. Using well-seasoned, varnished wooden rods, laying them end to end, he measured an 11.4-kilometer base line on the road from Paris to Fontainebleau. From this base line, he proceeded by triangulation to measure an arc from Malvoisine, near Paris, to the clock tower of Sourdon, near Amiens—retracing somewhat Fernel's carriage route. But Picard was much better equipped than Fernel. For measuring angles he used a stout iron quadrant fixed on a heavy stand, which was sturdy enough to reduce the possibility of error from vibrations. The usual pinhole sights were replaced by two telescopes with cross hairs; he was the first to apply the telescope to measure angles in surveying. Picard also attached a 3-meter-long telescope to the instrument he used for the star sightings he made to determine latitude. From the 13 triangles of his survey Picard calculated that a degree measured 110.46 kilometers, a remarkably accurate result.

But Picard's measurement, however accurate, had not settled the matter of a degree. The problem was not so simple, as scientists in France and England were beginning to suspect.

At the time Picard was laying out his triangles in the field, one of the greatest of all theoretical scientists was formulating in his mind the theory of universal gravitation. The idea came to Isaac Newton in 1666 while he was studying Galileo's work on the pendulum. (A falling apple may or may not have been the eye opener.) The circular pendulum led Newton to the conception of centrifugal force and that led to an idea that gravity holds planets in place while they swing around the Sun. In 1687 Newton finally published his comprehensive theory, which states: every particle in the universe attracts every other particle with a force that is equal to the product of their masses divided by the square of the distance between them.

Word of this new theory had spread years before its formal publication, and it caused quite a stir. Newton held that if there was such a thing as centrifugal force (the word comes from Latin words meaning "fleeing from the center," and anyone who has ridden a fast merry-go-round knows the feeling), it would mean that the Earth could not be perfectly round. And if Newton was right, a basic assumption of geodesy and cartography was wrong.

According to Newton, the Earth's rotation creates a centrifugal effect, a tendency to push material away from the center of rotation. Moreover, this tendency would not be uniform. As a result of the spinning, different points on the Earth's surface move at different speeds. The surface moves faster at the equator, where it must make a circle of nearly 40,000 kilometers in twenty-four hours; to the north or south of the equator, a spot on the Earth travels more slowly, since it makes a smaller circle in the same twenty-four hours; and at the poles the surface is motionless. Centrifugal force of the spinning Earth is thus strongest at the equator.

If that was indeed the case, Newton said, the Earth should tend to bulge at the equator, the result of the greater centrifugal force pushing material out, but never pushing with enough strength for the material to escape the pull of gravity. Consequently, the Earth would be an oblate spheroid, expanded at the equator and somewhat flattened at the poles. And if that should be found true, Newton further said, it would confirm the existence of centrifugal force.

So, just as Picard was perfecting his methods of measuring a degree, and the Earth's circumference therefrom, his essential premise of a spherical Earth was being challenged. If the Earth was a sphere, the value of a degree would be the same everywhere, for both longitude and latitude. If not, if the Earth was not perfectly round, it would not be possible to take the measurement of one degree in France and say that was that. How many kilometers to a degree would depend on where you were. If the Earth was an oblate spheroid, as Newton believed, a degree would get gradually longer as you approached the flattened poles.

Although Picard continued to believe the Earth spherical, the evidence from the outset seemed to favor Newton. Astronomers looking through telescopes could now see more clearly the globes of Jupiter and Saturn and discover that they were flattened at the poles and bulging at the equators. What had happened to distort those spinning globes could have occurred to the Earth. In 1673, a French expedition to Cayenne, a colony on the northeast coast of South America and near the equator, obtained some down-to-earth evidence. During his astronomical work there, Jean Richer noticed that the pendulum clock that kept good time in Paris lost two and a half minutes a day in Cayenne. Since it is the force of gravity that keeps a pendulum swinging, and since Newton postulated that gravitational force depended on the distance between the two bodies involved (the pendulum and the Earth, in this case), Richer could only conclude that the force of gravity was somewhat weaker in Cayenne than in Paris. Cayenne must, therefore, be further from the center of the Earth. If it had been a mountainous land, that might have explained the difference; but Richer's base was at sea level. To Newton, when he heard of this, it could only mean that he was correct: the Earth bulged at the equator.

But most French scientists were not willing to accept Newton's theories. They clung to the system of their own René Descartes, who had conceived of the universe as being full of matter whirling in vortices, with each star being the center of a whirlpool. To the French this seemed observably preferable to some strange force, gravity, that operated with mathematical precision but otherwise appeared intangible.

Late in the seventeenth century, as the Newton debate raged, the French Academy decided on a test involving an extension of Picard's meridian line to the north and south of Paris. Picard had died in 1682, and the task fell to Jean Dominique Cassini and his son, Jacques, both excellent astronomers. Following Picard's methods and using much the same instruments, they ran triangulations from Dunkerque on the English Channel all the way to the Pyrenees—and with surprising results.

The Cassinis divided the measured arc into two parts, one northward from Paris, another southward. When they calculated a degree of arc over the southern line, they reached a linear value slightly greater than Picard's degree farther north. This might have been due to an error, and so they said nothing. After his father died in 1712, Jacques Cassini resumed the survey to the north, and for him it confirmed his earlier finding. The length of a degree seemed to get shorter toward the pole, not longer. The difference was a matter of meters, not kilometers, but Cassini felt it might be important. In fact, Cassini became convinced that neither Picard nor Newton was right. The Earth was neither spherical nor an oblate spheroid. The Earth, Jacques Cassini said, bulged at the poles and was flattened at the equator—a spheroid, yes, but prolate and not oblate.

Cassini's announcement to the Academy fired an even more intense controversy between French and English scientists. If this had been an earlier age, the debate might have settled into a duel of dogmas, but this was the time of the Scientific Revolution, and it was decided to put the issue to a rational and decisive test.

With the authorization of Louis XV, the Royal Academy of Sciences laid plans for two geodetic expeditions. One would go to the equator, the other to the Arctic Circle. They would determine the different lengths of a degree of latitude at different latitudes, using an extension of the method of Eratosthenes. If the degree at the equator was found to be shorter than at the Arctic Circle, it would prove Newton and the English right, that the Earth was oblate or flattened at the poles and bulging at the equator. A reverse finding would prove Cassini and the French right, that the Earth was prolate, or drawn out at the poles like a lemon or an egg.

"The Question, for its Curiosity only, might well merit the Consideration of Philosophers and Mathematicians," the leader of the Arctic expedition, Pierre Louis Moreau de Maupertuis, wrote. "But the Advantages arising from the Discovery of the Earth's true Figure, go beyond mere speculation; they are real, and of very great importance."

Maupertuis had in mind the needs of navigators and cartographers. "Were the Positions of Places with respect to Longitude and Latitude even so exactly marked on our Globes and Charts," he continued, "it would signify little to the finding of their true distances, while we were ignorant of the length of the Degrees of the Meridians and of the Parallels of the Equator. And if the distances of Places are not very well known, to what dangers must the Ships be exposed that are bound for them!"

Although the first expedition left in 1735 for Peru, it was the Maupertuis expedition to Lapland, which embarked a year later, that produced the first definitive measurements leading to a resolution of the oblate-prolate issue.

Maupertuis was a gentleman of the salons and one of the most versatile scientists of his time. Born in St. Malo in 1698, he served as a soldier in Louis XV's army, studied mathematics in his leisure time, experimented in genetics, and was among the first on the Continent to appreciate the significance of Newton's laws of gravitation. While in his early thirties, he became a member of the French Academy, where he was the foremost Newtonian; later, he became president of Frederick the Great's Academy of Sciences in Berlin. Maupertuis enunciated a theory of his own, the "principle of least action." He conceived it out of a belief that the operations of nature conform to a rule of economy in the expenditure of energy. In the ensuing debate, Voltaire, an old friend who turned into his most biting enemy, attacked Maupertuis with such withering ridicule that his reputation never quite recovered. "Earth flattener" was one of the mocking phrases Voltaire directed at Maupertuis, an allusion to the heroic expedition to Lapland that stands as one of the undisputed classics of geodesy.

In 1736, at the age of thirty-eight, Maupertuis left France for Lapland. He headed a party of three other academicians and a priest who was a corresponding member of the Academy. At Stockholm, where a Swedish astronomer joined the party, they heard but were undaunted by dark and discouraging accounts of what lay ahead. "Nothing could retard us," Maupertuis reported, "neither the frightful stories they told us at Stockholm, nor the Goodness of his Swedish Majesty; who, notwithstanding the Orders he had given in our behalf, told us oftener than once, that it was not without a sensible concern he saw us pursue so desperate an Undertaking."

Maupertuis and his party reached Tornio (Torneå) in early summer. This bleak village at the northern end of the Gulf of Bothnia, not far below the Arctic Circle, would be their base. The spire of its church would be the starting point of their triangulation survey. The Torne River would be their avenue north, as it roughly followed the 24° E meridian and the Swedish-Finnish border. On the mountaintops rising above the marshes and dense forests they would set up their stations.

On July 6, the party proceeded up the river in boats laden with instruments and supplies. Too soon they had reason to recall those dire warnings back in Stockholm. Maupertuis complained of the "wretched Diet," of cataracts in the river, and most of all, of the torments of "great Flies with green Heads, that fetch blood wherever they fix." The flies, he further noted, were "so insufferable as to drive the Laplanders and their Rain-Deer from their Habitations to seek shelter on the Coasts of the Ocean."

From the river Maupertuis scanned the terrain for likely sites on which to establish surveying stations. They had to be high above the forests, and each one had to be visible to at least two other stations. For the first site he chose a steep mountain called Niwa, its bare rock summit rising above the river. From there Maupertuis selected for the next station another peak several kilometers to the north. He and another member of the party made their way to the mountain, Avasaksa, where they cleared the crown of trees and erected a "signal." This was a great hollow cone formed of many logs stripped of their bark. The whiteness of the stripped logs was, Maupertuis said, "enough to be visible 10 or 12 leagues." Surveyors at other stations could sight on this target with a quadrant fitted with a telescope and thus measure the horizontal angles between their stations and Avasaksa.

The expedition continued toward the Arctic Circle, selecting mountaintops, clearing and erecting signals, and taking angles. But they could not escape the flies. They were delayed several days by heavy fog. A forest fire, which "had been no doubt occasioned by our not taking sufficient care to extinguish our Fires," destroyed one of the log signal cones.

When they finally crossed the Arctic Circle, on July 30, they headed for Pullingi, the tallest mountain they would use. Its steep sides and mossy rocks made for "exceeding difficult access," and once at the summit Maupertuis and the others had to spend six days felling trees and fighting off more flies.

At the Finnish village of Pello, about forty kilometers north of the circle, Maupertuis saw what he decided would be the northernmost point in the chain of triangles. It was Kittis Mountain, and it could not have been more perfectly situated. As Maupertuis told the French Academy, "Our several Courses, in the 63 days we had past in these Deserts had furnished us as compleat a Set of Triangles as we could have wished for; and an Undertaking begun in some sort at random, without knowing if it was at all practicable, had turned out so much better than Expectation, that it looked as if the placing of these Mountains had been at our disposal."

From Kittis Maupertuis planned to shoot the angle of a known star to determine latitude. Later he would shoot the same star from the church back in Tornio. The difference between the two latitudes would establish the arc of the meridian between Kittis and Tornio.

But first Maupertuis and his party returned to Tornio to pick up a new angle-measuring zenith sector that had just arrived from England. They had been using a small quadrant for taking the angles of their triangles. The new sector, with a radius of three meters, was capable of the finer accuracies needed in astronomical observations.

On their return to Kittis, they built two observatories on top of the mountain, one for the smaller quadrant and other instruments, the other for the three-meter sector. "The Sector almost took up the whole room," Maupertuis said. "What difficulty we had in carrying up so many Instruments to the top of the Mountain, I shall not mention; it is sufficient that we carried them up. The Sector's Limb was placed exactly on the Plane of a Meridian Line which we had traced; and verified its Situation by the time of the Passage of a Star, several Altitudes of which we had taken."

The surveyors used the sector on October 1 to make several sightings of a star in the Dragon constellation. Preparations were then made to abandon Kittis as soon as possible. Winter was closing in; snow had already fallen. In a few days, the party might find themselves stranded, their boats locked in a frozen river. Maupertuis also had a compelling scientific motive for leaving. He needed to return to Tornio without delay to make similar observations of the same star before its position changed. As he explained, "This star must, during the time of Observations, either keep fixt to the same Point of the Heavens, or at least, if it does not, the Laws of its Motion must be known, that the difference of Altitude arising from its proper Motion may not be confounded with that which proceeds from the Curvature of the Arc of the Terrestrial Meridian."

The river was freezing behind them as the Maupertuis party sailed back to Tornio. They arrived on October 28 and three days later, a month after the Kittis observations, were able to measure the altitude of the same star to determine latitude at Tornio. Maupertuis found that the arc of the meridian between Tornio and Kittis was just shy of a full degree—57′27″.

One more measurement remained before Maupertuis could supply a definitive answer, from the polar regions at least, to the question of the Earth's shape. He had yet to measure a base line. Without a measured base line, the known length of one side of one triangle, all the other interconnecting triangles were meaningless. Without the base line Maupertuis could not calculate the linear distance between Tornio and Kittis and, in another step, find the length of a degree in the Arctic.

On December 21, the day of the winter solstice, the party set out on sledges to measure a line from the west bank of the river across to the station at Avasaksa. They used ten-meter rods made of fir to measure the straight line. By working across the frozen river they were assured of an easy-to-measure smooth plane. Otherwise, there was nothing easy about the operation. Since the pale Sun peeped above the horizon only briefly at noon, the surveyors had to struggle along in the gloom of long twilights. Only the whiteness of the snow and a steady shower of meteors, Maupertuis said, gave them light enough to work four or five hours every day.

Describing the "Fatigues and Dangers," Maupertuis wrote:

> Judge what it must be to walk in Snow two foot deep, with heavy Poles in our hands, which we must be continually laying upon the Snow and lifting again: In a Cold so extreme, that whenever we would taste a little Brandy, the only thing that could be kept liquid, our Tongues and lips froze to the Cup, and came away bloody: In a Cold that congealed the Fingers of some of us, and threatened us with yet more dismal Accidents. While the Extremities of our Bodies were thus freezing, the rest, through excessive Toil, was bath'd in Sweat. Brandy did not quench our thirst; we must have recourse to deep Wells dug thro' the Ice, which were shut almost as soon as opened, and from which the Water could scarce be conveyed unfrozen to our Lips; and must thus run the hazard of the dangerous contrast which iced Water might produce in our heated Bodies.

Despite the cold, the surveyors measured the line twice before they returned to Tornio to sit out the rest of the winter. Maupertuis made his calculations. The length of the degree in the polar regions, he found, was 111.094 kilometers. This was more than a half kilometer longer than a degree as measured in France.*

The Earth was definitely not a perfect sphere.

Early in the summer of 1737, after rechecking and verifying the entire network of triangles and taking new star sightings, the Maupertuis expedition left Lapland and returned to France. On November 13 Pierre Louis Moreau de Maupertuis stood before the Royal Academy of Sciences in Paris and delivered his account of ordeal and success, concluding with the declaration: "Whence it is evident, That the Earth is considerably flatted towards the Poles."

The Academy had yet to receive an equally definitive report from its expedition to Peru.

L apland was almost a picnic compared to Peru.

The expedition to measure a degree at the equator embarked from France in May 1735. At the time, with equatorial Africa still a wild, uncharted region, South America was the only large and

---

*Subsequent measurements showed that Maupertuis made a relatively large error—of more than 360 meters. His degree was too long. Had the error been on the short side, the question of the shape of the Earth might have gone unanswered for decades or longer.

accessible landmass in the southern hemisphere. And Quito, a city in the province of colonial Peru that is now Ecuador, lay just south of the equator, a convenient staging base for the survey. Permission was obtained from the Spanish king, and the expedition sailed for Peru by way of Martinique and Santo Domingo, arriving there in March 1736.

The leaders of the party, Pierre Bouguer and Charles Marie de La Condamine, were prominent young members of the French Academy. Bouguer, who was born in Brittany in 1698, was a professor of hydrography and a pioneering investigator of the absorption of light in the atmosphere. He joined the expedition at the last minute, despite, as he said, "a proneness for seasickness." La Condamine fit more the mold of scientist-explorer. Born in Paris in 1701, he had abandoned a military career for the study of science and geography. An earlier scientific expedition to the Middle East had whetted his appetite for adventure and given him some experience in leadership. At Quito, La Condamine assumed virtual command of the Academy's expedition.

Among the other members of the expedition were Louis Godin, another academician; a botanist; a surgeon; a French Navy captain, who was an engineer; a draftsman; an instrument-maker; a young boy, the nephew of the Academy's treasurer; and two Spanish lieutenants, who were able mathematicians. They were a large and diverse group of men and not a happy group, or particularly fortunate. They encountered, as Bouguer said, "difficulties not to be imagined."

All went well at first. Indians brought them gifts of fruit, helped them build a station, and provided them with horses. When they reached Quito, the welcome was even warmer; they were fêted as heroes. Soon, however, the Spanish and Peruvians grew suspicious. Why were these Frenchmen there? Could they be on the trail of hidden gold? The people of Quito were not at all satisfied with the Frenchmen's explanations. Who would cross the ocean and climb the Andes merely to measure the Earth's shape? Finally, La Condamine had to suspend the survey and make an eight-month journey to Lima and back to get papers from the viceroy that enabled them to work without interference.

The expedition's triangulations extended along a 3° line, from north of Quito and the equator down to the south of Cuenca. Nearly every step of the way was a struggle, every angle measured a testament to their endurance.

Eighteenth-century French maps of
Lapland (top) and South America (bottom)

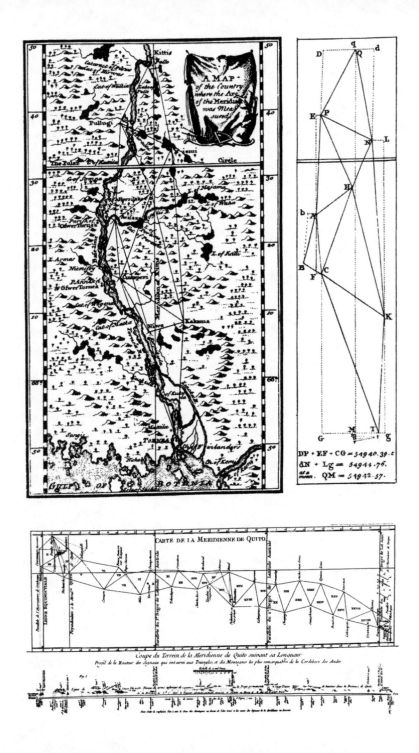

DP + EF + CG = 54940.39. t
dN + Lg = 54944.76.
at a
mean. QM = 54942.57.

CARTE DE LA MERIDIENNE DE QUITO

Coupe du Terrein de la Meridienne de Quito suivant sa Longueur
Profil de la Hauteur des Signaux qui ont servi aux Triangles, et des Montagnes les plus remarquables de la Cordelière des Andes

In his report to the Academy, Bouguer said: "We have sometimes been obliged to purchase, for a month and a half's patience, a single quarter of an hour of fine weather; and in one of these stations we have been longer detained, than we should have been toiling through a whole meridian in Europe. We were working, too, in a country, to which even its inhabitants themselves were strangers; and obliged continually to penetrate into deserts, where no paths but those made by wild beasts were to be discovered."

On the upward trail, the mules had to stop for breath every seven or eight steps. The party would sometimes spend an entire day getting across a treacherous ravine. High in the Andes, at altitudes upward of 3,600 meters, members of the party fainted, vomited, and were subject to "little hemorrhages" of the lungs. "We were continually in the clouds," Bouguer said, "which absolutely veiled from our sight every-thing but the point of the rock upon which we were stationed." Storms swept away sighting signals. Rain would fall for days on end, making it impossible to build a fire. "We had to live on bad cheese, and biscuit made partly from maize," Bouguer recalled.

At Pinchincha, a rocky summit outside Quito, Bouguer hit upon a surprising discovery. He noticed, by observing the slower swing of his pendulum, that the value of gravity in the mountains, even after he had accounted for the altitude (and thus the greater distance from the center of the Earth), was less than in the plains not far away. He had expected the opposite to be the case, thinking that the huge mass of Andean rock would exert considerable additional attraction on his pendulum. In reporting his observation, Bouguer expressed his opinion that the density of rocks in the Andes was less than that of the rocks in the plains.

Bouguer's observation was the first hint that the density of the Earth's crust is not homogeneous and, consequently, that the Earth's gravity is not fixed but varies from place to place on the surface and in the interior. This further complicated the work of surveyors and geodesists. It was becoming clear not only that the Earth is not a perfect sphere but that it also may be mined with gravitational booby traps for unsuspecting surveyors. Could they ever again trust their pendulums and plumb bobs?

It would be another century before surveyors and geodesists began to understand density variations. Even so, credit for the discovery goes to Bouguer. The differences between observed values of gravity at one place and the mean value are called Bouguer anomalies. Maps of Bouguer anomalies are increasingly important tools today in geophysics.

Toward the end of the Peruvian triangulations the expedition received word that the Lapland surveyors had already returned to France with convincing evidence in favor of an oblate Earth. It was a bitter blow to La Condamine's men. For all their efforts, all the suffering and risk, they would not taste the triumph of being first to bring back the answer to one of Europe's outstanding scientific questions. Although they still had work to do, they hardly had the spirit for it.

Instead, the party retired to Cuenca to rest. Tragedy followed. The surgeon became involved in a local scandal and was killed during a riot at the bullring; Bouguer and La Condamine barely escaped the stone-throwing mob, taking refuge in a church. The botanist, having lost his collection of specimens, suffered a mental breakdown from which he never fully recovered. The draftsman died in a fall from a scaffold while carrying out architectural work on a church. Even earlier, the young boy had died of fever.

Nonetheless, in 1742 and early 1743, Bouguer and La Condamine completed the survey, Bouguer working at the northern terminus of the arc, La Condamine at the southern. By taking several simultaneous star sightings, they determined the latitudes at each end of the arc and could thus translate the network of triangles into a measurement of a degree of meridian at the equator. The result was 109.92 kilometers—compared to the 111.094 in Lapland. The Earth was definitely more curved in Peru, at the equator, than in Lapland, at the Arctic Circle.

By all standards, except scientific, the expedition to Peru came to a sad and disorderly conclusion. Bouguer and La Condamine were no longer speaking to each other, and so they made their separate ways back to France. The two Spanish officers in the party brought a lawsuit against La Condamine for his failure to include their names on commemorative pyramids placed at either end of the base line near Quito. Godin had married a thirteen-year-old Peruvian girl, Isabella, and strange, romantic circumstances caused him to remain in South America for a quarter of a century.

Pierre Bouguer was the first to return to France. On November 14, 1744—nine and a half years after the expedition had left—he stood before the Royal Academy of Sciences and reported: "The experiments already made to ascertain the weight of bodies which are found to diminish therein in proportion as they draw towards the equator; the various operations undertaken in France to measure the extent of the degrees of both latitude and longitude; everything, indeed, concerns to satisfy us, that the earth is not completely spherical."

Three months later, Charles Marie de La Condamine arrived in Paris with incredible tales of his return voyage down the Amazon River by raft, the first such journey by a foreigner, and with a weary lament: "We returned seven years too late to inform Europe of anything new, concerning the figure of the Earth."

T he two French expeditions, Voltaire said, had been enough to "flatten both the poles and the Cassinis." They had proved beyond doubt that Newton was right about centrifugal force and the shape of the Earth. The Cassinis had been misled by an accumulation of small errors in their arc measurements. The Earth was not prolate, it was an oblate spheroid.

According to subsequent refinements, an east-west degree along the equator measures 111.321 kilometers—for an equatorial circumference of 40,075 kilometers. A north-south degree is 110.567 kilometers at the equator, gradually increasing in length until it measures 111.900 kilometers as it reaches the poles. The circumference of the Earth along any meridional circle is 67.2 kilometers shorter than the circumference along the equator.

The gathering of this new knowledge about the shape and size of the Earth marked an important turning point in geodesy and cartography. The history of geodesy is usually divided into three epochs. From Eratosthenes to Jean Picard, in the seventeenth century, was the spherical epoch, a time when the Earth was thought (except in myth and dogma) to be a sphere and was measured and mapped accordingly. Geodesy entered its second epoch, stimulated by the theories of Newton, with the introduction of advanced surveying tools and methods, particularly triangulation. The more precise surveys led to the new conception of the Earth as being a less than perfectly round body—a spheroid or ellipsoid. The ellipsoidal epoch extended until the twentieth century, when even more sophisticated instruments permitted geodesists to define the figure of the Earth, every variation and undulation, with incredible exactitude. This was to become known as the geoidal epoch.

For cartography the new precision of the ellipsoidal epoch was already sufficient to lay the foundations of scientific mapmaking. The framework of any map is derived from a knowledge of the size and shape of the entire Earth and from measurements of the length of a degree along a meridian. The more reliable the measurements, the more reliable the latitude and longitude coordinates of a map. This had been Ptolemy's message long ago. But it had not been, could not be, a systematic practice of mapmaking until the late seventeenth century. Only then, with the measurement of the degree, the development of triangulation, and the emerging knowledge of the Earth's shape, did it become possible to draw maps on which the relative positions of distant points were fixed with increasing accuracy.

The pursuit of accuracy transformed the character of mapmaking. Cartographers left the printing shops and cloisters and went into the field. They fanned out, soldiers and adventurers, navigators and scientists, to survey the land and the sea. Their surveys helped define the known and encompass the unknown.

The Family That Mapped France

In April 1669 Giovanni Domenico Cassini arrived in Paris and was presented forthwith to France's young and very ambitious monarch, Louis XIV. Louis coveted power and reveled in splendor, and to him this meant he had to have not only the mightiest army and most magnificent art and architecture, but also the most luminous science. Toward this end the Royal Academy of Sciences had been founded, and offers of generous salaries and research funds and a chance to work with some of the newest instruments attracted many of Europe's best scientists, Cassini among them.

The forty-three-year-old Cassini was a rising figure in European science. He was a surveyor of waterworks and fortifications, professor at Bologna, astronomer to the Pope, even an experimenter in blood transfusions from one animal to another. But it was his publication of tables of the eclipses of Jupiter's four major satellites that brought Cassini to the French king's attention and to Paris.

When he accepted the Academy's invitation and went before the king, however, Cassini had no intention of remaining in Paris for more than a few years. He wanted only to use the Academy's new facilities for some further studies of Jupiter. But Jean Baptiste Colbert, Louis's influential minister of finance and an amateur scientist himself, had other ideas. He wanted Cassini to work on one of his (and the king's) pet projects, the mapping of France.

Colbert had for years deplored the state of cartography. He had roads and canals to build. He had many schemes to invigorate the economy and thereby magnify his king's power and stature. But, when he had asked to see maps of the provinces where this work might be done, Colbert discovered that the maps were either nonexistent or woefully incomplete and inaccurate. He discovered, moreover, that no one seemed to know how to provide him with better maps.

The country's mapmakers still practiced cartography in a rather medieval manner. They compiled and reworked old material, reprinted from old plates, perpetuated misinformation, and rarely went into the field to observe and measure. When would they come down from their ivory towers and out of their printing shops? When would they concentrate less on decorative maps and more on those that would be useful and well surveyed?

Soon, Colbert vowed. He wanted the kind of maps that indicated man-made and natural features as determined by precise engineering surveys and measurements. They would portray the shapes and elevations of mountains, valleys, and plains; the network of streams and rivers; the location of cities, roads, political boundaries, and other works of man. In short, Colbert wanted topographic maps.

At the urging of Colbert, Cassini agreed to stay in France and help with the surveys by which such topographic maps could be constructed. In 1673, the Italian-born Cassini became a French subject and changed his name to Jean Dominique Cassini.

His decision marked the beginning of a remarkable dynasty in cartography. French mapmaking, it seemed, often ran in families. There were the Sansons, from Nicolas in the seventeenth century through three sons, a son-in-law, and a grandson, all publishers of maps and atlases. There were the Delisles, Claude the father and two sons, Guillaume and Nicolas. There were Philippe Buache and his nephew, eighteenth-century marine cartographers. But none of these families could match the innovative drive and scientific spirit of the Cassinis. Over a period of more than a century the Cassinis—father, son, grandson, and great-grandson—surveyed and mapped the whole of France. Through their efforts France became the first country to be mapped in its entirety on a basis of extensive triangulation and topographic surveying.

Cassini made his first major contribution to the mapping of France by doing what he did best and liked to do most; he studied Jupiter and its satellites. He looked to the heavens for a solution to one of cartography's most nagging problems: how to find the longitude of a place on Earth.

Surveyors and navigators had already become adept at finding latitude, the position of a place north or south of the equator. Cassini's contemporary Jean Picard was particularly skilled in using angle-measuring instruments and mathematical tables to fix latitude by determining the angular height of the noon Sun above the horizon or the height of the Polestar. Maps of the time could often be trusted on latitude, but not longitude.

Longitude is also an expression of angles. The longitude of a place is its angular distance east or west of 0° longitude, the prime meridian. (On most maps today this is an imaginary north-south line drawn through the Royal Observatory at Greenwich, England, but in Cassini's time it usually was shown to run through the island of Ferro in the Canaries.) The longitude of a benchmark in New York City, for instance, is 73°59′31″ W. That is its angular distance west of Greenwich. If you could slice a long wedge out of the Earth along the meridians at Greenwich and at New York, and if the wedge reached to the axis of the Earth, the exposed angle at the axis would measure 73°59′31″.

But longitude can be more readily understood and reckoned as a function of time. In fact, meridian means midday. Since the Earth rotates 360° every 24 hours, it turns 15° every hour, or 15′ every minute. When it is noon at Greenwich, it is an hour before noon at 15° W, six hours before noon at 90° W, and midnight on the opposite side of the Earth at 180°. Because of this correlation between longitude and time, the meridians on most world maps are counted by 15s, spaced to represent one hour's turning of the Earth. Likewise, time zones conform more or less with each 15° of longitude.

Since time and longitude go hand in hand, it should be a simple matter to determine the position of a place east or west of the prime meridian. Simple, that is, if you know simultaneously the local time and the time at the prime meridian.

In 1669, local time could be obtained from observations of the midday Sun or even with the available pendulum clocks, if they were regularly corrected. But there was no way of knowing the precise time at Greenwich or Paris when it was noon in some faraway, uncharted place or even somewhere within a country the size of France. Nowadays, a mapmaker can get an accurate time fix by short-wave radio or by having a precision clock set to the prime-meridian time—Greenwich Mean

Time. Neither was available in the seventeenth century, and this was the crux of the longitude problem.

If a solution was to be found, the prevailing assumption held that it would somehow be reached through astronomy. Much thought and effort were thus directed toward finding some method of celestial time-keeping for longitude to compare in accuracy with the celestial guides used in fixing latitude.

Galileo had first proposed using the moons of Jupiter. In the early seventeenth century, with the newly invented telescope, Galileo discovered the four largest Jovian satellites and studied their positions at various hours of the night and on various nights of the year. During twenty-four years of tireless observation he prepared tables of the satellites, plotting their motions and eclipses, and wrote detailed instructions for their possible use in finding longitude.

The idea was never tested during Galileo's time, which was probably just as well; his tables and telescopes were not adequate to the task. But after he was summoned to Paris, Cassini decided the time had come for the test. He and other astronomers at the Academy's observatory revised and enlarged his eclipse tables for Jupiter's satellites, completing the work by 1676. Cassini trained astronomers and surveyors and sent them into the field.

Observing Jupiter's satellites was not a job for the careless or impatient. Even through the best telescope the four largest moons were but faint pinpoints of light darting in and out of view. Bad weather often delayed observations for days or weeks. A moment's inattention could spoil a whole night's work. While one man recorded the time, another had to peer through the telescope and wait for a satellite to emerge from behind the brighter light of Jupiter or to reach some other predicted phase in its orbit. The time of the satellite's appearance had to be noted carefully. By referring to the tables, which predicted the time the satellite would appear that night, the observers could know what time it was in Paris. A pendulum clock set by star or solar sightings could tell them the local time at the same instant. The difference between the two times represented the difference in longitude between the observation site and Paris, or whatever prime meridian was being used.

Though slow and tedious in practice, the method began to produce results. Longitude fixes became a fairly routine part of the new surveys of France, which were now under way.

Picard's measurements of a meridian line north and south of Paris, though intended primarily to solve the degree question, served also as the starting point in the survey of France. His methods of triangulation were soon employed along the coast to determine more accurately the outlines of the country. In 1679 Picard and Gabriel Philippe de La Hire worked at Brest, Nantes, and points between, establishing everywhere they went the latitude by solar and star sightings and longitude by the eclipses of Jupiter's moons. This was followed the next year by a survey up the coast from Bayonne to Bordeaux and to the south of the Gironde. In 1681 the two men fixed the latitude and longitude of points around Cherbourg, then went their separate ways to survey Mont-Saint-Michel, Calais, and Dunkerque.

La Hire took the results of these and other coastal surveys and constructed a new outline map of France. When it was superimposed on the latest contemporary map, glaring discrepancies were exposed. Brest, for example, had heretofore been located too far west, out in the open sea. Marseilles had been placed too much to the south. Hardly any stretch of coastline had been properly located by either longitude or latitude.

Soon afterward, on May 1, 1682, Louis XIV paid a visit to the Paris Observatory. When he was shown the new map and told of the real cartographic boundaries of his kingdom, Louis exclaimed, "Your work has cost me a large part of my state!"

Cassini, meanwhile, had been corresponding with astronomers across Europe, instructing them in the Jovian method of finding longitude and encouraging them to make their own observations. This they did, and the results poured in to Cassini. Not surprisingly, the new data rendered all the maps of Europe obsolete.

The time had come, Cassini decided, to produce an up-to-date map of Europe and of the world. Since he expected an unceasing flow of data, Cassini conceived the idea of compiling a sort of open-ended map, one that could be updated with the arrival of each new bit of information. This would be a master map, which he called a *planisphère terrestre*.

Cassini cleared the third floor of the observatory's west tower. On the floor draftsmen inked in a circular map about ten meters in diameter. The north pole was placed at the center, with meridians radiating from it to the periphery at intervals of 10°; parallels of latitude were laid out at the same intervals. No place was included on the map unless it had been located to Cassini's satisfaction, which meant it had been fixed by careful astronomical observation. Several hundred cities and towns in Europe eventually found their way onto the planisphere, each being located with reference to the Ferro prime meridian and to all the other cities and towns.

The same high standard of precision was applied to the placement of coastlines and continents. Cassini was thus able to portray correctly the length of the Mediterranean—42°—for what is believed to be the first time.

Over the next few years the Academy sent astronomers on expeditions to Egypt, South America, and the West Indies. Their measurements fixed with greater precision the position of the New World relative to the Old. Jesuit missionaries trained to make latitude and longitude determinations traveled to Madagascar, Siam, and China. Their reports back to Cassini were scrutinized and, if the work seemed scientific and careful, the new data were duly recorded on the planisphere. A more correct outline of Eurasia began to emerge.

In 1696 Cassini transferred the information on the tower floor to paper, publishing a world map. The planisphere became the standard for many of the better world maps published in the early eighteenth century.

After the death of Picard in 1682, Cassini the astronomer had to devote more time to being Cassini the surveyor. It was not always easy. Colbert's death in 1683 cost him his most influential backer. Wars and straitened finances brought the surveys to a virtual standstill. Thus it would often be for the Cassinis, all four generations, as they struggled to map France.

By 1700, however, Cassini got support to resume work on Picard's meridian. Since he was then well into his seventies, he had to have the help of his twenty-three-year-old son, Jacques. The father and son ran triangles on a line through the south of France, with two objectives always in mind. There was the degree problem, of course, but also Picard's grand strategy. Picard had proposed that the mapping of France should proceed in two steps: first lay out a framework of triangulation, then fill in the topographic detail. He reasoned that until there was a carefully measured framework of points fixed by latitude and longitude (control points, they are now called), no one could be sure the towns, roads, hills, and other topographic detail were correctly placed on a map. The plan may seem commonplace today, since it is a practice generally followed in surveying, but at the time it was a new conception.

The elder Cassini died in 1712, at the age of eighty-seven. Jacques Cassini, though he continued to survey, became embroiled in the oblate-prolate controversy and felt that it was more important to define the length of a degree before going on with the mapping project. He was right, in the sense that the size and shape of the Earth had a bearing on the mapping of any particular part of the Earth. But for this and other reasons the younger Cassini's report to the Academy in 1733 was more a summary of what needed to be done than of what had been done. After more than sixty years, the Cassinis and their associates had produced an outline map of France, fairly accurate charts of the Atlantic coast, and an excellent topographic map of the Paris environs; the latitude and longitude of all major French cities were now plotted with considerable accuracy. But the framework for the greater map was little more than the one line of triangles from Dunkerque through Paris to the Pyrenees.

With the support of Louis XV, who had studied geography and considered himself an amateur scientist, Jacques Cassini set out to survey a line perpendicular to the so-called Picard line. From the summer of 1733 through the following year, he and his son, César François, ran a line of triangles from the Atlantic coast to Strasbourg. A second line, from Orléans to Brittany, was surveyed in 1735. During the rest of the decade, the two Cassinis, together or separately, surveyed the southern and western coasts and the northern frontier; other teams measured the eastern frontier, Normandy, and Brittany. César François Cassini even spent two grueling years resurveying the old Picard line. At the end of 1740, they had enclosed and crossed France by an uninterrupted chain of 400 triangles laid out from 18 fundamental base lines.

Never before had there been so much precision surveying concentrated in so few years. Out of this experience César François, the third-generation Cassini, emerged as the most persistent and accomplished mapmaker in his illustrious family. According to George Fordham, a historian of cartography, he was a man of "enthusiasm, honesty, and practical ability." In recognition of his many achievements he was made a count, Cassini de Thury.

Though an astronomer like his father and grandfather, and director of the Paris Observatory, Cassini de Thury devoted most of his time to cartography. In 1745 he presented to the Academy a map in eighteen sheets representing the results of surveys over the previous decade. It consisted of eight hundred connecting triangles, but only a minimum of topographic detail. To Cassini de Thury it marked the culmination of the first step in Picard's strategy, the framework for topographic surveys to come.

The following year Cassini de Thury went to Flanders to help map the Low Countries. Cassini ran triangulations as far north as Bergen op

Surveying by Henry Beighton, 1728

Zoom, a terminus of Snell's pioneering survey. Considerable topographic detail was added by military engineers. Then, when the triangulation and topographic surveys were combined, Cassini went to the king and showed him the results. Much impressed, the king immediately declared that the whole kingdom of France should be mapped in the same manner. He told Cassini to formulate the plan and had his comptroller-general make the necessary financial arrangements.

By 1750 the "infilling" survey had commenced. Royal engineers were working on the systematic mapping of the country's main roads. Cassini's surveyors, using plane tables, sketched in the rivers and canals, towns and châteaux, vineyards and other important landmarks, always locating them within the framework of the triangulation survey.

Six years later, when royal funding for the survey came to an abrupt end, Cassini was faced with the sad prospect of having to discontinue the whole operation. He hurried to Louis and showed him the most impressive of the new maps, but the king would not restore the financial support. Undaunted, Cassini hit upon another approach—a capitalist mapmaking venture. With the king's blessing, he rounded up a number of investors who agreed to buy shares in the project in return for proprietary rights to the published map, and so the survey continued.

In 1783, the year before his death, Cassini de Thury dispatched a message to the English government in which he proposed a cooperative mapping venture. He suggested that the two countries join in a triangulation survey extending across the English Channel. As an astronomer, Cassini cited the advantages of knowing more precisely the difference in latitude and longitude between the observatories at Paris and at Greenwich. As a cartographer, he recognized the necessity of laying a mathematical framework for tying the map of one country to the map of adjoining countries. Only in this way, triangle by triangle, country by country, could cartography move toward a scientifically based map of the world.

Cassini's proposal was referred to the Royal Society, the English counterpart to the French Royal Academy, where the initial reaction was somewhat frosty. Cassini had pointed out that, as a first step in the undertaking, the English would have to make a careful triangulation survey from London to Dover. The implication was unmistakable, and quite a few English scientists took umbrage. How could this Frenchman tell them that the land between London and Dover was not as well surveyed as that between Calais and Dunkerque!

But it was true. While the Cassinis had been mapping France, the English had neglected to initiate any such systematic surveys of their own.

No one knew this better than William Roy, the man eventually assigned by the Royal Society to direct the English side of the cross-Channel survey. Roy, a major-general, had learned surveying during campaigns on the Scottish border and had for years wanted to put his talent to work on a broader scale. He had tried to interest the government in a general survey of Britain, but got nowhere. Even so, he enjoyed scouting the countryside for likely triangulation routes and once, "for my own private amusement," measured a base line—all for a survey no one seemed to want. Then came the French proposal and his appointment. Roy plunged eagerly into the new project, hoping it would finally open the way to the fulfillment of his dream of a good general map of Britain.

No Frenchman could have had cause to criticize Roy's devotion to accuracy in the London-Dover survey. He spent nearly all the summer of 1784 measuring a base line of more than eight kilometers on Hounslow Heath outside London. He tried chains and wooden rods but finally settled on hollow glass tubing for the measurement; the glass did not contract and expand as much from weather changes.

One of the many visitors who came out from London to watch the work on the Heath was George III, whom Roy acclaimed as "the patron of the sciences." The king, now that the burden of the American war was lifted, promised his full support.

Nonetheless, it took another three years to proceed from the base line to the rest of the survey. This was how long it took Jesse Ramsden, the instrument maker, to build and deliver "the curious instrument" that Roy had ordered. Roy was driven by Ramsden's procrastination to the use of language so harsh that it had to be excised from his Royal Society paper before its presentation. But the theodolite was the most advanced surveying device yet developed, "rendered extremely perfect," Roy said, for measuring both horizontal and vertical angles to within fractions of a second of arc. The ninety-kilogram instrument consisted of a graduated brass circle nearly a meter in diameter, fitted with fine-adjustment attachments, telescopes, and lanterns for night observations.

The theodolite was immediately put to use in the summer of 1787. Roy began running a string of connecting triangles, each 20 to 30 kilometers to a side, from Greenwich toward the coast. But the going was slow, and with the "French gentlemen" waiting and the autumn rainy season approaching, it was decided to defer the rest of the intermediate triangulation. Roy proceeded to Dover, where on September 23 he met the French delegation led by Jacques Dominique Cassini, who had upheld the family tradition by taking charge of the project from his late father. After two days of cordial meetings, the plans were set.

Signal towers were erected at four stations—at Dover Castle and Fairlight Head on the English coast, at Cape Blancnez and Montlambert on the French side. Theodolite observations from the four stations would give the horizontal angles necessary to determine two principal triangles—Fairlight-Dover-Montlambert and Montlambert-Blancnez-Dover.

The angle measurements would be made at night. Roy was an early advocate of night surveying, which was to become a common practice, as a means of minimizing sighting errors due to "the tremulous motions or boiling in the air." A mirage of air boiling off a highway or desert is an extreme example of the tricks the atmosphere can play on the eye in daylight. By working at night, moreover, the observers could more certainly see the illuminated targets through their telescopes. On each tower there glowed a bright white light of burning lime—limelight.

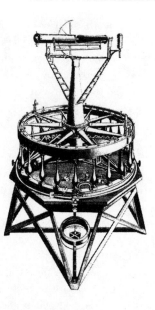

Ramsden's theodolite,
eighteenth century

No surveyors had ever before dealt with such large triangles. It was 75 kilometers from Fairlight, near Hastings, to Montlambert, somewhat inland and near Boulogne; 77 kilometers from Fairlight to Blancnez, below Calais. On many nights the fog and rain frustrated the surveyors. But on the few clear nights they were able to intersect their distant targets of flickering white light. The Channel had been crossed, as Roy said, "to establish for ever, the triangular connection between the two countries."

Afterward, Roy and other surveyors completed the triangulation from Dover to Greenwich. In February 1790, Roy read his formal report to the Royal Society—"An account of the trigonometrical operations, whereby the distance between the meridians of the Royal Observatories of Greenwich and Paris has been determined." He seized the opportunity to make another plea for a general survey of Britain. He was convinced that, with the development of Ramsden's theodolite and the experience of the cross-Channel survey, the country was ready. And, with a nod to the Cassinis, Roy sought to shame his countrymen into "having at least as good a map of this as there is of any other country."

A memorandum written by Roy for the Duke of Richmond, who

Part of a map of the triangulation of France,
Giovanni Maraldi and Jacques Cassini, 1744

later created the national survey of Britain, provides an excellent summary of surveying techniques as practiced toward the end of the eighteenth century. Entitled "General Instructions for the officers of Engineers employed in surveying," it reads in part:

> In every District to be surveyed, the first thing to be considered will be; What situations are the best, for the Base or Bases to be measured, as the foundation of the work, and for connecting the different Serieses of triangles together?
>
> These Bases should therefore be as long as the circumstances of the ground will permit; not less than a Mile, or a Mile and a half: and as often as possible, they should be measured on the sand of the Sea Shore; because in such cases no Reduction of any kind will be necessary on account of difference of level.
>
> Every Base should be measured at least twice; and oftner if there should be any remarkable disagreement between the first and second measurement. For this purpose, one chain should be kept as a standard, with which those in common use will be from time to time compared at least at the beginning and end of any operation, that a true mean may be taken for the ultimate length. And with a view to still greater accuracy it will be proper to observe the heat of the Air, as shown by the Thermometer at stated intervals, while the operation is going on.
>
> The principal Triangles, connected with the Base or Bases, will be such as are nearly equilateral, formed by the Church Steeples, Windmils, single trees, or other conspicuous objects and in each of these Triangles all the three Angles should be as often as possible actually observed with the large Theodelet, that the reduction to 180° may be properly made.
>
> In many cases it will be advisable and even necessary, to establish signals by Camp Colours or otherwise on the chief eminences, whose situation being permanently marked on the ground, so as to be referred to occasionally, will form so many auxiliary triangles for connecting the Survey, where other remarkable Objects may be wanting.
>
> With regard to profil or elevation, the relative heights of the angles of the great triangles, are such, as on all occasions should be first determined: Because these being once settled, the relative heights of all other chief commanding points of any general Range running parallel to the shore, or to a River, such as an Army would occupy to oppose the Descent of an Enemy on the coast, or his penetration into the country after he had effected a Landing, will be subsequently ascertained with respect to the first; and the whole should refer to Low Water Mark at spring Tides, by some permanent Mark, taken on a Quay or Wharff, or some other substantial Building situated near the Shore, to which reference may be had on any future occasion.

In certain cases, the smaller heights near the shore will be best determined by the accurate application of the Telescopic Spirit Level. But in general, the Business will be greatly expedited, by taking the Angles of elevation or depression, with Ramsden's best Theodelet, from some centrical point, whose distance from a number of others has been already ascertained by trigonometrical computation, and from which point all the others can be distinctly seen.

In leveling, if the Telescopic level be adjusted by inversion in its Ys at any intermediate point exactly half way between the two Station Staffs, the relative heights of the vanes, or their distance from the centre of the Earth will be obtained at once, without any allowance for curvature or Refraction. But by the Angles of elevation or depression, allowance, according to the distance, must be made for curvature and Refraction, at the same time, that great accuracy must be observed in adjusting the Instrument, and taking the Angles repeatedly, that a true mean may be obtained for the ultimate Result.

One method of keeping the Books must be adhered to by all the Engineers, that any one of them may be able to lay easily down the observations of the others. Perhaps for common Surveying, the best kind of Book would be one of the quarto size, with certain Columns ruled on the page towards the left hand to contain the Angles and measured distances of the stations, commencing at the bottom of the page; while the Right hand page, contained the corresponding Sketch or Eye Draught.

The Commanding Engineer on the spot will charge himself with the determination of the great triangles, and will register in a Book kept by himself everything concerning them; as well as what may relate to the relative heights, whether determined by the Level or by the vertical appartus of the Theodelet.

The filling in or surveying the interior part of the great triangles will probably be executed in the common manner by the Junior Engineers, with the small Theodelets and chains provided for the purpose. They will consequently proceed along the contours and Creeks of the shore; along the great Roads and lanes; and also along the courses of the Rivers, Rivulets. The Boundaries of Forests, Woods, Heaths, Commons or Morasses, are to be distinctly surveyed, and in the enclosed parts of the Country all the hedges, and other Boundaries of Fields are to be carefully laid down. Altho' the exact Turn of every one need not be surveyed, if frequent Cuts in different directions are made thro' the inclosures and the direction of the fences laid down where they intersect these cuts, the remainder may generally be taken by the Eye.

The Risings or irregularities of the ground are every where to be expressed with care; so as to render the plan truly topographical, by preserving that gradation or keeping which should distinguish at first sight the higher part above those that are lower; and these last above such as are quite flat. To do this in the best manner, the plan of the Lines, or great features of the Country, should be first laid down; which being done, the particularities of the surface, will be more readily and truly represented afterwards.

The first survey will be made from the Magnetic Meridian; But in every district it will be necessary by observation of the Sun or Stars to determine a true Meridian; by ascertaining the Angle that it makes with some one of the longest sides of the great triangles; whence the variation of the Compass will at the same time be determined.

Each Field cannot be represented on less than two Inches to a mile; which may be that generally made use of for the general plan. Particular Sea Ports of consequence, such as the Thames and the Medway &c, will require a scale of about six Inches to a mile. These may afterwards be reduced to a Scale of one Inch to a mile for the Island in general.

A Book of general miscellaneous Remarks should also be kept, wherein may be entered every thing that occurs relative to the nature of the Coast, such as, what parts of it are accessible, and what not; at what distance from the shore, ships of war may come to an Anchor to cover a debarkation from Boats; and what sort of communications there are leading from the Coast to the interior Country, in case an enemy had made his landing good: also the nature of the soil, how far the clay, chalk, or Gravel country Extends. Whether there is plenty of Timber and of what sort, and size. And as in every chalk country the Rivulets generally run underground & the Inhabitants are supplied by wells dug to a considerable depth; it will be proper to mention the depths of such wells.

Roy was not fated to live to see such a survey. England's first great surveyor died later in 1790. In the following year, George III was persuaded to launch a national survey as a military project. This was the beginning of the Ordnance Survey, the British government's agency for topographic mapping.

At the death of Cassini de Thury in 1784, all but a few sheets of the *Carte de Cassini,* as the map was being called, had been completed. The end of more than a century of work was in sight.

The map—182 sheets on the scale of 1:86,400—was eventually completed and published in 1793. The sheets, when all assembled, measured about 11 by 11 meters. If it lacked the esthetic appeal of earlier maps, being crammed with detail at the expense of artistic flourish, the *Carte de Cassini* fulfilled Colbert's requirements. It was scientific, complete, and useful. It showed roads, rivers, canals, towns and hamlets, abbeys, châteaux, vineyards, marshes, lakes and ponds, even windmills. The only serious flaw was its crude qualitative representation of relief.

At the time of the Cassinis, mapmakers had yet to figure out a satisfactory way to depict a third dimension on a two-dimensional map. Earlier cartographers simply drew tiny pictures of hills, rows of mounds like molehills or sharp peaks like dragons' teeth, and left it at that. Others drew "hatches," which were fine lines that through a shading technique made objects stand out in relief. If done poorly, as was often the case, the result looked more like hairy caterpillars than mountains. The French refined the method, calling it *hachure*. By making the hachures thick or thin, widely spaced or packed together, slanted, shortened or lengthened, the cartographer could give an impression of contour and slope. At its best, as in the G. H. Dufour atlas of Switzerland in the mid-nineteenth century, the effect could be strikingly realistic. But still it was only an impression; absolute elevations could not be shown by hachures alone.

Another method, hypsometric tinting, would soon be tried. Successive zones of elevation were colored differently, usually following the spectrum—starting with green for low land and proceeding through the buffs and oranges to dark red or brown or even purple for higher elevations. This made for handsome, easy-to-interpret maps (and it is still useful in many maps), but there were often not enough tints to depict more than gross differences in the lay of the land.

Most cartographers eventually settled on contour lines as the simplest and most informative method of showing relief. Those long, faint, undulating lines may be a puzzle to many map readers, but actually a contour is nothing but an imaginary line, every part of which is the same distance above sea level. The nearest approach to a contour line in reality is the outline of a lake; if the level of a lake dropped one foot each night for ten consecutive nights, leaving a different shoreline ring behind each time, you would see, in effect, contours of the sloping shore.

Contour lines provide a way of showing vertical measurements, as distance scales are a way of showing horizontal measurements. A look at contour lines and how closely they are drawn gives one a quick idea, for example, of a plain. On a map the vertical distance between lines—the

OVERLEAF: *Topographic map of Paris and environs, from the* Carte de Cassini, *published in 1793*

"contour interval," usually noted at the bottom of the sheet—can be anything from 1 to 100 meters, perhaps even more. A map of flat country, like a prairie or desert, would require a contour interval of perhaps 1 meter in order to accentuate whatever variations there are in elevation. A map of mountainous terrain would have to have much larger contour intervals, or else the lines showing a steep slope would stack one on top of the other—6-meter contours would be hard to see separately on slopes greater than 26°.

The earliest use of contours on wide-area topographic maps is attributed to a French engineer, J. L. Dupain-Triel, whose map was produced in 1791. Not until the mid-nineteenth century, however, did contouring become an accepted part of topographic mapmaking. The Cassinis made do with hachuring, which was better than hairy caterpillars but hardly as expressive as the later Swiss shading techniques, and the insertion of occasional spot elevations, which were usually derived from barometric readings of varying reliability.

Publication of the completed Cassini map coincided with the French Revolution. Jacques Dominique Cassini, though a royalist whose forefathers had been grateful beneficiaries of the Bourbons, saw no reason for this to make a difference. He was a scientist and cartographer, not a politician. And so he went before the revolutionary government and presented the sheets: here was a project initiated by the *ancien régime,* but for the benefit of all France. The new government liked the map so well, recognizing its military and political value, that it agreed to support further publication costs—and to back revisions as they became necessary.

Nevertheless, in 1794, during the Reign of Terror, the revolutionists arrested Cassini for his royalist sympathies. He was forced to spend seven months in prison, but left with his head still on. Later, Napoleon Bonaparte honored Cassini, proving no doubt that the mapmaker had been faithful to the inscription on the Cassini home in Paris: "Search for truth but follow the middle road."

With the completion of the map of France, the long line of Cassini cartographers came to an end; Jacques Dominique Cassini's son gave up astronomy for botany. But the *Carte de Cassini* became the basis for all future maps of France and the standard for national topographic mapping.

Today, topographic maps are the most versatile and useful type of maps. More than 10 million copies of such maps are distributed annually by the United States Geological Survey to engineers, scientists, industrial and urban planners, real-estate developers, hunters, fisher-men, and other outdoors people. They are also used as bases for more specialized maps—the aeronautical, coastal, census, flood zoning, geo-logic, soil, weather, and road maps.

Colbert saw the need. Picard devised the strategy. But it was the genius and industry of the Cassinis, all four generations, that proved topographic mapping to be achievable on a national scale.

John Harrison's Timepiece

An urgent petition reached Parliament on March 25, 1714. Certain "Captains of Her Majesty's Ships, Merchants of London, and Commanders of Merchantmen" wanted something done about the day's most pressing problem in navigation, the problem of longitude.

If the elder Cassini thought he had solved the problem with those moons of Jupiter, he should have heard the profanity and obscenities being uttered on the waterfront or on any of the ships at sea. Those long, heavy telescopes and cumbersome pendulum clocks might be all right for some astronomer in Paris, but no seaman could use them to track and time a tiny Jovian moon from the rolling, tossing deck of a ship. There had to be a better way of finding longitude at sea.

Until there was, ships could never be sure of their positions east or west of home port or of anywhere else. They could never be sure if they were a few kilometers or many kilometers from a particular landfall, which would, in either case, probably not be very well mapped, also because of the longitude problem. Neither the explorers who found a place nor the cartographers who put it on the map, having themselves had only a rough idea of longitude, could have ever known exactly the position of the place. Thus, as long as the problem of longitude at sea defied solution, both navigation and cartography suffered.

The longer the problem went unsolved, the graver became the situation. In 1707 four British naval vessels and two thousand men were lost in a wreck off the Scilly Isles as a result, in large part, of an insufficient knowledge of their longitudinal position. This was no isolated incident, as the petitioners pointed out to Parliament: "That the Discovery of the Longitude is of such Consequence to Great Britain for the safety of the Navy and Merchant-Ships as well as for the improvement of Trade, that for want thereof many Ships have been retarded in their Voyages and many lost. . . ."

What affected the navy and maritime trade was no trifling matter to Englishmen. So Parliament voted to offer a reward "for such person or persons as shall discover the Longitude." The offer: £10,000 for any method capable of determining a ship's longitude within one degree; £15,000, within 40′; and £20,000, within half of one degree. A permanent board of examiners, composed of scientists and admirals, was given responsibility for evaluating proposals and judging the results of accuracy tests. This became known as the Board of Longitude.

Parliament's action set in motion one of the earliest examples of goal-oriented, state-directed research. It and similar offers by other maritime nations stimulated astronomers, mathematicians, inventors, and not a few cranks to bring their minds to bear on the longitude problem.

Many approaches to a solution had already been tried. Mariners once thought that Earth's magnetism might hold the key. As early as Columbus and the Cabots, explorers had noted variations of the compass needle from true north. On their voyages they plotted the variation, or declination, by taking a bearing on the Polestar and then noting how many degrees the needle pointed east or west of the star. For a long time it was hopefully assumed that the variation changed at a uniform rate corresponding to changes in longitude. But research in magnetism during the seventeenth century reached a disheartening conclusion: the variation lacked uniformity.*

This left mariners with nothing better than dead reckoning, which often was little more than an educated guess. By keeping a careful record of compass courses steered and of the ship's speed, mariners

---

*Many years thereafter, in 1777, John Churchman of Maryland, a surveyor and self-taught student of magnetism, conceived a new theory of the variation of the magnetic needle whereby longitude could be determined, or so he said. According to him two natural satellites were revolving around the Earth, one around the North Pole and the other around the South Pole. They had never been seen by anyone, he said, because they were not visible from the lower latitudes. The satellites caused the variation in a predictable manner according to longitude. Of Churchman's theory, Francis Hopkinson said: "This learned Man understands as much of Philosophy as can be acquired by practical surveying; and as much of Navigation as can be obtained from paddling a canoe."

could estimate where they were east or west of their home port, or, for purposes of determining longitude, from the prime meridian. The ship's speed could be estimated only by the rather crude device known as the *logline*. This consisted of a long rope attached to a weighted piece of wood, often a chip off a log. When the chip was tossed overboard, it remained stationary in the water while, as the ship moved forward, the line was paid out. The operation was timed by a sand glass. When the logline was hauled in, a measurement of the length of line used provided an indication of the distance traveled in a specified time, which then gave the mariner an estimate of the vessel's speed per hour. These periodic measurements were duly recorded in a journal. This came to be known, after the type of measurement device, as the ship's log. Oftentimes, to facilitate calculations, knots were tied at measured intervals in the rope, giving birth to another seagoing expression— knots, for nautical miles per hour.

Dead reckoning and loglines were found wanting, particularly on long voyages. A pilot in Magellan's fleet proved to be almost 53° off in calculating the longitude of the Philippine Islands. Is it any wonder that mariners despaired of ever having a reliable means of determining longitude?

Astronomers never gave up, however. The Royal Observatory at Greenwich Park, outside London, had been established in 1675 for "the advancement of navigation and nautical astronomy." Charles II had made it clear that longitude should be an overriding concern. He is said to have informed the Royal Society that he was "graciously pleased" with the Society's work on useful problems such as finding the longitude at sea, but not their "childish diversions" such as weighing air.

It was at this time that Cassini was perfecting the Jovian method of finding longitude, at least longitude of places on land. But another approach favored by many scientists was based on the fact that the Moon changes its position very rapidly with respect to other heavenly bodies. The angular distance between the Moon and a prominent star changes 13° in every 24 hours. If the lunar motion could be observed by the new reflecting octant and perfectly tabulated, time by the Moon could be accurate to about two minutes, once atmospheric refraction and Earth's motion were allowed for. Two minutes of time is half a degree of longitude. If this method could be perfected, the £20,000 prize would be won.

The Moon was certainly easier to observe than Jupiter's faint satellites. But in 1744 the existing data on the motions of the Moon were too unreliable, being based on imprecise observations, to be of much help. The lunar method was not ruled out, but it was not yet prizeworthy.

Since dead reckoning, magnetic declination, and Jovian satellites had been discarded, and the lunar method remained an uncertain prospect, there seemed to be only one other possible approach—a highly accurate mechanical clock to be carried on ship. The clock would keep the time of the prime meridian, whether at Greenwich or Ferro or Paris. By comparing this time with the local time at sea or on some distant shore, it would be possible to know one's longitude relative to the prime meridian.

This idea had been proposed as early as 1530, when clocks were still a relatively new invention, crudely built and hardly accurate enough for longitude calculations. Even the addition of the pendulum, another of Galileo's ideas, had failed thus far to improve clocks sufficiently for this work. In tests at sea, the motion of the ship upset the regularity of the pendulum. Clocks run by springs proved unacceptable owing to the effects of temperature; the coiled spring loses strength when heated and gains it when cooled. Some clockmakers proposed keeping the timekeeper in a brass box over a stove or in a partial vacuum, neither of which proved practical.

A superior clock just might be the solution, though, as the eminent Isaac Newton, by then president of the Royal Society, said in 1714. But its invention would not be easy. Appearing before a Parliamentary committee convened to consider the longitude petition, Newton said: ". . . by reason of the Motion of a Ship, the Variation of Heat and Cold, Wet and Dry, and the Difference of Gravity in Different Latitudes, such a Watch hath not yet been made."

Could it ever be? This was the challenge that preoccupied a man named John Harrison for nearly all of his long working life.

Harrison was twenty-one years old when Parliament announced the prize. He was born in 1693 at Foulby in Yorkshire, the eldest son of a poor carpenter. An attack of smallpox at the age of six may have been decisive in shaping his life. While convalescing, he became fascinated by a watch his parents, seeking to amuse him, had laid on his pillow. He listened to its ticking for hours and studied its moving parts. He never forgot that watch.

In subsequent years young Harrison learned his father's trade, earned some extra money surveying land, and in the evenings read lectures on mechanics and physics. By 1714, he had learned to make and repair clocks, first as a sideline, then as his full-time trade. The more he worked with clocks, the more he applied his emerging inventive genius to improving the way they were built.

For one thing, Harrison decided that pendulums could be improved upon. They were made of iron or steel rods, which contracted in the winter, making the clocks run fast, and expanded in the summer, causing the clocks to lose time. Harrison all but eliminated this fault with the invention of a pendulum resembling a gridiron. It consisted of nine alternating steel and brass rods, so assembled that the different expansion and contraction rates of the two metals canceled each other out.

Another of Harrison's inventions was the "grasshopper" escapement, a new type of control device for the step-by-step release of a clock's driving power. His escapement was almost frictionless and required no oiling; it thus pointed the way to further improvements in clockmaking.

In 1728, when he was thirty-five years old, this self-taught country clockmaker felt himself ready to enter the longitude competition. Harrison packed full-scale models of the gridiron pendulum and the grasshopper escapement, to show what he had done, and the drawings for a proposed marine clock, to show what he planned to do, and went to London in search of financial assistance. It was suggested that he visit George Graham, the country's foremost horologist. Graham must have been impressed by Harrison. He loaned Harrison money out of his own pocket and advised him to return home and build the marine clock he proposed. Harrison, a methodical man and something of a perfectionist, spent the next seven years building his "Number One," as it was later called.

While Harrison worked away at Barrow-on-Humber, longitude became the rage. In conversation, finding longitude became synonymous with the impossible, no less than flying to the Moon. In *Gulliver's Travels,* Swift made longitude one of the great discoveries that would "enrich the minds." In a print for *The Rake's Progress,* Hogarth included in a madhouse scene of horror a comparatively serene lunatic busily working on a solution to longitude.

Every manner of proposed solution, many blatantly absurd, came before the Board of Longitude. The following entry in the board's minutes was not atypical: "A Memorial from Mr. Owen Straton was read, proposing a method of finding out the Longitude by means of an Instrument of his invention, and the said Mr. Straton, who was attending, being called in, and it appearing that the instrument proposed is a Sun Dial, he was told it could not be of any service, and then withdrew."

Harrison's clock was finished in 1735. Large and heavy, standing almost one meter tall, it was no thing of beauty. Harrison had eliminated as many moving parts as possible. Instead of using a pendulum, which had proven unreliable at sea, he designed a system of two large brass balances connected by wires. The motions of the two balances, of equal weight, were always opposed so that the effect of a roll of the ship on one would be counteracted by the other. A committee of the Royal Society, after examining the clock's mechanism, decided that it was sufficiently promising to be worthy of a trial at sea.

There is no record of how well the clock performed on the voyage, in 1736, to Lisbon. George Procter, captain of the *Centurion,* probably reflecting the seaman's skeptical attitude toward new-fangled instruments, wrote that "the Difficulty of measuring Time truly, where so many unequal Shocks and Motions stand in Opposition to it, gives me concern for the honest Man, and makes me feel he has attempted Impossibilities." On the return trip, aboard the *Orford,* Harrison seems to have made a more favorable impression. Relying on his clock, he estimated the ship's position to be one and a half degrees west of the dead-reckoning calculations. He was proved right on landfall.

Even so, the voyage to Lisbon and back was made along a well-known trade route following a roughly north-south course, which was hardly a definitive test of the clock. Harrison received a subsidy from the board and went home again to build Number Two. But at the time it was completed England was at war with Spain, and the Admiralty, fearing that the clock might fall into enemy hands, vetoed a sea trial.

Harrison retired to work on yet another version, Number Three, which took him seventeen years to build. Then, without even offering it for a sea trial, he turned to the construction of Number Four. The first three were all heavy, boxlike instruments, but not Number Four. It was more like a large pocket watch, about twelve centimeters in diameter, with a jeweled mechanism that was the product of years of exquisite craftsmanship. Into the clock had gone "fifty years of self-denial, unremitting toil, and ceaseless concentration," Harrison said, but he was finally confident that the prize was within grasp. "I think I may make bold to say," Harrison wrote, "that there is neither any other Mechanism or Mathematical thing in the World that is more beautiful or curious in texture than this my watch or Time-keeper for the Longitude."

A sea trial for Number Four began in November 1761. Harrison, now sixty-eight years old and not up to a voyage to the West Indies, entrusted

John Harrison's Number Four timepiece, which won him the £20,000 prize in 1773. This is an early version of the modern marine chronometer.

the clock to his son and co-craftsman, William. By arrangement with the Board of Longitude, the timepiece was placed aboard the *Deptford,* bound for Jamaica, and secured in a case with four locks. The four keys were given to William Harrison, Governor Lyttleton of Jamaica, who was taking passage on the ship, Captain Dudley Digges, and his lieutenant. All four had to be present, with their keys, to unlock the case at each winding.

The run from Spithead to Madeira put the timepiece to its first test. William Harrison wondered if Number Four would really meet the Board of Longitude's standards. Captain Digges was willing to bet on his dead reckoning. All hands aboard took a more than usual interest in the proceedings, after they found that the ship's beer had spoiled and they would have to drink water until they reached Madeira. They were most eager to make the island on the first approach, and woe betide the navigator who missed the island altogether and left them to cross the Atlantic without their beer.

Nine days out the *Deptford's* longitude, by dead reckoning, was 13°50′ W, but according to Harrison's calculations, based on the timepiece, it was 15°19′ W—a difference of nearly 160 kilometers. Harrison told a skeptical Captain Digges that if Madeira was properly placed on the chart they should sight land the next day. Despite his reservations, Digges held to Harrison's course. At 6 A.M. the next day, the lookout reported seeing land dead ahead. It was Porto Santo, the northeastern island of the Madeira group.

The landfall at Jamaica was equally successful. The *Deptford*, following Harrison's predictions throughout the voyage, arrived three days before another ship that had sailed for Jamaica ten days earlier. Number Four was taken ashore and checked against Jamaica's longitude, as determined by astronomical observations. After allowing for a predetermined rate of error (2⅔ seconds a day), the clock was found to be only 5 seconds slow. This meant an error in longitude of 1.25′—or 1.25 nautical miles, 2.3 kilometers. Number Four had more than complied with the requirements for the great prize.

On the return voyage, Number Four was put through the shocks and stresses of extremely stormy weather. Nevertheless, when it was checked back in England, its total error in longitude for the five-month round-trip voyage was 28.5′. The timepiece was within the limit of half a degree. But, if John Harrison had earned the prize, he had yet to win it.

Harrison, when he sought to collect the money, encountered stubborn resistance from the Royal Society and the Board of Longitude, the very body that had encouraged him during the long years of labor. The board decided that the clock's performance could have been a fluke. Harrison was awarded £2,500 as an interim prize and was told that the full prize would be his only if he submitted his clock to an examination of its mechanism and to another and more exacting trial at sea.

Although the board may have been justified in its caution, its motives were suspect. These men of the scientific establishment, learned and well-born, could not readily admit that a Yorkshire clockmaker, the son of a poor carpenter, had solved the problem of longitude. To admit as much would be a bitter blow to many of their friends and to some board members, who had themselves sought the prize. The board was under particular pressure in this regard from Nevil Maskelyne, whose bias in favor of the lunar method of finding longitude was deep and unshakable.

While Harrison was building and testing Number Four, considerable progress was being made toward the lunar solution. The established scientists had always favored this method and had pointed most of their own efforts in that direction. It may have been their conservative nature, their stubborn desire to be consistent. If latitude was determined through celestial means, it stood to reason, in their minds, that longitude could somehow be found in the same way. But this required that the Moon's motions be thoroughly known and charted. The Moon's position had to be calculated for many years in advance if it was to be useful in determining time and, in turn, longitude. These tables, not available in 1714, were now taking shape.

Johann Tobias Mayer of Göttingen devoted the last fifteen years of his life, from 1747 to 1762, to working on such tables. Mayer, a mathematician and cartographer, was drawn to the longitude question not from any nautical experience (he never even saw the sea) but from a desire to improve land surveying in Germany. He read all about the current theory of the Moon's motions around a spheroidal Earth and began preparing the tables by which the Moon's motions could be predicted. In time the usefulness of his tables was demonstrated, and Mayer applied for the longitude prize.

Maskelyne tested the tables with success on a voyage to St. Helena in 1761, but Mayer died before the board could act on his application. There was general agreement that Mayer deserved the prize, or at least a part of the prize, and this put the board in a quandary. Harrison and his timepiece seemed also to have a just claim. James Bradley, the Astronomer Royal, had also worked on a compilation of lunar tables, and he thought he should receive the prize. In the end Mayer's widow received £3,000. Mayer had come closer than anyone else to determining longitude by astronomical means, but he had not fared much better than Harrison at the hands of the Board of Longitude.

Under the circumstances, with Maskelyne succeeding to the post of Astronomer Royal and more influential than ever, Harrison's chances of receiving the full prize seemed dim. He had little choice but to submit to another trial at sea. His son, William, embarked this time in March 1764 aboard the *Tartar,* bound for Barbados, with John Lindsay as captain. Once again, Harrison correctly predicted the landfall at Madeira and also at Barbados. And who should be at Barbados when Harrison arrived but Nevil Maskelyne, who had been sent out to compare Harrison's results with observations of Jupiter's satellites. The board was taking no chances.

Number Four, however, had performed even better than on the voyage to Jamaica. The timepiece showed an error of 38.4″ over a period of seven weeks, or 15.36 kilometers of longitude. Further calculations showed that, after the round trip, the error came to less than one tenth of a second a day.

In a resolution in early 1765, the Board of Longitude unanimously concluded that "the said timekeeper has kept its time with sufficient correctness, without losing its longitude in the voyage from Portsmouth to Barbados beyond the nearest limit required by the Act 12th of Queen Anne, but even considerably within the same." But it still withheld the full prize, declining to pay more until Harrison had disclosed the secrets of his timepiece, so that it could be reproduced, and had turned over all four machines to the board.

This might seem, on its face, to be a reasonable request, since the whole idea was to produce many instruments for solving longitude. But the board revealed its abiding reluctance to acknowledge Harrison's genius by throwing in a hooker. Even if he did all this, Harrison would be entitled to only £7,500, which would make a total of £10,000, or half of the full award. To earn the rest, Harrison would have to build two more timepieces and submit them to exhaustive testing.

Harrison was an old man by this time, and his eyesight was failing. "I cannot help thinking," he wrote to the board, "but I am extremely ill used by gentlemen who I might have expected different treatment from. . . . I hope I am the first, and for my country's sake, shall be the last that suffers from pinning my faith on an English Act of Parliament."

After an inspection of his machine by watchmakers appointed by the board, Harrison received the £7,500. Maskelyne, of all people, was given Number Four to test at the Royal Observatory. He put the clock through the most rigorous tests, placing the clock dial up and dial down, and then produced an unfavorable report.

Harrison was not without friends, and with their help he published broadsides attacking the "priests and professors" and "men of theory" who supported the lunar school at the expense of his timepiece. He also decided to enlist the aid of one who might have seemed an unlikely ally in a battle against the establishment. He went to George III, a move that set the stage for a story-book dénouement to the whole proceedings.

George III had an informed and sympathetic interest in technology. He had been so impressed by the accounts of the *Tartar*'s voyage that he had granted Harrison and his son an audience at Windsor. Harrison's son, William, now asked for, and received, another audience. Upon hearing the story, George III lost his patience. "By God, Harrison, I'll see you righted," the king declared, and so he did.

Harrison's Number Five was tested at the king's private observatory at Kew. George III took a daily interest in checking the clock's performance. In ten weeks the clock was in error by 4½ seconds, an average of just under ¼ second per day. The king needed no further proof, and instructed Harrison to petition Parliament for the full prize.

Everyone in Parliament knew the petition had the backing of George

III and that he was willing, if necessary, to testify in Harrison's behalf. This did not prove necessary. In 1773 Harrison received the balance of the £20,000 prize.

"Longitude Harrison," as he was often called, died at his house in Red Lion Square in London on March 24, 1776, in the eighty-third year of his life. He had provided the instrument for bringing a new dimension to navigation and mapping, the dimension of longitude. The instrument was the earliest version of the modern marine chronometer.

In the year of Harrison's death, James Cook sailed again to the Pacific, where he demonstrated beyond any doubt the utility of the chronometer in marine mapping. Nearly two centuries later, the honored guest at a dinner at 10 Downing Street was an American, who rose to propose a toast to John Harrison. His invention, the American said, enabled men to explore the Earth with precision and, when most of the Earth had been explored, to dare to build navigation systems for voyages to the Moon. "You, ladies and gentlemen, started us on our trip." The speaker was Neil A. Armstrong.

Surveyors of Sea and Shore

CHAPTER TEN

In cartography, as in all human endeavor, tradition dies hard, and some traditional ideas unsupported by fact can have the longest and most resolute life spans.

One such idea so misled explorers and cartographers that for the longest time they could not conceive of Earth as having room enough for its greatest ocean, the Pacific. For as long as the Ptolemaic or Columbian estimates of Earth's circumference were accepted as being approximately correct, learned men could only assume that Asia lay at the other side of the Atlantic. There could not possibly be another intervening ocean, certainly not one stretching across 160° of longitude and covering an area larger than that of the entire land surface of the globe. Little wonder that Vasco Núñez de Balboa and his men "Look'd at each other with a wild surmise" when, in 1513 at Darien on the isthmus of Panama, they first cast eyes upon the Pacific.*

With its discovery by Europeans, the Pacific eventually took its rightful place on the map and in the minds of men. This came about through the considerable influence of two other ideas conceived more out of wishful thinking than observed fact. One was that hardiest of geographic illusions, *Terra australis incognita,* the unknown southern land

---

*John Keats mistakenly credited "Stout Cortez" with the discovery, not Balboa. The lines: "Stared at the Pacific—and all his men / Look'd at each other with a wild surmise— / Silent, upon a peak in Darien."

whose existence had been postulated by mapmakers. The other was the Northwest Passage, the navigable short cut from Europe to Asia through North America, a dream that became an irresistible and sometimes fatal lure to many a post-Columbian explorer.

There *is* a southern land, Antarctica. But this is not the "golden province" the early believers in Terra Australis inferred from a careless reading of Marco Polo. This is not the vast continent they had expected to be stretching far above the South Pole and dominating much of the southern hemisphere.

The idea goes back at least to the ancient Greeks, who, being vaguely aware of lands near the Arctic Circle, reasoned that comparable landmasses should exist far to the south. Hence the name Antarctica, from the Greek word *antarktikos*, the antipode of the Arctic. The Greeks held out little hope of ever reaching those distant shores, for they assumed the equatorial heat imposed an impenetrable barrier between the northern and southern worlds. But to Aristotle the southern land had to exist, reachable or not. It was a logical necessity to preserve the symmetry that had been observed elsewhere in nature and in man. The simplicity and neatness of symmetry, like sphericity, never failed to impress the Greek mind.

In the fifteenth century, maps based on Ptolemy's geography depicted the southern continent with growing authority. The continent often bore the optimistic label *Terra australis nondam cognita,* the "southern land not *yet* known," and was usually shown as a long rectangle in the southern part of the Indian Ocean. But it grew in size and legend with each new voyage of discovery.

Magellan, while navigating the strait now bearing his name, saw Tierra del Fuego to the south of South America. So deeply ingrained was the idea of the continent that mapmakers disregarded the opinion of Magellan's men, who believed Tierra del Fuego to be only an island, and drew it as a northern tip of Terra Australis. And so it remained on maps for almost a century. Finally in 1616, the Dutch navigator Willem Schouten sailed round the south of Tierra del Fuego, discovered and named Cape Horn, and proved that Magellan's men, not the mapmakers, had been right.

But the mapmakers saw no reason to alter their inflated vision of Terra Australis. During the sixteenth century the continent's outlines grew larger and more detailed with each new map. In 1531, Oronce Fine, a French mathematician, drew the first map showing Terra Australis as something more than a shapeless blob; he endowed the supercontinent with capes and bays. The first modern world atlas, by Abraham Ortelius in 1570, contained even more details, including rivers. He gave names to some of the more prominent features—the Land of Parrots, Cape of the Good Signal, River of Islands, and the Sweetest River. Someone studying the map would never suspect that Terra Australis was not yet known.

On such maps the continent extended up from the pole well into the temperate zone, covering an area larger than all the Americas. Mercator's map of 1569 reflected the contemporary thinking. There must be, Mercator wrote, "under the Antarctic Pole a continent so great that, with the southern parts of Asia, and the new India or America, it should have a weight equal to the other lands." Greek symmetry was not dead.

This view held for two more centuries. In 1767, Alexander Dalrymple, a Scottish geographer, declared that such a continent "was wanting on the south of the Equator to counterpoise the land to the North, and to maintain the equilibrium necessary for the Earth's motion."

At the time of Dalrymple the British harbored a vague hope of realizing in Terra Australis a new and more docile colony than the rebellious thirteen in America. But another persistent geographic idea, the Northwest Passage, held promise as an even more real and immediate boon to British fortunes.

The two known sea routes to Asia were round the Cape of Good Hope and through the Strait of Magellan. Either way it was a long, arduous voyage from the ports of northern Europe, and it meant plying waters under Spanish and Portuguese control. If only they could find the Northwest Passage, the British thought, it would revolutionize world commerce.

The quest began soon after Columbus returned with the news of his discovery. At the time, and for many years thereafter, America was not a land to be explored for its own sake; it was a barrier in the way to Cathay that had to be skirted or somehow penetrated. Columbus's fourth voyage was a search for a strait from the Caribbean to the Indian Ocean, and John Cabot lost his life trying to find a strait through North America.

Giovanni da Verrazzano, the Florentine navigator, searched the American coast from Florida to Newfoundland "in continual hope of finding some strait or northern promontory at which the land would come to an end, in order to penetrate the *quelli felici liti del Catay*"— "those happy shores of Cathay." In 1524, looking across the narrow Outer Banks of North Carolina, he saw a great expanse of water (Pamlico Sound) and was convinced he had found *el mare orientale,* the Eastern or Pacific Ocean. Though he failed to find an inlet or make further investigations, Verrazzano's error was perpetuated for more than a century by cartographers, whose maps showed the Carolinas and Virginia to be a narrow isthmus separating the Atlantic and Pacific oceans. On the Bailly Globe of 1530, the "Sea of Verrazzano" inundates much of the present United States.

Verrazzano must have been an overly cautious man, and for an explorer rather incurious. Sailing up the coast, evidently keeping a respectful distance from shore, he missed entirely the entrances to Chesapeake and Delaware bays. If he had seen them, he would surely have gone in to explore for possible straits to Cathay. When he dropped anchor in the Narrows of New York Harbor, the first European to glimpse the site of the future city, he remained only long enough to decide this held slight promise as a route to the Pacific. His backers were silk merchants who would be satisfied with nothing less than a passage to Cathay.

From the reports of Verrazzano and others it became increasingly clear that there was no passage in the middle latitudes. The first person to promote a search in the higher latitudes may have been Robert Thorne, an English merchant living in Seville. In a letter to Henry VIII in 1527, Thorne urged that three possible northern routes be explored: eastward along the northern coast of Europe, westward round the north of America, and directly across the polar region. "There is no land unhabitable, or Sea innavigable," Thorne asserted, volunteering "to attempt, if our Seas Northward be navigable to the pole, or no."

Although Thorne never made the attempt, word of his proposition spread. The English and Dutch tried first to find the Northeast Passage, but were blocked by ice; they had to settle for the discovery of new fishing grounds and a profitable new trade with Russia. The English turned more hopefully to the idea of a Northwest Passage.

The maps of the day were most encouraging. Mercator drew open navigable seas right up to the North Pole; he probably subscribed to a contemporary theory that sea ice formed from the shoreline outward and could not exist in the open sea. A chart by Humphrey Gilbert in 1576 showed a strait traversing the continent from the St. Lawrence to the Gulf of California. His chart was, however, little more than a piece of propaganda. It was based partly on Spanish tales, partly on belief in a northern passage equal to the open seas south of Africa and South America, but mostly on a desire to promote a company Gilbert had formed to send out expeditions of discovery.

The search for the Northwest Passage thus became an obsession, and the modern maps of Canada are sprinkled with the names of those who tried and failed to find the passage:

Frobisher Bay—Martin Frobisher, three voyages in 1576–78, for Gilbert's Company of Cathay. On a 1578 map by George Best, "Frobussher's Straights" cuts across Canada to a western sea.

Davis Strait—John Davis, three voyages in 1585–87. "I have bene in 73 degrees, finding the sea all open. . . . The passage is most probable, the execution easie." He had not ventured far enough. Davis Strait, between Baffin Island and Greenland, leads to the entrance by which the Northwest Passage was eventually to be navigated. The globes prepared by Emery Molyneux in 1592, the earliest English terrestrial globes, include considerable information furnished by Davis, with whom Molyneux may have sailed.

Hudson Strait and Hudson Bay—Henry Hudson, 1610–11. Before reaching Canada, he had sailed up the Hudson River in New York as far as present-day Albany, also in a vain search for the passage. In Hudson Bay, mutineers seized his ship and marooned Hudson, who was never heard from again.

Baffin Island and Baffin Bay—William Baffin, 1612–15, 1616. He sailed as far as 78° N, discovering and naming Lancaster Sound, which he mistakenly assumed came to a dead end. The sound is the true beginning of the Northwest Passage.

Foxe Channel and Foxe Basin—Luke Foxe, 1631. After he charted the western shore of Hudson Bay, the supposed water link between the bay and the Pacific disappeared from most seventeenth-century maps. Foxe's course was drawn on a map made from a piece of rough-hewn ship timber.

To the south, meanwhile, fond hope continued to cloud men's vision. Such was the power of the idea that in 1621 Francis Billington, who had arrived on the *Mayflower* with the Plymouth Colony settlers, sighted from a treetop "a great sea, as he thought." It was soon found to be a lake, the source of the town brook of Plymouth, but to this day local residents call it "Billington Sea." An early map of Virginia, published in London in 1651, showed the Pacific Ocean separated from the Chesapeake Bay by only a narrow strip of land. Verrazzano may have been long dead, the victim of cannibals in the Caribbean, but not his mistakes.

The French, in their explorations of the Great Lakes, kept thinking they would find the passage. When Jean Nicolet reached a town of Winnebago Indians near the head of the Green Bay of Lake Michigan, he was not sure but what he had come upon an outpost of Cathay. And so he put on a robe of fine Chinese damask and advanced to meet the waiting villagers.

Success eluded them all, but their failure on the Atlantic coast did not deter those who sought the passage from the Pacific side.

In their case, hope was buoyed in part by an Ulster landowner, Arthur Dobbs, who revived for an eager audience the story of a Greek pilot, Apostolos Valerianus. This man had supposedly sailed for the Spanish under the name of Juan de Fuca. As the story goes, de Fuca was dispatched up the northwest coast of America by the viceroy of Mexico. In 1592, between latitudes 47° and 48°, he claimed to have sailed into a broad inlet, discovered a land rich in gold and silver, and in due course reached the Atlantic Ocean. Convinced the story was true, Dobbs used it in 1741 to persuade the Admiralty to mount another expedition, which was unsuccessful, and to move Parliament to offer a reward of £20,000 for the discovery of the passage. Maps of the time dutifully showed an immense body of water penetrating the interior north of California, de Fuca's legacy to a wishful world.

Just as there *is* a southern continent, there *is* a Northwest Passage. The believers of the eighteenth century were right in their assumption that the American landmass could not extend all the way to the North Pole, but they did not imagine that the passage would be so far north and be blocked much of the way by a thick sheet of ice.

The actual passage runs from Baffin Bay through Lancaster Sound, thence through either Barrow Strait or Prince Regent Inlet and Bellot Strait, through Franklin Strait and James Ross Strait to Queen Maud Gulf, across the northern mainland coast to Beaufort Sea. The first successful voyage through the passage was made in 1903–1906 by the Norwegian Arctic explorer Roald Amundsen. In 1969, the *Manhattan,* a huge tanker with reinforced hull, became the first commercial vessel to go through the passage. The nuclear-powered submarine *Nautilus* finally proved in 1958 the feasibility of the third route proposed in 1527 by the English merchant Thorne. The American submarine went directly across the North Pole—but under the ice, a short cut now frequently used by submarines. But this is hardly the route Frobisher sought, Hudson died in search of, or Juan de Fuca conjured up.

In the late eighteenth century, however, the lure of Terra Australis and the Northwest Passage remained sufficiently compelling to attract to the Pacific two of the most accomplished surveyors ever to put out to sea.

Beginning in 1768, James Cook led three expeditions that remain a model of reconnaissance mapping. His instructions were to survey the vast ocean, to search for the southern continent, and to make yet another attempt to find the Northwest Passage. George Vancouver followed Cook and, between 1792 and 1795, conducted a hydrographic survey of much of the west coast of North America. His charting of coastline, harbors, bays, inlets, and islands set a high standard for thoroughness and accuracy.

O f few men can it more aptly be said that he was the right man at the right time and the right place.

Born in Yorkshire in 1728, James Cook was, as the baptismal record stated, "ye son of a day labourer." He was eighteen, in those days a relatively late age to be embarking on a naval career, when he went into the North Sea coal trade. At twenty-seven he joined the Royal Navy as an able seaman, but in two years he passed the examination to become a master "of any of His Majesty's Ships from the Downs thro' the Channel to the Westward and to Lisbon."

Being a young man without influential connections, Cook had to make his way up through the ranks by the sheer force of ability. Advancement came slowly, the reward of competence and steadfastness, not manifest brilliance. The few portraits of Cook reflect those characteristics. He was a tall, rawboned, solid man with a strong face: set jaw, determined mouth, large nose, and steady, rather patient eyes. This was a commanding but unpretentious man. This was, as he revealed himself in his career, a thorough and absolutely dependable man. What better attributes for a man who must lead others into the unknown and seek to chart that unknown with an efficiency hitherto unattempted? What better attributes for any pioneering surveyor?

Cook's introduction to surveying suggests equally laudable characteristics—a ranging curiosity and a readiness to learn.

One day in July of 1758, while master of the *Pembroke* off Canada during the Seven Years' War, Cook went ashore and caught sight of a man carrying a small square table and an attached tripod. Every so often, the man would set the table down, squint along its top, and then make some notes. This was a plane table, a standard piece of surveying equipment in that day, and with it the man was observing angles. The man was Samuel Holland, a British Army surveyor who would later become surveyor-general of Quebec. Holland and Cook fell into conversation, during which Cook asked for instructions in the use of the plane table.

Holland obliged, and in the next few days, as he recalled, he made

Cook "acquainted with the whole process." In the autumn, whenever his naval duties permitted, Cook studied trigonometry and astronomy and practiced his new skills in a survey of Gaspé Bay. The result was the first engraved and printed chart bearing the name of James Cook. Other charts of the Canadian coast followed, and Lord Colville, Cook's superior, was so impressed that in 1762 he wrote to the Admiralty the following commendation:

> Mr. Cook late Master of the Northumberland acquaints me that he had laid before their Lordships all his Draughts and Observations, relating to the River St. Lawrence, part of the Coast of Nova Scotia, and of Newfoundland.
>
> On this Occasion, I beg leave to inform their Lordships, that from my Experience of Mr. Cook's Genius and Capacity, I think him well qualified for the Work he has performed, and for greater Undertakings of the same kind. . . . These Draughts being made under my Eye I can venture to say, they may be the means of directing many in the right way, but cannot mislead any.

This was a rare man, Cook. Not only could he sail a ship, but he could chart a coast. Most captains left chartmaking to others, few of whom, Cook once remarked, "are capable of drawing a Chart or Sketch of a Sea Coast."

Cook soon got his chance to demonstrate how it should be done. With the end of the war, the victorious British had much new territory in need of mapping. Newfoundland, with its 9,600 kilometers of rugged coastline, presented a particular challenge, to which Cook applied himself between 1763 and 1767.

With characteristic diligence, Cook went well beyond the traditional method of marine surveying, which was the running traverse off the coastline. This entailed the careful noting and plotting of the ship's course as she sailed along a coast and the fixing of prominent coastal features from cross bearings taken from the ship. In *The Practical Navigator and Seaman's Daily Assistant,* first published in 1772, the instructions for such a survey were: ". . . take with the Azimuth Compass, the Bearings, in Degrees, of such Points of the Coast as form the most material Projections or Hollows; write down these Bearings, and make a rough sketch of the Coasts. . . . Then let the ship run in a direct line, which must be carefully measured by the Log or otherwise for one, two or three miles."

Then, according to this method, the compass bearings of the same land features were to be recorded again. When the bearings of a given feature were plotted from successive positions of the ship, the point at which the bearing lines intersected fixed the position of the land feature with respect to the ship's track. While this was repeated down the coast, an experienced draftsman would sketch in the outline of the shore, particularly noting inlets, islands, and harbors. At the same time, a member of the crew would take occasional depth soundings.

Depth sounding is as simple, though time-consuming, as it is ancient. As practiced in the time of Cook, it involved lowering a hemp line with a lead weight attached to the end. When the weight was felt to hit the bottom, the depth could be roughly determined by the length of the line that had been paid out. Herodotus wrote of the practice in the fifth century B.C., and the Apostle Paul described the use of sounding line in Acts, 27:27, 28: "But when the fourteenth night was come, as we were driven up and down in Adria, about midnight the shipmen deemed that they drew near to some country; and sounded, and found it twenty fathoms: and when they had gone a little further, they sounded again, and found it fifteen fathoms."

Surveyors using the traditional method took it for granted that there would be errors. It was difficult to log the ship's track or fix positions with real precision; these errors, in turn, threw off all other calculations, whether it was the soundings of submarine landscape or the placement of coastal features. From a distance offshore it was not always possible to be sure if an inlet was a bay, a strait, or a river, or to know always if some stretch of shore was the mainland or an island masking the mainland.

For these reasons Cook decided upon a more thorough survey. His intention became clear by his choice of instruments. He took with him a newly-developed theodolite for measuring angles between lines of sight and a brass telescopic quadrant for determining latitudes from celestial altitudes. Since neither instrument could be used on board a rolling ship, Cook obviously planned more than a running traverse.

His technique can be seen from some typical lines in the ship's journal: ". . . went into Bay Sacre, Measured a Base Line and fix'd Flaggs on the Different Islands. . . . At Noon took the Suns Meridian altitude on shore and found Cape Norman to be in Latitude 51°39′ North. . . . At 6 A.M. the Master [Cook] with the Cutter went ashore to Continue the Survey, Stood to the westward about a League off shore, brought too and sounded every mile."

Cook was combining land-surveying techniques he had learned from

Holland with the traditional offshore marine surveying. Going ashore nearly every day, Cook measured a base line by chain and marked the ends of the line by implanting flags. He fixed the latitude of the line by taking quadrant sightings on the Sun. Then, using the theodolite, he determined angles between each end of the line and some distant point, a tree or rocky crag or implanted flag. This was the same technique of triangulation used with such impressive effect by the Cassinis of France. In time Cook obtained enough angles to plot a network of triangles whereby it was possible to know the correct relation between many land features along the coast. A draftsman then sketched in the detail between the measured points.

After running a series of triangles, Cook then took theodolite sightings on the mast of his ship, thereby linking his shore survey with any offshore soundings. It was tedious work: day after day of measuring base lines, fixing flags, sighting, sounding, plotting, and sketching.

Although not the kind of work that could be hurried, Cook's "first employment" in the Newfoundland Survey was performed under tremendous pressure to do the best job possible in the shortest time. The problem was diplomatic. Under terms of the Treaty of Paris, England was to cede to France two small rocky islands, St. Pierre and Miquelon, off the entrance of Fortune Bay. England was willing, but only after the islands could be mapped. It might be the last chance to survey this piece of alien, and potentially enemy, territory.

While Cook was engaged in the task, a French frigate came onto the scene bearing the islands' governor-designate, M. d'Anjac, and a company of soldiers and settlers. They fully expected to take possession, as agreed, on June 10, 1764. But somehow, in a diplomatic *tour de force* that has never been fully explained, Captain Charles Douglas of the Royal Navy persuaded d'Anjac to remain on board his ship until July 4, by which time Cook had removed himself—chains, theodolite, new charts, and all—from St. Pierre and was hastening to survey Miquelon.

The islands were handed over to the impatient French on July 31, and a plan of the islands, 3½ inches to a mile, was turned over to the British Admiralty. As Douglas explained in a letter to the Admiralty, "I procured him all the time I could by staying at St. Peter's under various pretences, until towards the 17th, and then went to the Road of Miquelon—where we made shift to keep the Commandant in some sort of temper, until

Captain James Cook's chart of the south coast
of Newfoundland, published in 1767

A

Chart of Part

of the SOUTH COAST of

NEWFOUNDLAND

INCLUDING the ISLANDS

St. Peters AND Miquelon

with the Southern Entrance into the Gulph of St Laurence

from actual Surveys taken by Order of

COMMODORE PALLISSER

GOVERNOR of NEWFOUNDLAND, LABRADORE, &c.

BY

James Cook

SURVEYOR.

—Larken sculp. 1767.

Scale of English and French Leagues, 20 to a Degree.

CONNOIRE BAY

WHITE BEAR BAY

BURGEO ISLES

RAMEA ISLES

NB. The Longitude of the Burgeo Isles, by an observation of the Eclipse of the Sun on the 5 of Aug 1766 is of 51 W:9t from the Meridian of London.

15r. 20' West

the beginning of August; when, thro' the unwearied assiduity of Mr Cook, the survey of that Island too, was compleated." In conclusion, Douglas asked the Admiralty for £50 "for the extraordinary expenses I was put to." The Admiralty thought it a fair price for a good map.

After each of the five surveying seasons, which ran from about May to November, Cook returned to England and published his newest charts. Since the Admiralty had no hydrographic department as such, a surveyor like Cook had to resort to commercial cartographers for publication, at his own expense, of the fruits of his work. Some of the charts were drawn at a scale of an inch to a mile, sometimes 3½ inches to a mile. They included far more inland topography than was customary on hydrographic charts of the day. A chart by Cook was oriented to magnetic north, had a generous number of soundings, included latitudes in the "remarks," and was dressed up with the use of brushwork in brown and green to show relief. Somewhere on the map was always the statement many contemporary chartmakers could not make—based on "actual surveys."

Of the Newfoundland surveys, J. C. Beaglehole, Cook's modern biographer, stated: "Cook was to carry out many accomplished pieces of surveying, in one part of the world or another, but nothing he ever did later exceed in accomplishment his surveys of the southern and western sides of Newfoundland."

Cook, the diligent surveyor of Newfoundland, the experienced seaman, was now ready to embrace the Pacific and his destiny. In 1768, this son of a poor Yorkshireman finally won a Navy commission and embarked as commander on the first of three historic voyages into the Pacific.

Cook was an ideal choice for the assignment. Beneath that stolid exterior beat the heart of an explorer, for he confessed to an ambition which "leads me not only farther than any man has been before me, but as far as I think it possible for man to go." And he set out for the Pacific with every intention of combining precision mapmaking with ranging discovery. The mapmaking of earlier explorers in the Pacific had been so unreliable that some islands remained, in effect, undiscovered; the Solomon Islands, for example, were so poorly plotted that they had to be rediscovered two centuries later. "The world," Cook said, "will hardly admit of an excuse for a man leaving a coast unexplored he has once discovered."

On the first voyage, from 1768 to 1771, Cook headed for Tahiti by way of Cape Horn. His ship, *Endeavour*, was a three-masted Whitby collier, a small, almost flat-bottomed ship selected for its stability and its maneuverability in coastal waters. There were gales to weather and the monotony of uneventful days to endure, weeks without sight of land or any other European ship, the uncertainties of fate in an unknown sea far from civilization, all this before Cook reached Tahiti, the recently discovered island paradise.

At Tahiti Cook went about his first order of business—the observation of a transit of the planet Venus across the face of the Sun. Using a telescope and a pendulum clock, scientists with Cook attempted to record the times Venus's dark shadow "touched" the solar disk, crossed it, and then disappeared. From these and other data collected at other points around the world scientists eventually reached a rough estimate of the distance between Earth and the Sun.

From Tahiti Cook sailed southward, where, as the Admiralty had advised him in sealed orders, "there is reason to imagine that a Continent or Land of great extent, may be found." He was to search for Terra Australis.

This search took Cook south as far as the fortieth latitude, well beyond where some early cartographers had imagined the continent to be. He thus proved that the still-putative southern continent had to be smaller than had been assumed. Next, he sailed west to New Zealand, which he circumnavigated—the first to do so—and thereby showed that, contrary to general belief, it was not part of Terra Australis.

Lacking the time he had in Newfoundland, Cook had to settle for a running survey of New Zealand's entire coastline. But he made the most of the six months he had. He used the more precise reflecting sextant as well as a compass for taking bearings on coastal landmarks. He plotted the ship's track from intersecting rays on landmarks, not from the generally unreliable dead reckoning, and adjusted it from time to time with astronomical observations. On a few occasions, as in Queen Charlotte Sound, he took the time to go ashore for triangulations. The result was the first chart of New Zealand, notably exact in its latitudes and outline and, even without benefit of a chronometer, not more than one half of a degree off in its longitude.

Indeed, the accuracy of Cook's chart was soon to be confirmed. A French navigator, Julien Crozet, sailing on the northern New Zealand coast in 1772, wrote: "As soon as I obtained information of the voyage of the Englishman, I carefully compared the chart I had prepared of that part of the coast of New Zealand along which we had coasted with that prepared by Captain Cook and his officers. I found it of an exactitude and of a thoroughness of detail which astonished me beyond all powers of expression, and I doubt much whether the charts of our new French coasts are laid down with greater precision. I think therefore that I cannot do better than to lay down our track off New Zealand on the chart prepared by this celebrated navigator."

A mapmaker could ask for no higher accolade.

With much the same care, and with equal accuracy, Cook charted the east coast of Australia, covering 3,200 kilometers of coastline in four months, courting disaster among the coral reefs. On August 17, 1770, the *Endeavour* was driven toward a reef "so that," Cook wrote, "between us and destruction was only . . . the breadth of one wave." Yet the three men taking observations for longitude remained at their instruments, presumably undaunted. The observations, according to the ship's astronomer, "were very good."

More than a quarter of a century later, in 1801, the English naval officer Matthew Flinders charted more of the Australian coast and was the first to establish that Australia was not made up of several islands but constituted a sixth continent. After this discovery was recognized by geographers, William Faden was one of the first cartographers to name the continent Australia, after the mythical southern landmass.

Cook returned to England in 1771, but after barely a year at home he set out again for the Pacific, this time with two ships, *Resolution* and *Adventure,* and a chronometer built to Harrison's design. His instructions were to search for the southern continent by circumnavigating the globe as close to the South Pole as possible. Cook's second and most arduous voyage lasted from 1772 to 1775.

Cook sailed deep into southern waters in November, at the beginning of the summer season in the southern hemisphere. At latitude 50°40' the crew came upon the first of their "islands of ice." Some of the icebergs they saw were "as high as the dome of St. Paul's." Their rigging froze over with ice, snow coated the sails, and fog enshrouded them, separating the two ships. On Christmas Day, in this icy wilderness, they all got drunk.

In the course of the voyage Cook crossed the Antarctic Circle three times. He proved that no continental land lay north of 60° S, once more diminishing the possible size of Terra Australis. On the second penetration of the circle, in 1773, Cook reached as far south as 71°10' without sighting land. If he had been able to keep going south for another 500 kilometers, he would have reached the Walgreen Coast of Antarctica. But Cook saw nothing but icebergs, fog, and snowy mists. He could only conclude that if there was a continent it must be closer to the

South Pole than he could reach. Thus came to an end all the more romantic notions about a rich and extensive Terra Australis. Cook's undiscovery forced cartographers into retreat, erasing all those imaginary shorelines as they went. They would have to wait another half century before the discovery of Antarctica, considerably diminished, would give them something real to draw at the bottom of their world maps.

But Cook's second voyage added, as well as subtracted, land from the world map.

After each of the first two Antarctic penetrations, Cook rested his men in New Zealand or some sunny island, gathering fresh food and charting landfalls. (He made his third Antarctic penetration, a more cursory survey, en route to Cape Horn, homeward bound at the end of 1774.) At one time or another, they cruised by the Marquesas, the Society Islands, the Friendly Islands, the New Hebrides, New Caledonia, and Norfolk Island. At each island Cook took careful sextant and quadrant readings for latitude and checked his chronometer to determine longitude. It would never be said of an island Cook discovered that it could not be found again.

Cook was, nonetheless, somewhat apologetic about these reconnaissance surveys, realizing that, however reliable, they could not match in accuracy and detail his work in Newfoundland. In his journal, Cook wrote: "The word Survey, is not to be understood here, in its literal sence. Surveying a place, according to my Idea, is takeing a Geometrical Plan of it, in which every place is to have its true situation, which cannot be done in a work of this kind."

The objective of Cook's third voyage, beginning in 1776, was a search for the Northwest Passage along the Pacific coast of North America. This took Cook and his two ships, *Resolution* and *Discovery*, into northern Pacific waters for the first time. Cook discovered the Hawaiian Islands, where he paused for food and fresh water, and then continued on to the American mainland, reaching the coast of what is now Oregon.

During most of 1778, spring, summer, and autumn, Cook conducted a remarkable reconnaissance survey of 5,000 kilometers of coastline, from Oregon north along much of the Alaskan coast. Since he worked mostly from shipboard, Cook was not always able to distinguish islands from promontories, inlets from rivers; he mistakenly assumed Cook Inlet in Alaska to be a river. He somehow missed the mouth of the Columbia River and the fact that Vancouver Island was an island. Yet, even though Cook would not have called it a survey "in its literal sence," the result was the first fairly detailed map of the American northwest

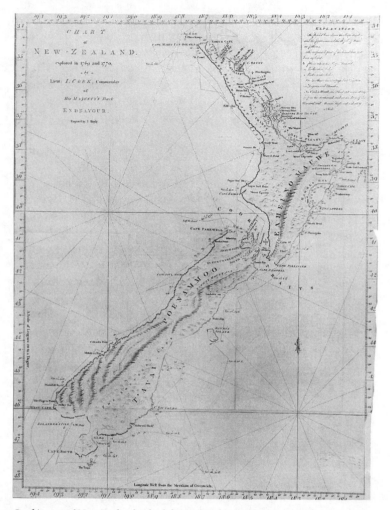

Cook's map of New Zealand, which he explored in 1769–70

coast, and it was firmly based on accurate coordinates of latitude and longitude. He traced in considerable detail both the Asian and American sides of the Bering Strait. He reached as far north as 70°44', only 200 kilometers from Point Barrow, which meant that he had come almost as close to the North Pole as he had to the South Pole.

Nowhere did Cook find any waters that he considered likely to lead to the Northwest Passage. Before he faced west for a winter in Hawaii, however, Cook drew up plans to make another and final attempt to find the passage during the summer of 1779. "But I must confess," he said, "I have little hopes of succeeding."

He was never to make the attempt. While in Hawaii, gathering supplies and repairing the ships, Captain James Cook was murdered by natives on February 14, 1779. His dismembered body was found on shore the next day. It was identified by a scar between the thumb and forefinger of his right hand, the result of an exploding powder horn on the Newfoundland coast back in the days when he was learning and perfecting his great surveying skills.

It has been said of Cook that the modern map of the Pacific is his epitaph. When Cook's three voyages were over, only a handful of Polynesian islands remained undiscovered. This sturdy, persistent, no-nonsense man had dispelled two longstanding geographic illusions, the ideas of Terra Australis and the navigable Northwest Passage. This first-class surveyor, Beaglehole concluded, "put in the place of dispelled illusion so many positive discoveries, and laid the foundations of Pacific geography as a science, of Pacific anthropology, of antarctic hydrography."

Cook's discoveries also opened the European mind to a new world of ideas, some romantic and disquieting, others profound and revolutionary. No major new entry on the map, certainly not one covering one-third of Earth's surface, could leave the mind unaffected.

While Europeans were not slow to see the Pacific as a new arena for trade and colonization, it was the philosophical aspects of the discoveries that more immediately intrigued people. "The Europeans went into the Pacific in search of a physical Antipodes," writes Wilcomb E. Washburn, in an essay on the intellectual assumptions and consequences of exploration. "They found an intellectual and moral Antipodes."

Tales of the South Sea islanders, especially the Tahitians, seemed confirmation of Rousseau's ideal of the Noble Savage. The naked, beautiful Tahitians, simple and guileless, living happily off nature on an idyllic island, seemed to embody the virtue of man before the Fall. Theirs must be the lost innocence, this must be the earthly Paradise.

The narrowness of traditional European culture stood exposed. Self-doubt became common. If the pagan Tahitians had learned to live a serene communal life in which no one starved, what case could be made for the superiority of Christian morality? If carefree Tahiti was such a perfect society, what of the European idea of progress through hard work?

People could not read or hear enough about the South Seas. In London, Captain Cook's journals were best-sellers, and variations of Tahitian dances became the rage. The enthusiasts, of course, usually overlooked published references to such Tahitian practices as thievery, infanticide, and human sacrifice, for they were wishing so much to believe in and perhaps learn from this new Eden. In 1785, reflecting a hope of somehow bringing the two cultures together, the Theatre Royal in Covent Garden produced a play about Omai, heir to the Tahitian throne, wooing an English girl, Londina.

Some of the sophisticates scoffed at all this. Horace Walpole grumbled that he had not read anything about Cook's voyages and never intended to. Samuel Johnson, to whom London was world enough, wondered why the expedition had brought back so many insects when, as everyone knew, there were already plenty in England—a reaction that would be repeated another day in connection with the return of rocks from the Moon.

This indifference, however, did not discourage the growing curiosity among other learned men who wanted to know more about the rest of the world, a curiosity aroused by the discoveries and the new maps. This directly influenced the heroic scientific travels of Humboldt and Darwin and Cook's immediate successor in Pacific surveying, Vancouver.

On Cook's second voyage, when the *Resolution* reached its southernmost penetration of the Antarctic, a "young gentleman" of the ship rushed to the head of the bowsprit before the turn was made. Waving his hat in the air, he cried out, *"Ne plus ultra."* The exuberant young man was George Vancouver, who would later claim that he had been closer to the South Pole than any other man. Yet his fame rests more securely on his exhaustive survey of the Pacific coast of North America.

Vancouver was born in 1757 at the North Sea port of King's Lynn. He was the son of a well-to-do customs official, and it was perhaps through his father's connections that he secured a coveted berth on Cook's second and third expeditions, where as an aspiring naval officer he served first as an apprentice seaman and then as a midshipman. Little is known of Vancouver's performance, except that he learned astronomy and navigation from William Wales, the astronomer on the second voyage, and was described in the journal of one shipmate as a "Quiet inoffensive young man." But the incident on the bowsprit suggests a youthful eagerness to make his mark in the unknown.

Following service with Cook, Vancouver passed examinations for a commission as lieutenant and saw extended duty in the Caribbean, where he charted Kingston harbor. Then, in 1791, he was given command of the *Discovery* for an expedition to chart the Pacific coast of North America and, taking up where Cook left off, make one more search for the Northwest Passage.

The *Discovery*, a three-masted, full-rigged ship, and her consort, the two-masted *Chatham*, sailed from Falmouth on April 1, 1791. Along the way Vancouver ran a 560-kilometer survey of the southern coast of Australia and stopped for repairs in New Zealand, at Dusky Sound. While he paid profuse tribute to Cook in his journals, Vancouver seemed to take delight, whenever possible, in making improvements upon Cook's work. In Dusky Sound, for instance, Vancouver went to an inlet which was described on Cook's chart as "Nobody Knows What," and after a survey Vancouver could write upon his chart, "Somebody Knows What."

After a month's stopover at Tahiti and a visit to Hawaii, Vancouver's expedition reached the coast of North America, north of San Francisco Bay, a year after its departure from England. Vancouver's men made eighty-five sets of lunar observations there to establish accurately the starting point for their survey. This suggests the high standards of surveying Vancouver was to practice over the next three years. He applied the same methodical procedures Cook had used in Newfoundland, but had to cover considerably more ground.

Vancouver used small boats to fullest advantage. In a typical operation, the two ships would drop anchor offshore, while Vancouver and several of his officers and men would make for an inlet or for shore in two or three small boats, usually a yawl and a launch from the *Discovery*, a cutter from the *Chatham*. One party might go ashore, establish a camp, and begin measuring for triangulation. Another party or two, with at least five days' provisions, would navigate the small craft in and out of the many coves on the rugged shoreline. It was slow, tedious work, from daybreak to dark, under sail when the wind blew, at the oars for long hours when the wind did not blow.

Whenever they came upon a prominent point, the officers would land and take compass bearings of all other prominent points in sight. At every opportunity, they would observe the Sun at noon to get latitude and check their chronometer for longitude. And they would keep careful records of the boat's track and make sketches of the shoreline. After the boats returned to the ships, Vancouver would supervise the drawing of a finished chart from all the observations. The new chart would then be tied into the mapping that had been done before.

These techniques were first applied in the inland waterways around present-day Seattle. In April 1792, Vancouver entered the Strait of Juan de Fuca, named for the Greek pilot of Northwest Passage legend, and set about to determine if it could lead to a navigable short cut across the continent. A camp was established at Port Discovery. A two-boat expedition under Lieutenant Peter Puget and Joseph Whidbey, with two weeks' supplies, was dispatched to trace the shoreline of one promising arm of water. For this feat Vancouver honored the lieutenant by naming the body of water after him—Puget Sound. Other features in the area also were named after Vancouver's officers: Port Orchard, Mount Baker, and Whidbey Island.

In like fashion, Vancouver's men surveyed waterways to the north, exploring the Strait of Georgia past the site of the present-day city of Vancouver and proving the insularity of Vancouver Island. There were groundings on the unknown shoals, tense encounters with Indians, countless hours spent in the small boats.

Archibald Menzies, botanist of the expedition, had high praise for the exertions of the surveyors: "Such an intricate & laborious examination could not have been accomplished in so short a time without the cooperating exertions of both men and officers whose greatest pleasure seemed to be in performing their duty with alacrity & encountering the dangers & difficulties incidental to such service with a persevering intrepidity & manly steadiness."

In 1793, after wintering in Hawaii, Vancouver's expedition returned to the coast for a survey extending from 56° N, in southeastern Alaska, down to 30° N, the planned southern limit of the survey, in Baja California. The most arduous and time-consuming phase took place along the coast of British Columbia. The surveyors had to enter each of the many fiords and channels, making sure none led to the Passage. On one such boat expedition, the men spent 23 days and covered 1,100 kilometers, yet managed to map no more than 100 kilometers of the continental shoreline.

During the third surveying season, in 1794, Vancouver cruised the Alaskan coast. He sent two boat expeditions through the treacherous tides of Cook Inlet. Cook had examined its muddy waters and assumed it was the mouth of a river, thus encouraging the lingering myth that it was actually the western entrance to the Northwest Passage. But Vancouver and his men quickly came to mountainous dead ends. Afterward, Vancouver wrote that if Cook had "dedicated one more day to its further examination, he would have spared the theoretical navigators . . . the task of ingeniously ascribing to this arm of the ocean a channel, through which a north-west passage existing according to their doctrines, might ultimately be discovered."

By now Vancouver, in failing health, seldom ventured out in the boats himself. Puget and Whidbey led most of the expeditions, sometimes staying out sixteen to eighteen days at a time as they charted the last stretch of coastline. Once Whidbey kept his crew at their oars all night to escape a band of hostile Indians. It was thus a happy occasion when, in August, the expedition reached a point near Petersburg, Alaska, where the 1794 survey joined the end of the 1793 survey. They called the place Port Conclusion.

Vancouver could then head back to England with justifiable pride of accomplishment.

In three visits to Hawaii, Vancouver produced the first complete and accurate map of those islands. As a self-appointed diplomat, a role he enjoyed greatly, Vancouver acted as a peacemaker between warring tribes of Hawaii, negotiated the cession of Hawaii to England, and even arranged a reconciliation between Kamehameha, the king, and his young wife.

In three long seasons, he produced the first complete chart of the Pacific Coast of North America, which Vancouver felt confident would be found "tolerably correct." A comparison of Vancouver's chart with modern charts of the same region shows only one error of any consequence, according to Bern Anderson, Vancouver's biographer. The error involved some features of islands such as the Queen Charlotte Islands and Kodiak, which he had copied from Spanish and Russian charts.

Vancouver's reports were as meticulous as his surveys. In describing every measurement and every sighting in such minute detail, he explained that it was his intention to make "the history of our transactions on the north west coast of America, *as conclusive as possible,* against all speculative opinions respecting the existence of a *hyperborean* or *mediterranean ocean* within the limits of our survey."

After a voyage of 100,000 kilometers, probably a record at the time, Vancouver returned to England in 1795. Though he was only thirty-eight years old, his health was sinking rapidly, owing probably to a hyperthyroid condition known as Graves' disease. In the brief time remaining to him, Vancouver wrote his journals and supervised the engraving of his charts. He died May 12, 1798, a relatively obscure and unappreciated man.

Recognition came later. Charles Wilkes, who headed a United States naval exploring expedition in the Pacific, resurveyed Puget Sound in 1841 and marveled at the accuracy of Vancouver's work. As late as the 1880s, Vancouver's charts were still the most trustworthy descriptions of Alaskan waters. And, in tacit recognition of his thoroughness, no one after Vancouver made another serious attempt to find a western outlet to the Northwest Passage anywhere below the Arctic Circle. Vancouver had exposed de Fuca's marvelous lie.

Despite the successes of men like Cook and Vancouver, at the turn of the century hydrographic mapping was an old but woefully neglected branch of cartography. The discoverers had gone to great lengths to fill in the marine charts of the globe, but the burden of hydrography extended beyond the reconnaissance of discovery. As in the days of the portolan chartmakers, the task was to provide a graphic guide (as up-to-date as possible) to the location of channels and anchorages, dangerous shoals and reefs, and navigation aids and important land features. But no one—not even the seafaring British—was charting the coastal waters and shipping lanes with any of the systematic care and thoroughness of contemporary topographic mapping as practiced, say, by the French. Writing of that time, George H. Richards, a British admiral, said that "there was scarcely what could be termed a correct chart of any portion of the globe in existence," even of the coasts of the British Isles, excepting on the Channel.

One of the problems was that, with few exceptions, maritime surveying was never the sole purpose of any of the ships afloat. Vessels of war or exploration sometimes took time out to make "running surveys" of a coast, the accuracy and detail of which seldom measured up to the work of a Cook. The consequences could be dire. During the Napoleonic Wars British losses by shipwreck, caused by bad charts as well as bad weather, were eight times as great as those inflicted by the enemy.

Another reason for the deficiency in nautical charts was institutional. Private publishing houses had traditionally sold all nautical charts, including those produced from surveys commissioned by the Admiralty. Cook, as has been noted, had to make his own chart-publishing arrangements. Commercial considerations thus dictated what charts would be available to mariners: only charts deemed readily salable would be published, and important revisions might not see print until existing editions were sold out. It was to expedite the publication of more and better charts that the British Admiralty created the post of hydrographer in 1795.

At first the hydrographic office was little more than a chart depot, for its charter spoke of "selecting and compiling" information for charts but nothing about conducting any surveys of its own. This began to change, tentatively in the beginning and vigorously after Francis Beaufort became hydrographer in 1829. A modern biographer, Alfred Friendly, called Beaufort the "greatest of British hydrographers," who in twenty-five years at the post, longer than any person before or since, created "the world's finest maritime surveying and chartmaking institution."

Francis Beaufort was born in Ireland in 1774. His father was a gentleman farmer, an Anglican cleric, and an accomplished topographer, who in 1792 published the first really good map of Ireland.* When Francis was fourteen years old, he went to sea for a time and had a harrowing experience that seems to have determined the career toward which he would struggle for many long years.

Beaufort was a young assistant on the *Vansittart,* a cargo vessel of the East India Company, bound for China. One of the ship's duties was to survey the Gaspar Strait between the islands off the Sumatra coast, the shortest route for a ship coming from the Cape of Good Hope in the direction of China. Because depth measurements in the strait were few and unreliable, the *Vansittart* spent three weeks making soundings by lead line. Beaufort, who had studied mathematics and astronomy before dropping out of school, took an active part in all this, making sightings and taking bearings. On the last planned traverse, the ship went aground on a shoal she had set out to chart. All hands abandoned ship. They were rescued after two days adrift, but Beaufort would never forget what could happen for lack of an accurate chart.

Wherever he went, as a midshipman and officer in the Royal Navy, Beaufort seized any opportunity to take latitude and longitude observations and sketch the shoreline. His surveys of the River Plate provided the first reliable charts of the approach to Montevideo. Finally, in 1810, he received orders to make a full-time job out of maritime surveying. He

---

*The map was misleading in one interesting respect, reflecting the cartographer's prejudice. As Friendly noted in *Beaufort of the Admiralty*: "Besides the expected geographical features of rivers and towns, mountains and bays, it listed every religious establishment from bishops' seats to curates' lodges—Church of Ireland ones, that is: a stranger would never discover from map or *Memoir* that there was one Roman Catholic in all Erin."

was given command of the *Fredericksteen* and instructed to chart the southern coast of Turkey, the existing map of which was hardly more than an outline.

In two seasons of surveying, Beaufort ranged from Izmir (Smyrna) south to the Gulf of Iskenderun. He often went ashore for triangulations and, more and more as he went along, to explore classical ruins. By copying inscriptions and studying classical literature, he managed to identify the classical names for many of the places he mapped, an exercise in mapping both the past and the present. Beaufort recorded his geographical and archeological discoveries and other experiences in his book *Karamania*, published in 1817 to much acclaim. The book, according to Friendly, "remains to this day the most charming *vade mecum* for any seafarer along the southern coast of Turkey."

On one surveying excursion ashore, Beaufort and his party were set upon by a mob of Turks. It happened in the Gulf of Iskenderun (Alexandretta of classical times), and not without some warning. "An old blackman" in one of the villages, Beaufort wrote in his journal, "advised us to beware of the inhabitants to the Eastward [saying], 'Tell your Captain not to trust himself on shore, for he will not be the first that they have killed or carried away.'" Two days later, the mob rushed the surveying party. Beaufort had his men fire over the attackers' heads to frighten them off. But one of the "dastardly Dervishes," as Beaufort called them, had time to fire and hit Beaufort in the groin.

The wound became infected, and for a time Beaufort despaired of ever recovering. But after a convalescence at Malta, he was able to resume command and bring his ship back to England. It proved to be his last voyage.

For the next two years Beaufort confined himself to the preparation of his Turkish charts—eleven basic sheets, at a scale of two miles to an inch. The quality of the work is attested by the fact that until 1972 the Beaufort charts continued to be cited by the Admiralty as the principal authority for surveys of the southern coast of Turkey.

For the next fourteen years after that, until 1829, Beaufort enjoyed the honors that came his way, including membership in the Royal Society, and advised others on matters of science and exploration. But the job he really wanted eluded him. He wanted to be hydrographer to the Admiralty.

At the age of fifty-five, after being passed over at least once, Beaufort was finally appointed hydrographer. His energy and ambitions for an enlarged chartmaking operation coincided with the expansion of the

British Empire, whose existence depended on operations at sea—and a well-charted sea, preferably. More ships were assigned to surveying duty. What Beaufort expected of each survey can be seen in his instructions to each captain, such as the following example of 1831:

> The first and principal object will be the hydrographic contour of the Gulf [of Arta, in Greece] with its islands, roads, shoals and soundings . . . an accurate examination of the entrance . . . and judicious marks for passing over the bar in the deepest water. These marks are always best conveyed to the seaman by addressing them to his eye, that is by sketches of the objects when in the desired position. The height of all headlands, isolated hills and remarkable peaks should be trigonometrically determined and inscribed on their summits on the charts—as they afford the seaman a ready means of ascertaining his distance by the dip table.
>
> The nature of the shore, whether high cliff, low rock or flat beach, is of course inserted on every survey—but much more may be easily and usefully expressed—for instance the general elevation of the cliffs and their colour, the material of the beach, mud, sand, gravel or stones, etc.
>
> Advantageous landing places and their nearest point of approximation to the known roads should be distinguished.

At one time Beaufort's fleet of survey ships numbered twenty, the best known of which was the *Beagle*. On his famous second voyage (1831–36) the captain, Robert FitzRoy, charted major sections of the South American coast from the River Plate around Cape Horn and as far north as Ecuador. But the expedition is remembered because of the naturalist on board—Charles Darwin.

One of Beaufort's most flamboyant surveyors was Edward Belcher, who produced data for more than sixty charts of the west coast of Africa, the China Seas, East Indies, and the west coasts of North and Central America. And he seems to have had time to engage in Britain's first Opium War with China and in some secret missions.

As much as Beaufort admired Belcher's work, the man sometimes exasperated him, as can be seen in one letter from Beaufort to Belcher:

> Your last letter is really all Hebrew to me; ransoms and dollars; queens; treaties and negotiations? What have I to do with these awful things; they far transcend my limited chart-making facilities. . . . The harvest I look for at your hands does not stretch beyond the reach of a deep sea-line and all the credit I crave for you, and through you for myself, must be won in the Kingdoms of science and reaped in hydrographic fields.

Some of Beaufort's surveyors died of fever in the West Indies and on the African coast, of exhaustion in New Guinea, of drowning off Panama, and at the hands of murderers in Malta and Africa. One survey ship with all aboard was lost in a North Sea gale. More often than not, the hardships and risks of those who went to sea to map the world were far greater than of those who mapped the land.

By the time Beaufort retired in 1855, the hydrographic office had produced some 1,500 new nautical charts, a record 130 in his last year. He was never too busy to spend days fussing over each chart before it went to the engraver. His standards were the highest, so much so that the expression "safe as an Admiralty Chart" entered the language as a simile for integrity.

Francis Beaufort died in 1857 at the age of eighty-three. His contributions to hydrography had earned him a knighthood, the rank of admiral, and a place on the world map, Beaufort Sea in the Arctic being only one of several geographic features named for him. A historian of the hydrographic office, L. S. Dawson, a surveyor himself, credited the "master mind of Beaufort" with doing "more for the advancement of maritime geography than was effected by all the surveyors of European countries united."

It would be another century before marine cartography could extend its reach with any authority beyond coastlines and coastal waters to the topography of the deep-sea floor. Meanwhile, the exploration and mapping of the Pacific by Cook and Vancouver and by Beaufort's surveyors elsewhere meant that the essential outlines of Earth's lands and seas were now known, with the major exception of Antarctica. The important blank spaces remaining on the world map lay in the interiors of Africa, Central Asia, and the Americas, and in those directions the explorers and cartographers turned their attention throughout the nineteenth century.

Soldiers, Pundits, and the India Survey

So far as Kim could gather, he was to be diligent and enter the Survey of India as a chain-man. . . .

"Yes, and thou must learn how to make pictures of roads and mountains and rivers—to carry these pictures in thy eye till a suitable time comes to set them upon paper. Perhaps some day, when thou art a chain-man, I may say to thee when we are working together: 'Go across those hills and see what lies beyond.'"

With these instructions from Colonel Creighton, Kim, the self-possessed orphan of Rudyard Kipling's creation, found himself drawn into the adventurous, sometimes clandestine operations of the Survey of India. Kim had what it took: enterprise, daring, devotion to duty. He learned all about rods and chains and angles, compass bearings, the "simple remedies of the Road," and how to "pace all his distances by means of a bead rosary." And those about him—Creighton, who dreamed of becoming a fellow of the Royal Society; Mahbub Ali, the Pathan horse dealer and secret agent; Hurree Babu, the fat Bengali adventurer and secret surveyor—were, though fictional, true to the life of British India in the most colorful era of the Survey, a mapping service with one of the longest and proudest traditions.

The Survey of India was founded by James Rennell, a young Navy officer with a passion for maps and experience as a surveyor, a man to whom, it was once said, "the blanks on the map were eyesores." In 1765 Robert Clive, governor of the East India Company, assigned Rennell the task of making a general survey of Bengal. Rennell took a detachment of sepoys and went about the countryside for seven years, fixing latitudes, plotting productive lands, and filling notebooks with information on rivers, villages, and terrain. On his first project, mapping the Ganges delta, Rennell made this entry in his journal: "We have no other Obstacles to carrying on our Business properly than the extensive thickets with which the country abounds, and the constant dread of Tygers, whose Vicinity to us their Tracks, which we are constantly trampling over, do fully demonstrate."

A survey assignment in those days was often tantamount to a death sentence, and this came fearfully close to being the case for Rennell. A tiger did attack the party and carried off one of the soldiers. On another occasion, when a leopard sprang from a low bough and mauled five sepoys, Rennell, a slight, sinewy man, seized a bayonet and thrust it into the beast's snarling mouth. In combat with a band of human marauders, he sustained deep saber wounds that, as Clive reported to the company directors, left the surveyor with "a shattered constitution."

At the age of thirty-five, a fatigued Rennell returned to England, where he published the *Bengal Atlas* and was acclaimed the "father of Indian geography." But he had never crossed the hills to see what lay beyond. He had reached the foothills of the Himalayas, only to be turned back by fierce tribesmen standing guard at the passes.

After Rennell a motley procession of men passed into and out of the Survey, gentlemen and scholars, adventurers and lost souls—the flotsam of empire. They included a restless parson who happened to be an ardent astronomer; a veterinarian and a mathematics teacher; a shipwrecked master of art from Aberdeen; Hanoverian soldiers seeking respite from regimental duties; and even the bassoon player in a ship's band, who took up drafting.

According to the ample records of the Survey, one of this breed was a Francis Wilford, whose activities for obvious reasons cannot go unnoted. He left a clerk's job in London to take a chance on finding a more exciting life in India. He earned a commission in the Bengal Engineers, was attached to the Survey, and in his spare time became a student of ancient Hindu geography. From 1788 to 1794 he mapped much of the country around Benares, where the work once thrust him into the middle of a local boundary dispute. "I really thought that a stop would have been put to our progress," Wilford wrote, "when a Snake springing from the ground between the Arbitrators, to their astonishment and terror, ran away towards the Boundary. The Arbitrators, and the parties themselves, concluding this was a singular interposition of Providence, considered themselves now obliged to abide by the decision of the Snake, and agreed that the Line the snake had described in his flight would be for ever their mutual Boundary."

In the early decades of the nineteenth century, before the time of *Kim*, surveying parties like Wilford's were a fairly common sight south of the Himalayas. Outside the cities, along riverbanks and roads, in the downpours of the monsoon and the searing heat, they went about their methodical work, everywhere a tableau set against the sweep of the land. There were the perambulator wallahs, half-naked Indians taking turns pushing the measuring wheel; coolies bearing the tents and sacks of supplies; a company of armed sepoys, barefoot, turbaned, and red-coated; and the officers, sweating profusely in their full regulation dress, operating the theodolite and other instruments.

Their instructions were generally simple. According to the surveyor-general's orders for an 1806 route survey, they were to "observe everything on the Road, or that is visible from it, which can be considered as of any importance, but particularly . . . Forts, Hill Forts, remarkable peaks, mountains or Hills, Ghats or Passes, Towns . . . villages, etc.; Rivers or Nullahs, with their names, and noticing the way the stream runs, whether right or left, at the crossing place; their breadths and directions as far as visible, up and down the stream. . . . The bearings of the Road shall be observed as frequently . . . as possible."

Increasingly, however, the orders called for more scientific surveys. Triangulation was introduced, and the Great Trigonometrical Survey was initiated. A major discovery in geodesy lay ahead.

W illiam Lambton was the prime mover behind the Great Trigonometrical Survey. A veteran of the American Revolution, he had gone to India to serve as a regimental officer under Arthur Wellesley, the future Duke of Wellington. While studying existing maps of India, Lambton concluded that there was a clear need for "a mathematical and topographical survey of the greatest accuracy."

Lambton's first step was to order a special theodolite from London. It arrived in 1802 after a most eventful voyage. The ship carrying the theodolite was captured by the French and taken to Mauritius, but fortunately—this being before the era of total warfare—the French sent the theodolite on to India with a complimentary letter. Then Lambton could start the survey, working virtually from first principles. He set about determining the linear value of a degree of latitude and of degrees of longitude at different latitudes. Toward this end he measured a base line near the city of Madras and then ran a line of triangles westward across southern India.

To some in the East India Company, practical men used to the more straightforward route surveys, a string of triangles seemed an outrageous waste of time and money. Scoffed one member of the finance committee: "If a traveler wished to proceed to Seringapatam, he need only say so to his head palankeen bearer, and be vouched that he would find his way to that place without any recourse to Colonel Lambton's map."

From his survey, however, Lambton learned that the width of the Indian peninsula along the 13° parallel was 575 kilometers—70 shorter than given on Rennell's map. But Lambton had his mind on more than his critics or the revision of old maps. He essayed to construct a framework for a really scientific map of India. This would, he said, "accomplish a desideratum still more sublime, viz., to determine by actual measurement the magnitude and figure of the earth."

In the pursuit of this objective, Lambton proceeded to the southernmost tip of the Indian subcontinent, Cape Comorin. There, in 1806, he commenced laying out the triangles for the Great Arc of India. No more ambitious arc-of-the-meridian measurement had hitherto been attempted, in Europe or anywhere else. According to Lambton's plan, the arc would run almost 2,900 kilometers, from just east of Cape Comorin north along the 78° meridian through the mountains of Mysore, the plateaus of Hyderabad, the tip of the Ganges plain to the snow-capped peaks of Kashmir and Jammu. It would be the reference line for all future Indian mapping.

Lambton devoted the remaining seventeen years of his life to the monumental task. He drove himself and his men. Once when the great theodolite was being hoisted to an observation station on top of a building, the guide ropes broke and the instrument crashed against the wall, damaged apparently beyond repair. Lambton was undeterred. He retired to his tent with the theodolite and for six weeks worked without interruption. He rebuilt the theodolite with his own hands. With Lambton there was only one thought: the survey must go forward, the map must be finished.

His young chief assistant, George Everest, wrote the following description of Lambton in the field: "This great and extraordinary man, who when he aroused himself for the purpose of adjusting the great theodolite, seemed like Ulysses shaking off his rags; his native energy appeared to rise superior to all infirmities, his eyes shone with the lustre and his limbs moved with the full vigour of manhood."

When Lambton died of tuberculosis in 1823 at the age of seventy, he was out in the field in his survey tent. Less than half of the projected arc had been measured.

Everest, Lambton's successor, was no less committed to the task. A lean and wiry man with deeply sunken eyes and bristling side whiskers, Everest was a stickler for detail and accuracy and had an extremely short temper. He displayed these attributes as he pushed the survey northward to the mountains. It was slow and hard work, as Everest's reports make clear:

> From the station of Kundagutt which I had visited previously to crossing the Kistna, I sought for a station to the east of Hydershahipett and the only eminence which offered itself was a long black-coloured range, strongly resembling in shape the back of an elephant. There seemed to be on this two places about sixty miles off, which bade fair to answer my purpose, and I accordingly despatched four of my most skilful flagmen, with an efficient guard, to seek these spots out, and occupy them with my signals.
>
> It took me about three weeks to run southward along one side of the series, and to return northward by the other side of Hydershahipett. Nothing whatever having been heard of my detached parties, great apprehensions were entertained by me for their safety; but at last a gap began to break open in the black mountain . . . and after a fortnight's further waiting I had sufficient daylight behind to distinguish the colours of the Great Trigonometrical Survey flying on the one spot, and a signal mark on the other. The secret of the delay now came out. The station of Hydershahipett was on the very verge of the great forests of teak and ebony, far into the depths of which was situated this elephant mountain, called Punch Pandol. The access to it was by a circuitous route, unknown to any but the few straggling natives who lived in those forests, in a state closely bordering on savage life.
>
> The nearest village was Poomrarum, about five miles from the summit, from which it was necessary to cut a road for the instruments and tents; . . . and how my unfortunate flag-men could have had perseverance enough to go through with such a task, how they could have coaxed any uninterested persons to accompany them; how, after having pierced through a forest of teak trees, seventy, eighty, and even ninety feet high, thickly set with underwood, and infested with . . . tigers and boa-constrictors, without water or provisions, and with the jungle fever staring them in the face, they could have wandered through such a wilderness until they selected the most commanding point for a station, utterly . . . passes my comprehension.

By 1834 the survey came within sight of the Himalayas. A visitor to Everest's camp, Godfrey Vigne, described the scene 3,600 meters above sea level close to the summit of a mountain known as Chaur. Everest lived in a tent heated with a stove and closed tight against the cold air. Not far away on the huge granite rocks that formed the apex of the mountain, Vigne remembered, workers "had formed a platform of loose stones, and in the centre they had planted a mast as a mark for the survey." Off in the distance, visible only through the theodolite, were several other marks that they had previously raised on other summits. And down at Saharanpur on the plains, almost one hundred kilometers away, a powerful heliotrope reflected sunlight in brilliant flashes. Vigne had seen a typical setup in the survey's mountain operations.

Everest relied heavily on the heliotrope in triangulation. This was simply a circular mirror twenty centimeters wide mounted on a staff so that it could be turned in any direction. A flagman would set it up on a mountain peak or some other point where it could be seen by surveyors at several other locations. At night or in a haze, the flagman would have to burn a blue light as a signal for the surveyors making their triangulation sightings.

By the time he retired in 1843, Everest had extended the Great Arc well into the high foothills of the Himalayas. But it is not known if, before he returned to England and knighthood, Everest ever saw Peak XV.

This majestic peak rises from the Nepal-Tibet border, which was forbidden territory to the men of the Survey of India, although some of them had seen the peak from a great distance. In 1849 surveyors climbed lesser summits and took theodolite measurements of Peak XV from six different directions, with no two measurements in agreement. The distances, compounded by the antics of light reflected off snow fields and glaciers, made it all but impossible to get the same theodolite angle twice in succession. By averaging out the measurements, however, a Bengali clerk determined that the peak was 29,000 feet (8,700 meters) high. When he rushed into the office of Andrew Waugh, Everest's successor, with this information, the clerk supposedly blurted out: "Sir, I have discovered the highest mountain in the world."

But would anyone believe the measurement? When surveyors in India had reported measurements of other high Himalayan peaks, their figures were dismissed as preposterous by learned Europeans, who maintained categorically that it was impossible for a mountain to be higher than, say, about 25,000 feet. They would certainly be skeptical of such a seemingly rounded-out figure as 29,000 feet. So the surveyors arbitrarily added 2 feet to the calculation and for decades so it read in the geography books—29,002 feet.

Peak XV was known in Tibetan as Chomolungma, "Goddess Mother of the World." The surveyors had made it a practice to give each newly mapped peak an identifying number instead of a name, on the principle that it would be presumptuous to add new European names to peaks which were already known to and revered by the Asians; a laudable policy, and an exception to the usual practice of explorers and surveyors. But a peak as imposing as XV proved irresistible; it had to have a name, not a mere number. And so Waugh eventually ordered the mountain placed on the map and named after his predecessor. This is Mount Everest.

Subsequent measurements of Everest in the twentieth century produced a revised height of 29,149 feet, or 8,744 meters, and even that may be slightly off the mark. When it comes to measuring mountains, Colonel S. G. Burrard of the Survey once remarked from experience: "All observations are liable to error; no telescope is perfect; no leveling instrument entirely trustworthy; no instrumental gradations are exact; no observer is infallible."

The fallibility of trained observers and the perversity of nature were brought home to the Survey of India in the early 1850s. As the surveyors ran the Great Arc north into the mountains, they became more and more uneasy about a discrepancy in their data. The distance between Kaliana, at the base of the Himalayas, and Kalianpur, about 600 kilometers south, was first measured by triangulation and then checked by direct surface measurement. The two measurements failed to agree by 150 meters. This may seem minor, but it was an intolerable discrepancy in surveying of the kind being conducted in India. Something was seriously wrong, in their equipment or in their methods. Scientists in many countries, hearing of the error and the ensuing consternation, considered possible solutions to the so-called "Indian puzzle."

The surveyors in India suspected that the error was due to the plumb bob. They had used a plumb bob, a weight suspended by cord, in aligning instruments during the astronomical observations to establish latitude. If Earth's mass were homogenous, thus making its gravitational attraction uniform, the plumb bob would hang truly perpendicular to the surface. But Earth's density varies, as Bouguer had discovered in the

Andes, and this causes gravity anomalies. These variations in gravity, in turn, play tricks on plumb lines. The angle between the direction a plumb line takes and the direction of a reference perpendicular is called a deflection of the vertical. Knowing this, the surveyors in India had sought to compensate for such deflections when they calculated their angles. In particular, they made adjustments for the mass of any mountains nearby. They assumed that the enormous mass of the Himalayas would be sufficient to deflect the plumb line by at least 15″ of arc.

This proved to be the source of their error. Clearly, the Himalayas deflected the plumb line, but not nearly to the extent that their extra mass above sea level should require. The deflection rate, it turned out, was only 5″ of arc. This discovery confirmed Bouguer's suspicion of a century before. Mountains are made of material less dense than lowlands. Moreover, the discovery led to a new understanding of the nature and structure of Earth's crust.

From the Survey of India findings John Henry Platt, an English archdeacon and amateur scientist, surmised in 1854 that mountains must rise from the crust like a fermenting dough, the density of which would be less the higher they rose. The imagery was appealing, but not necessarily correct.

A year later, George Biddell Airy, a respected English scientist, conceived of an explanation that is more acceptable to geodesists today. If a mountain range were simply a protuberance of rock resting on the continental platform and wholly supported by the strength of that platform, then a plumb line should be deflected from the true vertical by an amount proportional to the gravitational attraction of the mass of the mountain. Since this was not the case in the Himalayas or the Andes, the surprisingly lower deflection angle could be explained only if the mountain's mass below sea level reaches deeper but is of lesser density than the material underlying plains and the ocean floor. Airy concluded, therefore, that mountains have "roots" and the oceans have "anti-roots."

A new and important idea in geophysics was emerging. It would later be called *isostasy,* from two Greek words meaning "equal standing."

According to the theory, each unit column of Earth, from surface to center, has approximately the same mass. A level of compensation exists at some depth below the surface—a sort of imaginary surface on which the lighter crust of Earth floats on the denser plastic material of the mantle underneath. Wherever the crust rises above sea level, as a continent or mountain, an appropriate amount of material must extend deeper into the mantle, just as high icebergs extend deeper into the water than low ones. This is the case because floating bodies displace their own weight. As a result the bottom of the crust, where it meets the mantle, has an undulating shape something like a mirror image of the surface above. Beneath mountains the crust bulges down into the mantle; beneath valleys and under the oceans, the crust is thin and the mantle reaches closer to the surface.

Subsequent gravity measurements bore out the conclusion that under mountains there is indeed a deficiency in mass, a root of lighter material. Along the shores of South America, for example, a pendulum swings away from the Andes and toward the abyssal floor of the Pacific. Seismic measurements showed that mountains can have roots extending as deep as 70 to 75 kilometers.

These were the kinds of discoveries, sparked by geodetic surveyors in India, that Henri Poincaré must have had in mind when he wrote: "When the geodesist finds that he has to turn his glass a few seconds of arc in order to point it upon a signal he has erected with much difficulty, it is a very small fact, but it is a fact giving a great return, not only because it reveals the existence of a little hump upon the terrestrial geoid, for the little hump would of itself be of small interest, but because this hump gives him indications as to the distribution of matter in the interior of the globe, and, through that, as to the past of our planet, its future, and the laws of its development."

Geodesists of the nineteenth century drew increasing satisfaction from their new insights into the shape of the interior as well as the surface of Earth. But they could also see that their task was not as simple as they might once have imagined. The Survey of India had impressed upon them the importance of something as simple as the deflection of the vertical. More precise calculations of the deflection would depend on charting the gravity anomalies all over the globe, a task of gravity surveying that remains unfinished. Using pendulums and the more modern and portable gravimeter, a spring-balance device about the size of a Thermos jug, surveyors have spent years collecting gravity observations. These have been collected and transferred into one uniform system related to the gravity value at Potsdam, which is internationally accepted as the principal gravity reference point. Even so, gravity anomalies continue to complicate the work of those who seek to map Earth according to its true size and shape.

In the last half of the nineteenth century, the time of Kim, Creighton's command reverberated throughout the Survey of India. Go across those hills and see what lies beyond. It was the pride of the Survey that most of India had been mapped fairly well, often better than other less extensive lands, but there was the lingering challenge before the Survey to penetrate the mountain barrier to the north and put those forbidden lands on the map.

The barrier was more than physical, it was political and traditional. The 2,400-kilometer border of Tibet was closed by order of the emperor of China—"no Moghul, Hindustani, Pathan or Feringhi [Eurasian] shall be admitted into Tibet on pain of death." Some travelers did get through, but no surveying party, certainly not one that included Europeans.

Thomas G. Montgomerie, a captain in the Survey, would not let himself be frustrated. He was an intrepid mapper. Pictures of Montgomerie show a man with full whiskers, high forehead, and deep-set eyes that seem focused on far horizons. He and his parties had suffered the cold and falling rocks to scale peaks of 4,500 meters or more to set up measurement stations. He sighted the elevation of K-2 from eight other stations and averaged them out at only slightly less than Everest. But each year Montgomerie had to turn his back on those distant peaks and forbidden lands.

After one such experience, Montgomerie returned to Dehra Dun, the pleasant little north India town where the Survey had its northwestern headquarters, and he had an idea that just might work. If Europeans were not permitted to cross the Himalayas, why not train Indians, give them disguises, and send them across to get the mapping data?

"When I was in Ladakh I noticed the natives of India passed freely backwards and forwards between Ladakh and Yarkand in Chinese Turkestan, and it consequently occurred to me that it might be possible to make the exploration by that means," Montgomerie explained. "If a sharp enough man could be found, he would have no difficulty in carrying a few small instruments amongst his merchandise, and with their aid good service might be rendered to geography."

It was a form of spying, to be sure, but these were desperate times for the British in India. The Great Game was afoot. This was the sporting term Kipling applied to the British diplomatic, military, and clandestine struggles with Russia for political ascendancy in Central Asia. It was a deadly game: several British surveyor-spies who entered the contested territory had already been tortured and murdered. So Montgomerie set about to get permission from his superiors to use Indians in disguise to survey the playing fields of the Great Game.

Montgomerie had already had some experience using native agents in Kashmir and to the northwest, with uneven results. The problem was recruiting the right persons. His original notion, to search the bazaars looking for natives from areas he proposed to explore, had not worked, for the reasons he once noted: "Any number of men are willing to volunteer for such a service, and if their own accounts are to be believed, they are all well fitted for the task. But a very little enquiry, however, reduces the number of likely men nearly down to zero; many cannot write, others are too old, and have no ideas beyond those of trade, and nearly everyone has special ideas as to what pay and rewards they are to get, and generally have special stipulations to make; all, however, apparently thinking nothing of the risks and exposure involved."

For his operations across the Himalayas Montgomerie decided on a different source of agents, or "intelligence workmen," as he called them. He selected eight candidates who had been recommended to him by the Indian Education Service. Only two of them mastered the two years of rigorous training, which included instruction in the use of the compass and sextant, in what features to look for when making a survey, and in taking an absolutely uniform stride as they paced off distances. The two who completed the course were cousins, Nain Singh and Mani Singh. Nain Singh was a young schoolmaster in one of the upper valleys of the Himalayas—a pandit, the appellation for a learned man. Anglicized, it was pronounced pundit. He came to be the first and greatest of the pundit explorers.

Montgomerie sent the Singh cousins on their first mission in 1865. They were to proceed through Nepal and cross the border into Tibet. They were to try to reach Lhasa, the capital, whose position on the maps, Montgomerie said, was "only a matter of guess."

As the Chinese emperor's ban on nearly all outsiders remained in force, the Singhs traveled disguised as Bashahris, people who were known to enjoy certain travel privileges into Tibet. If anyone should ask, the Singhs were to say that they intended to buy horses and pay homage at the Lhasa shrines.

In Nepal, the two men became separated. Mani Singh eventually got into western Tibet on his own, collected some mapping data, and soon returned to Nepal. Nain Singh headed for the border nearer to Lhasa. At a frontier village below the pass of Gya La, Nain Singh negotiated with a party of traders to take him across as part of their caravan. Soon after the crossing, the traders slipped away at night, taking most of Singh's money and leaving him alone in a strange land.

Fortunately, the traders did not take Singh's most valuable possessions, his tools for the surreptitious survey. The tools were concealed in a box with a false bottom. They included a sextant, thermometer, compass, chronometer, and a container of mercury. Singh also carried a prayer wheel, a cylinder on the end of a stick, which was more than merely part of his disguise. Tibetan Buddhists believe that prayers written on a roll of paper and placed on the cylinder will, each time the wheel is turned, go to heaven. But in his hollow cylinder Singh had placed not prayers but paper slips on which he would record distances and compass bearings. In addition, Singh carried a special rosary. Instead of the 108 beads of the usual Buddhist rosary, his had 100, for obvious mathematical reasons, and every tenth bead was slightly larger than the others. He planned, as Kim was also taught, to measure distances by dropping a bead on the string at every 100 paces, a large bead for every 1,000 carefully measured paces.

With the prayer wheel and rosary, Singh gave the appearance of being what he said he was, a poor, devout lama on a pilgrimage to Lhasa. A picture of Singh on the frontispiece of *Records of the Survey of India*, volume VIII, shows a slight man with a strong jaw and piercing eyes, all in all a most determined visage. He must have been that kind of man to do what he did.

Singh trudged through the strange, desolate country in the summer of 1865. He begged for food from the occasional passing caravan. When he could be sure no one was around, he took compass bearings, fixed his position by sextant, and recorded all this on the paper rolled inside his prayer wheel. Finally, in January 1866, Singh entered the forbidden city of Lhasa. He took, at first, all the steps appropriate for a pilgrim, including a visit to the Great Lama of Tibet. In his report Singh described the visit:

> I accompanied the Ladakh merchant, called Lopchak, on the 7th of February, to pay homage to the Gewaring-bo-chi [Great Lama] in the fort, ascending by the southern steps. A priest came out to receive us, and we were conducted into the presence of the Gewaring-bo-chi, a fair and handsome boy of about thirteen years, seated on a throne 6 feet high, attended by two of the highest priests, each holding a bundle of peacock feathers. To the right of this boy, and seated on a throne 3 feet high, was the Raja Gyalbo-Khuro-Gyago, his minister. . . . We were ordered to be seated, and after making offerings of silks, sweets, and money, the Lama Guru put us

three questions, placing his hand on each of our heads: "Is your king well?" "Does your country prosper?" and "Are you in good health?" We were then served with tea, which some drank and others poured on their heads.

Singh, to support himself, arranged to teach merchants the new Hindu methods of keeping accounts. But the delicacy of his situation was brought home to him with a shudder the day he saw the public beheading of a Chinese man who had arrived in Lhasa without permission. "Owing to my alarm," Singh recalled, "I changed my residence and seldom appeared in public again."

The little inn where he stayed became his secret observatory. At night he would climb out of his window onto the roof and with the sextant take the angular altitude of stars to determine latitude, timing his observations with the chronometer. He poured the mercury he had brought with him into his beggar's bowl. This level reflecting surface served as an artificial horizon from which he could measure the altitude of the stars he was observing. Whenever he could, he would apply the boiling-point method of determining elevation above sea level. Though not very precise, it was the best he could manage under the circumstances. He would boil water on the stove and check its temperature with the thermometer he had brought along—the higher the altitude, the lower the water's boiling point. Singh thus calculated that Lhasa was 3,420 meters above sea level. Today's measurements put the elevation at 3,540 meters.

Singh repeated these observations several times over the next few weeks. One day in the market, however, he was questioned by two Muslim merchants, who seemed friendly enough but a little suspicious of him. So Singh decided that the time had come for him to pack up his surveying gear and leave Lhasa, while he still could. On April 21, he left with a Ladakh caravan, heading west along the great Tibetan river, the Tsangpo. This enabled him to trace the river's course for more than 800 kilometers, ever the prayerful lama and secret surveyor, furtively noting the distances and bearings. After two months with the caravan, Singh stole away one night and struck out alone for the Nepal border. He reached the Survey's headquarters at Dehra Dun on October 27, 1866.

Nain Singh had been away 21 months. He had surveyed a 2,000-kilometer trade route from Nepal to Lhasa, taken 31 latitude fixes, determined elevations at 33 places. He had returned with vivid descriptions of the Tibetan capital and the first reasonably correct position for its placement on the map.

And he proved the feasibility of Captain Montgomerie's covert survey of the trans-Himalayan frontier. On Singh's retirement, after several other missions, he received the Royal Geographical Society's gold medal as the "man who has added a greater amount of positive knowledge to the map of Asia than any individual of our time."

Over the next two decades Montgomerie and his successors trained and dispatched many other pundit explorers. In 1868 a Persian named Mirza Shuja undertook such a mission into the north of Afghanistan, surveying the Oxus River and the Pamirs; he was murdered in his sleep at Bokhara. In 1871 Hari Ram made his way to Lhasa and returned with descriptions and data on nearly 48,000 square kilometers of previously unknown territory.

One of Nain Singh's nephews, following the family tradition, set out on a secret mission for the Survey in 1878. He was Kishen Singh, known in Survey records by the code name A-K, and his assignment was to reach Lhasa and then make a route survey northward across Tibet and into China. He affected the customary pilgrim's disguise, carried the usual concealed instruments, and was accompanied by two assistants.

Two years passed without news of A-K or his assistants. Four years. Five. The Survey gave them up for dead. After an absence of nearly six years, A-K arrived back in India, emaciated but successful. He had been robbed by nomads, deserted by one assistant. But he held on to his concealed instruments. He traversed trackless plains in Tibet and crossed the Nyenchintanglha Mountains, which Nain Singh had observed and reported on one of his missions. As instructed, A-K mapped a caravan route to China. On the return from China, his survey took him to the headwaters of the Mekong, Salween, and Irrawaddy rivers.

Even more incredible was the travail of Kinthup.

Nain Singh's accounts had raised one particularly mystifying geographical question. Could it be, as Singh suggested, that the Tsangpo flowed through the Himalayas to become the Brahmaputra? It seemed unlikely, perhaps impossible. Only a gorge of majestic depths could enable the Tsangpo, which runs the length of Tibet, to flow south into the Brahmaputra and thence across the Indian plain to the Ganges delta. Could there be such a gorge somewhere in the Himalayas? The question was an intriguing one to a generation that had just witnessed the race to find the source of the Nile.

H. J. Harman, a captain in the trans-Himalayan section of the Survey, decided in 1880 on a method of obtaining a scientific answer to the question. It was a straightforward method, but it required careful timing and coordination. He would have something identifiable floated downstream in the Tsangpo. If it emerged in the Brahmaputra, then he would have a definitive answer. The plan called for slipping someone, once again, into Tibet.

For the assignment Harman selected a Chinese lama living in Darjeeling and a native surveyor from Sikkim, who would be disguised as the lama's servant. The native surveyor was a man named Kinthup. He could neither read nor write. The lama and Kinthup were instructed to travel to Tibet, follow the Tsangpo as far downstream as possible, and then to throw into the river 500 logs. Each log was to be cut a foot long. Fifty such logs were to be tossed into the river each day for ten days. A watch for the logs would be maintained where the Brahmaputra flowed out of the Himalayas.

But Harman had misjudged one of his men. The lama dawdled for eight months before reaching the northern side of the frontier, more intent on enjoying himself than anything else. He stopped at villages and got drunk, staying four months in one place. As Kinthup reported later, "The detention was owing to the lama falling in love with his host's wife. Finally, the state of affairs between the wife and the lama became known to the host and the lama had to pay Rs 25 in compensation."

The lama had from the start behaved brutally toward Kinthup. When they reached one Tibetan village, he sold Kinthup as a slave to the village headman. The lama was never heard from again. But Kinthup carried on.

For eleven months, from May 1881 to March 1882, Kinthup worked as a slave. When he got a chance, he escaped to Marpung and sought refuge in a monastery, where he became a novice monk, biding his time. After four months, Kinthup was granted permission to go on a pilgrimage. It turned out that he was in the area where the Tsangpo approached the highest Himalayas. Kinthup, therefore, went straight to the forest near the Tsangpo and worked for days cutting logs—five hundred of them. He hid them in a cave above the river and returned dutifully to the monastery.

After two more months, Kinthup asked to go on another pilgrimage. This time he walked to Lhasa to find a professional letter writer. And this is what he had written for him and sent to officials of the Survey of India:

Sir: The Lama who was sent with me sold me to a Djongpen [headman] as a slave and himself fled away with the Government things that were in his charge. On account of which the journey proved a bad one; however, I, Kinthup, have prepared the 500 logs according to the order of Captain Harman, and am prepared to throw them 50 logs per day into the Tsangpo from Bipung in Pemake, from the fifth to the fifteenth day of the tenth Tibetan month of the year called Chhuluk, of the Tibetan calculation.

Kinthup did as he promised. He threw the logs into the Tsangpo, fifty a day for ten days. But the watch on the Brahmaputra had long since been abandoned, as the lama and Kinthup were assumed lost, and Kinthup's letter from Lhasa would arrive too late. The logs must have floated through, however, for it was afterward proved that the Tsangpo and the Brahmaputra were indeed the same river.

Kinthup's mission was, in a sense, a failure. He did not receive the honors that had been bestowed on earlier pundit surveyors. He lived out his life as a humble tailor in Darjeeling. But his devotion to duty became a legend in the Survey of India, a legend that not even Kipling could have surpassed. Soon after Kinthup's mission, the practice of sending out disguised native surveyors came to an end.

In 1911 the Survey of India extended its triangulation, begun by Lambton and Everest, across the mountains to join with surveys carried out by the Russians from the north. India and the Himalayas became a part of the mapped world.

Mapping America:
The Boundary Makers

CHAPTER TWELVE

S ome of the most important lines on a map are those superimposed by society: boundary lines. Here and there, naturally enough, they follow a river or mountain barrier; elsewhere, they square off into the geometry of expectations, the grids of royal grant and political destiny, of speculator and settler; and then they zigzag and meander in ways bespeaking past battles and bargains, compromises by faraway diplomats, and the failings of not a few wayward surveyors.

The running of boundary lines, defining the framework of ownership and citizenship, is a staple of surveying and mapmaking. Compared to measurements of an arc of the meridian or the mapping of a vast ocean, land surveying may seem a rather pedestrian occupation. But most people who are surveyors spend most of their time surveying boundaries of one kind or another—the lines of property, province, or country. Theirs is a skill, more formally called cadastral mapping, dating back at least to ancient Egypt and Babylon.

At no time is a land surveyor cast in a more indispensable and exciting role than during the settlement of a new territory. This was certainly the case in the British colonies of North America. In the seventeenth and eighteenth centuries the emphasis in North America was on property lines and provincial boundaries, the laying out of new towns and the opening of roads. The services of land surveyors (as distinguished from topographic surveyors and mappers of unexplored regions) thus became very much in demand in colonial America.

Colonial surveyors were generally men who possessed a rugged constitution, a basic understanding of mathematics, and the essential tools of the trade, usually a plane table, chains, and a surveying compass known as the circumferentor. Their standard textbook, after 1688, was more often than not John Love's *Geodaesia, or The Art of Surveying and Measuring of Land Made Easie*. Love, who had served as a professional surveyor in North Carolina and Jamaica, understood the problems of American surveying. For instance, although the theodolite was already in use in England, he recommended against its employment in America. The early theodolites, heavy and bulky, were more suited to the cleared lands of England and Europe. The colonial surveyors found the circumferentor handier out in the dense forests. As Lewis Evans wrote in the "analysis" accompanying his *General Map of the Middle British Colonies in America* (1755), many European surveying practices were impractical in a country almost "every where covered with woods. . . . Here are no churches, Towers, Houses or peaked Mountains to be seen from afar, no Means of obtaining the Bearings or Distances of Places, but by the Compass, and actual Mensuration with the Chain."

The remuneration could be generous. A county surveyor in Virginia could by law charge 40 pounds of tobacco for every 100 acres surveyed of areas more than 1,000 acres, or not more than 400 pounds for areas that were less. Out on the frontier, surveyors were paid in money or shares of land, or both, so that an industrious surveyor could wind up a big landowner. Daniel Boone, who took up surveying on the frontier of Kentucky, claimed and surveyed new lands for himself and for land-hungry easterners on a fifty-fifty basis.

The early surveys were often haphazard, unavoidably inaccurate, and open to dispute. In Virginia, consequently, it became the practice, following an English custom, for all persons in each neighborhood to assemble every four years at a given point to retrace the boundary lines of all plantations in the area. The landholders were accompanied by surveyors, who redrew any disputed lines. This was known as land processioning.

Other surveying errors went unchallenged, and are incorporated in the maps of today. A common source of error was deviations of the compass. The needle of a compass never points precisely to true north, and the declination between magnetic and true north can vary somewhat from place to place. This may be inconsequential in surveying a small plot of land, but not in the case of long boundaries.

In his *Natural and Civil History of Vermont,* Samuel Williams, a nineteenth-century surveyor, told that, when the northern boundary of Massachusetts was run in 1742, the magnetic declination was less than the surveyor had been instructed to assume. The boundary line he charted, therefore, ran farther to the north of the parallel of latitude it was supposed to follow. This, Williams calculated, occasioned a "loss of 59,873 acres to Newhampshire; and of 133,897 acres to Vermont." If the boundary line had been correct, the northwest corner of Massachusetts, including Williamstown and North Adams, would now be in Vermont.

When the Virginia–North Carolina boundary was extended across the mountains toward the Mississippi River, along the present border between Kentucky and Tennessee, an unruly compass caused a variation in the line that has persisted through the years. Daniel Smith, one of the surveyors, noted in his journal that the deviation occurred because "there was some iron ore in that vicinity, which deflected the needle of the compass." This deflection caused a shallow V in the state line near the town of Portland, Tennessee. In running the line further westward the surveyors struck the Tennessee River some 18 kilometers north of the true line, the consequence of an accumulation of small errors. The survey later made from the Tennessee to the Mississippi River was on the true line, creating the offset at the Tennessee River that is still shown on maps.

N o one could have been surprised, in the atmosphere of colonial Virginia, when young George Washington decided on surveying as his first profession. Land was the way to affluence and social position, and where there was land to be claimed, parceled, and settled, there had to be surveyors.

Washington's father had done some surveying, as had most landowners, and his half-brother, Lawrence, was making bold plans to survey and develop the colony's western frontier.* As Douglas Southall

---

*Thomas Jefferson was also the son of a surveyor, Peter, who with Joshua Fry in 1751 produced an authoritative map of the southern colonies. Another Founding Father, Benjamin Franklin, was associated with cartography. He was responsible for the first accurate chart of the Gulf Stream, prepared in 1769–70 in collaboration with Timothy Folger. The chart was thought to have been lost for nearly 200 years, but in 1979 copies were found in Paris and London libraries, according to a report in *Science*, 207, Feb. 8, 1980.

OVERLEAF: *Recently discovered chart of the Gulf Stream, by Benjamin Franklin and Timothy Folger, 1769–70*

NEW

SOUTH WALES

NORTH

C. Monread
Maria
Mauell X.
P. Morning
Equam

Bare I.
The Cubbs
Iames
Saleisons
Tiderlos
Shepera

C. Iones

A
N
CAN

Viners Isa

Bay
Charlton

Peters R.

English Factory
Charles fort
Rupents R.
I. de Sal

Albany R.
Albany Fort
North
Shoald

NEW FRAN

L. Abittibis
Labrinth L.
Temiscaming L.

Monreal

Simangami

To find the Diſtance of two
Places in this CHART
Firſt if the two places differ only in Lattitude the Diff
Lat is the Diſtance 2 If y two places are both in the ſame
Lat take the diſtance between them, & apply the Compaſſes to
the Graduated Meridian, one foot ſtanding as much above y Lat
as y other below the Degrees interſepted is y Diſtance. 3 If y
two places differ both in Lat & Longitude. take y Diff of
Lat between them in Degrees from y Equinoctiall Line. & lay
ing a ruler upon y two places. apply one point of y Com
paſſes ſo to y edge of the ruler that y other turned
about may juſt touch ſome Eaſt or Weſt line croſſed
by y ruler. then take y Diſtance by y edge of y ruler from
the place where y Compaſs reſted to y place where y
ruler croſſes y ſaid Eaſt & Weſt line that ſpace mea
ſured on y Degrees of the Equinoctiall is the
diſtance ſought

NEW

NEW YOR
Albany
Hudſons
Burlington
Philadelphia

PENSILVANIA

Newcaſtle

Baltimore

MARYLAN
R. Patuxen
R. Pacomeck
Rapahanock R.
York R.
Iames Town
Elizabeth
Linhaven

VIRGINIA

Albemarle Coun

RICA

Naratake R.
Pameco R.
Nus R.
Claredon C.

Lockwood folle
Craven C.
Charles T.
Stono R.
Edeſto R.

C. Fear
C. Cartaret
Aſhly and Cooper R.
Port Royal

FLORIDA

N London

Wilton Iſland
Freſhes

Bay de Spirito
Sancto
Canaveral R.
Flores R.
C. Florida
Mexes
R. Grande
S. Maria

C. Cruce

B del Spirito
Sancto
B. S. Auguſtin
Bay Man
tancas
Bay S. Ioſephe
I. Iuſt R.
C. Iampo
Coſta
Fuegos

May R.

Iſle Marais
caſculos

Mouth of the
Miſſiſipi River

C. Carlos
R. Carlos e
B Iuan poa

ANA

Gulf of Florida

Membre
Bahama Banck
Abaco I.
Alden
Rock
Harbor I.
Eleuthera
BA

BAB

GULF of MEXICO

Freeman, his biographer, wrote of Washington's youth: "Everywhere the talk was of surveys—of the boundary stones that were set, of the tracts that were being opened in the region West of the Blue Ridge, and of the designs . . . formulating for a company to develop the Ohio country."

Though he had little formal education, Washington acquired a knowledge of arithmetic, or ciphering as it was called, and the elements of surveying. These were the practical sciences, and the Americans were already thinking of themselves as, above all, practical people. In 1745, when he was only thirteen years old, Washington found some surveying instruments that had belonged to his father—a brass surveying compass, a Jacob's staff (probably of the type used as a mount for instruments, not for astronomical sightings), and a chain and rods—and began running lines at his home and the neighboring plantations. One of his boyhood efforts, in 1747, was "A Plan of Major Law. Washington's Turnip Field as Survey'd by me." It was initialed "GW."

After further training from local surveyors, Washington set out in 1748, at the age of sixteen, on his first journey to the frontier. He accompanied George William Fairfax, eldest son of Lord Fairfax and a close friend, and James Genn, a veteran surveyor. They crossed the Blue Ridge into the still largely unsettled "Valley of Virginia." Washington, tall and husky, a superb horseman, liked the open life and, before the journey was over, made his decision to be a full-time surveyor.

He never lacked for work. His first job was to assist a surveyor who was laying out the town of Alexandria. In July 1749 he was sworn in as the official surveyor of Culpeper County. His style of mapping, largely concerned with running property lines by chain and compass, was generally neat but seldom ornate. He was conscientious, but probably no more or less accurate than any other experienced surveyor of his day. More than a hundred maps either surveyed or annotated by Washington are extant.

Washington's diaries attest to the hardships and problems of colonial surveying. He complained of traveling the "worst road that was ever trod by man or beast." He noted seeing an Indian war party and a "rattled snake." He could be matter-of-fact: "We run off four lots this day." Or he could be exasperated: one rodman, he wrote, "either stupidly or maliciously, cut his pole only fifteen feet long, instead of the sixteen and a half feet, as required by law."

By 1753, when he was only twenty-one, Washington had been commissioned by the Virginia militia and was beginning a second career. For his first important assignment as a soldier, however, he had to draw upon his skills as a frontier surveyor. Washington was instructed to proceed to Fort Le Boeuf near Lake Erie to deliver to the French commander there a warning from the Virginia governor that he would not tolerate any invasion of Virginia territory. Today, such a warning could be handled by a telephone call on the hot line or a televised speech by the secretary of state. But for Washington it meant leaving Williamsburg on October 31, 1753, leading a small party on a round-trip journey of more than 1,208 kilometers through mountains and snows, and returning on January 16, 1754. He found his way by compass and, ever the surveyor, kept a careful record in his notebook of the distances and directions along the traverse. And, when he filed his official report of the journey, it was accompanied by a map from the Potomac to Le Boeuf. Freeman wrote: "Even today, with no other paper than this, a traveler who used a compass probably could journey from Cumberland, Md., to the site of Fort Le Boeuf."

As his interest in military matters grew, Washington applied his surveying and cartographic skills to military maps. He drew several plans of Fort Loudoun in Virginia: a "Sketch of the Situation &c. of Fort Cumberland" in Maryland, and a map of a "Plan of a line of march in a forest country."

In the 1760s, after he resigned his militia commission, Washington traveled extensively and purchased large tracts of land, much of which he mapped himself. He continued to order instruments from London, including, as his records show, an "18 inch Circumferentor w. Sights to let down . . . £4.10.0, 1 Strong Chain . . . 0.7.6, 1 Loadstone Compt."

When the surveyor became a general, Washington had many occasions during the American Revolution to lament the poor state of topographic mapping in the colonies and to try to do something about it. "I have in vain endeavoured to procure" accurate maps, he once complained to Congress, but was "obliged to make shift, with such sketches, as I could trace out from my own observations and that of Gentlemen around me."

It would be for others, many decades later, to map the breadth of America in true topographic fashion. In his later years, Washington resumed the kind of surveying he did best, running boundaries by compass and chain. At Washington's death in December 1799, his possessions included a full range of surveying instruments—a tin canister of drawing paper, a case of instruments and parallel rules, compasses, plane tables, and a theodolite.

The first thoroughly professional and, for its day, sophisticated boundary survey in the American colonies was the work of two Englishmen, Charles Mason and Jeremiah Dixon. Their surveyed line, the Mason-Dixon line, is familiar today not for itself but as a symbolic division between North and South. As a figurative boundary, it was a line running between Pennsylvania and Maryland, along the Ohio River, and thence westward, the division between the free and slave states, a line drawn by political events that provoked the Civil War. In actuality, the line surveyed by Mason and Dixon defined only the boundary between the British colonies of Pennsylvania and Maryland.

For decades this had been a hotly disputed border and a source of enmity between the Penns of Pennsylvania and the Calverts of Maryland. George Calvert, the first Lord Baltimore, received the lands of Maryland as a royal grant in 1632. William Penn obtained the lands to the north by a royal grant in 1681; a subsequent grant gave Penn land to the south on the Delmarva (Delaware-Maryland-Virginia) peninsula. But it was not made clear where one grant ended and the other began. As for the northern boundary, the Maryland charter had read: "that Part of the Bay of Delaware on the North, which lieth under the Fortieth Degree of North Latitude, where New England is terminated." The fortieth parallel would have put Maryland's northern border within the present city of Philadelphia, which the Penns would not tolerate, and so a critical issue in the dispute was how far "under" the fortieth parallel the boundary was to run.

The dispute reached such an impasse in 1761 that the Penns and Calverts finally agreed to seek outside, independent help. Local surveyors had proved unable to conduct the type of astronomical and mathematical operations required to fix and connect the north–south and the east–west boundaries of the two provinces.

Thomas and Richard Penn, sons of William, were advised by their representatives "to send over from England some able Mathematicians with a proper set of Mathematical instruments." These persons, moreover, should be "of Great Integrity and totally unbiassed and unprejudiced on either side of the question." The Penns and Frederick, the sixth Lord Baltimore, appealed to the Astronomer Royal at Greenwich for recommendations. Mason and Dixon were his choices.

Charles Mason was an astronomer who had worked closely with the Astronomer Royal at Greenwich on a catalog of lunar positions and on improvements in astronomical instruments. Jeremiah Dixon was a

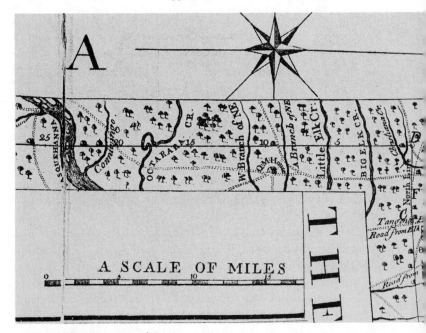

A section of the Mason-Dixon line

mathematician and surveyor from Durham. The two men had worked together as a team sent by the Royal Society to the Cape of Good Hope to observe the transit of Venus across the face of the Sun. It was when they returned to England in 1762 that they learned of their new assignment.

Mason and Dixon arrived in Philadelphia in November 1763. They brought with them two transits and two reflecting telescopes "fit to look at the Posts in the Line for ten or twelve miles." They also had a newly improved zenith sector, a graduated arc of a vertical circle used in conjunction with a telescope and plumb line. The sector would be employed to determine latitude by repeated observations of a number of fixed stars near the zenith as they cross the meridian at differing hours.

By this time the Penns and Lord Baltimore had reached an agreement without which the survey could not have begun. They decided that "under the fortieth parallel" would mean a line 15 miles, or 25 kilometers, south of the southernmost limit of Philadelphia.

First, therefore, Mason and Dixon had to fix the latitude of Philadelphia's southernmost boundary. They unpacked their instruments and established an observatory at a spot agreed upon by both sides. Using the zenith sector to make some 60 observations of stars over a period of three weeks, the two surveyors determined that the southern limit of Philadelphia was at 39°56′29.1″ N latitude. Later observations showed them to be off by a mere 2.5″.

Since a point 25 kilometers due south of Philadelphia would have put them in New Jersey, Mason and Dixon had to move about 50 kilometers westward on this line of latitude to the farm of John Harland on the Brandywine River. It was a two-day journey by horse and wagon, with the telescopes and sector cushioned against damage by a feather-bed. At the Harland farm they checked their position and then ran a direct line 25 kilometers south to "a plantation belonging to Mr. Alexander Bryan."

By their calculations Mason and Dixon determined that a spot in front of Mr. Bryan's house should be the starting point for the survey. The commissioners for the Penns and Lord Baltimore concurred.

But work on the east–west line had to wait. Between June and the end of September in 1764 Mason and Dixon surveyed the north–south line, the one between Maryland and the counties of Delaware, later to be the colony and then the state of Delaware. The two surveyors led a party of thirty-nine, including tent-keepers, cooks, chain-carriers, and axemen. As they measured off the line, they planted waist-high limestone markers at one-mile intervals. The 135-kilometer line was accepted by the commissioners, and a resurvey in the 1970s generally confirmed the accuracy of Mason and Dixon's work.

After the north–south line and a winter layover, Mason and Dixon returned to the Bryan farm in the spring of 1765 and prepared to run the line that would immortalize their names. There they placed a zero milestone, which they often described in their notes as the "Post mark'd West."

It was an arduous and hazardous venture. In their journal, primarily written by Mason, who assumed the senior role in the project, the two surveyors reported deep snows and extreme cold, tales of "border ruffians" and encounters with threatening Indians. Axemen would keep ahead of the surveyors, clearing a rough corridor about ten meters wide for the observations and measurements. Days would go by when the only entry in their journal was a single phrase: "Continued the Line."

Their work was meticulous. Horizontal measurements were taken with a Gunter's chain of 66 links on level ground and with a triangular-shaped surveyor's level on the slopes. "To prove that the Chain Carriers had made no error," Mason wrote in his journal, "I took a Man with me, a few days after, and measured it myself; and made it within a link of the same." Several times the party retraced their steps to check measurements. Every 10′ of a degree, or about every 17½ kilometers, they checked their positions by astronomical observations.

Mason and Dixon crossed the Susquehanna and continued the survey to the summit of the Blue Ridge Mountains, which they reached in October 1765. After spending the winter back on the East Coast, the surveyors returned along their measured line, replacing temporary posts with stone markers. Most of the summer and autumn of 1766 was spent in extending the "West Line to the approximate limit indicated by the Pennsylvania charter, or five degrees in longitude westward from the starting point."

Their primary task was completed. But during the winter of 1766–67, Mason and Dixon took a surveyor's holiday on the Delmarva Peninsula. They had remembered its smooth terrain and decided it would be an ideal place for a geodetic determination of the exact linear measure of one degree of latitude, a measurement that had been made in Europe, Lapland, and Peru but not in North America. The Penns and Calverts approved the project, since the results would be important in calculating the precise length of a degree of longitude along the West Line.

Meanwhile, colonial officials had been negotiating with the Indians to arrange for Mason and Dixon to undertake a further extension of the West Line. Upon the payment of £500, the Six Nations agreed to the survey and provided an escort of Indians for the party. Mason and Dixon were admonished by the colonial authorities to treat the Indians well because "the public Peace and your own Security may greatly depend on the good Usage and kind Treatment." The agreement also stipulated that the Indians be given daily rations of liquor, though the surveyors were urged to dilute "Spiritous Liquors" and pass them out not more than three times a day.

The extended survey proceeded slowly during the summer of 1767, amid rumors of Shawnee war parties prowling in the vicinity. Most of the survey's Indian escort stole away. The axemen threatened to leave. Mason and Dixon prevailed upon them to continue the work, until on October 9, 1767, they reached the mouth of Dunkard Creek about 50 kilometers east of what is now Pennsylvania's southwest corner. The axemen refused to cross the creek, and the remaining Indians in the party told the surveyors, "You go no further."

Mason and Dixon were not inclined to argue. They were not

unaware of the fact that the creek was named after a small town of white settlers which had been burned by Indians twelve years earlier. So the two English surveyors took their final observations: "233 Miles, 3 Chains and 38 Links from the Post mark'd West." This is a point on what is now the Pennsylvania–West Virginia border, about 25 kilometers southwest of Uniontown and 72 kilometers due south of Pittsburgh.

Their survey completed, Mason and Dixon prepared a map of the boundary and a full report based on their field notes. One entry, concerning the longitude measurement, suggests the scientific care and sophistication that were hallmarks of their work. They wrote:

> By comparing our measurement of a Degree of the Meridian with that made under the Arctic Circle, supposing the Earth to be a Spheroid of an uniform Density: a Degree of Longitude in parallel of the West Line is 53.5549 Miles. But as the Earth is not known to be exactly a Spheroid, nor whether it is everywhere of equal Density, and our own Experiment being not yet finish'd; we do not give in this as accurate. [The modern value is 53.2773 statute miles.]

In September 1768 Mason and Dixon sailed home to England. Little was heard again from Jeremiah Dixon, who died in 1779. Mason participated in other expeditions for the Royal Society, but in the 1780s he fell into financial and physical decline and returned to Philadelphia in 1786 for reasons that are unclear. Charles Mason died a few weeks later and was buried in an unmarked grave in Christ Church Burying Ground.

Thomas Hutchins arrived in Pittsburgh in early September 1785 with instructions to begin a survey. The survey he began set the pattern of American land development for more than a century, an unprecedented example of boundary surveyors shaping the look of the land.

At the time Pittsburgh stood at the threshold of the American West. Site of a strategic fortress in the frontier wars, it was a rude village of log cabins and some three hundred inhabitants on the fringes of settlement. They were linked to the seaboard by a wagon road, but had their eyes fixed west. They had the Ohio River and the beginning of a flourishing river traffic to the southwest. Across the Ohio, to the west and northwest, lay a vast domain of public land, unsettled Indian country that the young nation had fallen heir to after its victory over England.

As a military officer earlier in life, Hutchins had served at Fort Pitt and knew the region as did few others. He had led expeditions down the

Ohio to the Mississippi and traveled overland to Lake Erie and Lake Michigan. The map he drew established him as an authority on the region and a cartographer of note. When the Revolution broke out, Hutchins joined the Revolutionary Army as a surveyor on the southern front, and later was appointed geographer to the Continental Congress.

The lands west of Pittsburgh and north of the Ohio River were seen by the Continental Congress as a source of quick revenue. The idea was to sell tracts of land for settlement and with the proceeds pay off the national debt. Thus, Congress passed the Land Ordinance of 1785, providing for the surveying and auction of public lands. This was a modified version of a proposal by Thomas Jefferson and Hugh Williamson* that adopted the principle of rectangular survey, in contrast to the more "indiscriminate" metes-and-bounds surveys (unconnected to any overall system of coordinates) that prevailed in many of the seaboard states. The land was to be surveyed and divided into townships 6 miles square, each containing 36 sections of 640 acres each, 1 mile square. A rectangular survey, the Congress felt, would be "attended by the least possible expense, there being only two sides of the square to run in almost all cases and that there would be "exemption from controversy."

As James Monroe described the ordinance, the territory "is to be survey'd in townships containing abt. 26,000 acres each, each township mark'd on the plat into lots of one mile square, and one-half the country sold only in townships and the other in lots. 13 surveyors are to be appointed for the purpose to act under control of the Geographer, beginning with the first range of townships upon the Ohio and running North to the lakes, from [a point due north of] the termination of the line which forms the southern boundary of the State of Pennsylvania, and so on westward with each range."

This meant that the public-land survey would be tied to the Mason-Dixon line—a good starting point, since it was still the most accurately surveyed boundary in the nation. David Rittenhouse, a self-educated Philadelphia astronomer and instrument maker, and Andrew Ellicott, a Maryland surveyor who would later measure off the ten-mile square for the District of Columbia, had already extended the Mason-Dixon line to the southwest corner of Pennsylvania. In June 1785, a month after the ordinance was enacted, Rittenhouse and Ellicott set out with a team of

---

*Jefferson's original proposal, in 1784, called for slicing the federal lands into fourteen new states with an assortment of names: Cherronesus, Assenesipia, Illinoia, Michigania, Polypotamia, and others.

Pennsylvania and Virginia surveyors to run a line from the southwest corner due north to the Ohio River. Their methods were those of Mason and Dixon: axemen clearing a wide swath, measurements by chain, compass bearings, astronomical observations each night to correct the magnetic compass readings to true-north bearings.

In this plodding, careful way, the surveyors moved 100 kilometers north and on August 20 reached the south bank of the Ohio. Their immediate purpose was to mark off the territory of Virginia (present-day West Virginia) from that of Pennsylvania. But, with the new ordinance in mind, the surveyors also sent a party across the river to drive a stake west of where Little Beaver Creek flows into the Ohio. This was, as the public-land surveyors came to call it, the Place of Beginning.

Hutchins stayed in Pittsburgh most of September. He hired axemen and chain carriers, bought supplies and horses, and waited for the thirteen state-appointed surveyors to show up. Only eight did. He obtained assurances from the local colonel that he could "very safely repair with the surveyors" to the starting point. Hutchins would have reason to question the colonel's judgment, for the Indians across the river had their own ideas about which public had rightful claim to the public domain. As a Shawnee chief protested to some white emissaries, "We do not understand measuring out the lands—it is all ours. . . . Brothers, you seem to grow proud because you have thrown down the king of England."

Be that as it may, Hutchins, the eight surveyors, and about thirty helpers proceeded to the Place of Beginning and from the stake began running a base line west. This constituted the young nation's initiation of official mapmaking.

Hutchins's instructions from Congress were: "The line shall be measured with a chain; shall be plainly marked by chaps on the trees and exactly described on a plat, whereon shall be noted by the surveyor, at the proper distances, all mines, salt springs, salt licks and mill seats, that shall come to his knowledge and all water courses, mountains and other remarkable and permanent things over or near which such lines shall pass and also the quality of the land."

This, in other words, was to be more like a map than the simple plats drawn by property surveyors. But the survey got off to an inauspicious start.

The base line was supposed to run westward along a precise parallel of latitude. This would serve as the north boundary for the first "ranges," or tiers, of townships. From a mean of several observations of the Sun and Polaris, Hutchins reckoned the latitude running west from the stake at the Ohio to be 40°38′02″. He must have used a sextant rather than the more reliable quadrants or sectors, for he erred by nearly one kilometer. This was indicative of the early standards of the public-land surveys.

Hutchins was not destined to carry the line, however imprecise, very far on his first attempt. On October 8, after completing only 6½ kilometers, Hutchins received "disagreeable intelligence" concerning nearby Indian raids. He immediately left the line and fell back across the river. Efforts to obtain an escort of American soldiers or friendly Indians were to no avail. Hutchins felt that he had no other choice but to suspend the survey and return to Pittsburgh.

With the urging of Congress and a promise of military escorts, Hutchins resumed the survey in August 1786. This time 12 of the 13 state surveyors showed up, and 150 troops were placed at Hutchins's disposal. The geographer went to work, cautiously still, but with higher hopes of success.

Other errors, however, were allowed to creep in as the surveyors proceeded westward. The surveyors apparently did not pay the strictest attention to one of the complications of measuring a line of parallel, a problem caused by the Earth's curvature. If you start at a predetermined latitude, and run a straight line across the land, the line will in the northern hemisphere gradually drift below the latitude, and the longer the line, the greater the divergence. For an illustration of how a true line of parallel should look one need only glance at a map of the western United States; on a flat map the north boundary line across Oklahoma, New Mexico, and Arizona, for instance, is slightly curved in order to conform to its set latitude.

To adjust their line to the Earth's curvature, Mason and Dixon had followed a practice still employed by surveyors—the secant method. Every few kilometers Mason and Dixon would measure carefully calculated distances north or south from their straight surveyed line. From these calculations, based on mathematical formulas, they would correct the straight line to a curve and check again their position of latitude by astronomical observations. But it is believed that Hutchins must have simply approximated how the curved line should run.

At a point 10 kilometers west, the surveyor from New Jersey, Absalom Martin, peeled off from the party to run a line southward to mark off the first township. According to the plan, each of the twelve survey-ors was to do likewise and be responsible for measuring off an entire range of townships south of the base line.

All went well through August and into September. On the morning of September 18, the camp awoke to find that a pole marking the con-clusion of the previous day's surveying had been broken during the night. Everyone feared the worst, particularly when word arrived later in the day of Shawnee war parties gathering to the southwest. Hutchins prepared to retreat and dispatched messages to that effect to the other surveyors who were working independently on the ranges of townships.

The warnings never reached most of the surveyors. But, fortuitously, an expedition of Kentucky militia suddenly crossed the Ohio and descended on the Shawnee country on a mission of retribution, spread-ing terror and destruction. This diversion permitted the surveyors to keep working into November, when a small band of marauders attacked the surveyors' camp and drove off the horses. All of Hutchins's surveyors repaired across the Ohio for the winter, where they worked on their maps and reports.

This brought to an end the first effective season of national survey-ing. Only four ranges of townships had been completely surveyed, though parts of three others were subdivided. Because of errors in the base line, the township lines generally failed to meet, as they should, at clean right angles. Nonetheless, the first and only public auction of land in the Seven Ranges, as the area came to be called, took place in New York City in the autumn of 1787. It yielded little more than $100,000, hardly what the government had expected.

An impatient Congress decided that the public-lands survey was hopelessly slow and instead authorized a reversal of plans. Large tracts of land would be sold to speculators and then surveyed and subdivided.

Hutchins was supposed to survey the boundaries of the first two of these tracts, the one around present-day Marietta purchased by the Ohio Company and the other around Cincinnati, the so-called Miami Purchase of John Cleves Symmes. But Hutchins had grown weary of dealing with a parsimonious Congress, which had reduced his salary and delayed payment of expenses, and of the methodical routine of boundary surveying. He seemed to want to return to his real love—exploring and mapping, as he had done in his early career on the Ohio.

This, at least, would explain the letter Hutchins wrote in 1788 to the Spanish emissary to the United States, proposing to abandon his position as geographer to Congress for a comparable office under the Spanish government at New Orleans. He had in mind an assignment to map the Mississippi Valley and the river's many tributaries. But Hutchins, the nation's first official surveyor, died in Pittsburgh at the age of fifty-nine before his offer to Spain could be acted upon.

For all the problems and inadequacies, Hutchins's base line, which came to be known as the Geographer's Line, pointed the way west—both symbolically and in reality. Lessons were learned from his mistakes. Several of the men who worked under Hutchins became instrumental in laying out the rest of the Ohio country. And the concept of the rectangular survey was followed territory by territory, state by state, across the country.

One of the long-recognized problems with such a system of public-land surveys was the conflict between rectangularity and convergency. North-south lines in surveying are not in the long run parallel. They cannot be because they converge as they run northward toward the pole, as the meridians do on a globe. Therefore, without adjustments, it would be impossible for a six-mile-square township to be perfectly square or to be uniform in size compared to an adjacent township to the north. Jared Mansfield, one of the surveyors who took up where Hutchins left off in Ohio, devised a compromise between rectangularity and convergency that became the practice. By his concept the principal meridians and parallel base lines, the broad framework of the survey, would remain uninterrupted lines. But, within that framework, all the subordinate meridians would take a slight jog at each subordinate parallel and then start northward again the same distance apart as they were at the previous parallel. Such a map of township boundaries took on the appearance of off-line masonry.

In 1796, Mathew Carey published a map of the Seven Ranges showing in admirable detail a drainage pattern developed from survey records. This and the many other hastily produced maps found a ready market among would-be settlers back East. Later, Rufus Putnam, a veteran of the Hutchins survey, was able to produce a map of the newly created state of Ohio from a compilation of public-land surveys. This started a tradition in which, for another century, most of the mapping of the states was based on the public-land surveys and records.

For this reason the General Land Office (predecessor to the present Bureau of Land Management) is considered the oldest federal surveying and mapping bureau. It was established in 1784, Hutchins made the first surveys in its behalf, and everyone is now familiar with the expression "land-office business"—derived from the rush to claim and homestead the Land Office's newly surveyed and subdivided land on the frontier.

But the most visible influence of the rectangular survey, as initiated by Hutchins and practiced across the land by succeeding generations, is written on the land. It was, according to William D. Pattison, an authority on the history of the rectangular survey, "a striking example of geometry triumphant over physical geography."

A look at the map of the world illustrates what Pattison meant. Where else on the Earth are there so many right-angle boundaries, such as those of Colorado and Wyoming, for example, or the "Four Corners" where Utah, Colorado, New Mexico, and Arizona share a common boundary corner?

Or fly across the heartland of the United States and see that not only are the state lines run by a compass, but also the fields and roads: squares and rectangles broken only occasionally by a meandering river. The vast checkerboard laid out in precise north-south and east-west patterns is peculiarly American. It is the legacy of the thousands of public-lands surveyors, from Hutchins on, the boundary makers who staked out, measured off, and thus shaped the look of the land.

However, as Pattison wrote: "The lay of the land within the Seven Ranges has in the long run won out over the surveyed grid, as a determinant of the 'pattern of occupance.' A view of the Seven Ranges from the air shows fencelines, roads and buildings arranged in general conformity with the terrain, usually to the disregard of the surveyed grid. It is ironic that in this cradle, so to speak, of the rectangular surveying system, an influence which has shaped the greater part of the entire American landscape should have yielded almost completely to the dictates of Nature."

Mapping America:
Westward the Topographers

The young republic probed and pushed its way beyond the Seven Ranges of Ohio, beyond the Mississippi, across the plains and mountains to the western sea. In the person of trappers and traders, explorers and adventurers, soldiers and scientists, the republic made its imperial way west—a bold enterprise, the stuff of romance and legend, and an oft-told story. A vital role in the story was played by the explorers whose influence derived in large part from the maps they made or had made from the knowledge they risked their lives to gain.

It would be impossible, and unnecessary for the purposes of this book, to recall the entire cast of mapping explorers. Suffice it, then, to tell of a few men, the great and the obscure, and how their ambition, daring, resourcefulness, and dedication to duty put the American West on the map. Why they went west involves a complex of human drives, glory and greed, curiosity and escape, but most of them plunged into the unknown because they believed that there was something there worth knowing. Their expectations may have been inflated, which was often the case, and sometimes they deceived themselves and others in their efforts to square what they saw with their expectations. But such are the human strengths and weaknesses that produce maps of new lands.

How the best of the mapping explorers in the American West went about their work tells us much about the nature of reconnaissance mapping in the nineteenth century, when the thrust of cartography drove deeper into the world's vast and largely uncharted continental interiors. Their work encompassed the broad range of exploratory mapping—from discovery and pathfinding to the charting of rivers and railroad routes, the filling in of spaces on the map that had been blank, and the searching for resources. Their maps projected a hitherto unknown world on the minds of the known world. Their best maps replaced geographical lore with geographical reality.

The first of these mapping explorers were Lewis and Clark.

In 1803, with the purchase of Louisiana from Napoleon, the United States doubled its territory in one stroke, but had only the dimmest conception of the breadth and value of its new possession. It extended north from New Orleans, the Creole port, to Canada and west from St. Louis, the fur-trading river town, to the distant mountains. That much was known. It comprised the present states of Arkansas, Iowa, Missouri, and Nebraska, and parts of Louisiana, Minnesota, Oklahoma, Kansas, Colorado, Wyoming, Montana, and the Dakotas. That was not known in 1803, because the boundaries were poorly defined, the land virtually unexplored, and the maps, such as they were, of no real help.

There was, for example, only a vague knowledge of the "Stony Mountains." Samuel Hearne, a British explorer of the Canadian West, had in 1773 advanced the idea of the Continental Divide, the range of mountains in the western part of the continent beyond which all rivers ran westward. Using sketches of Peter Fidler, a surveyor for the Hudson's Bay Company, Aaron Arrowsmith, a prominent British cartographer, drew a single ridge of western mountains on his influential map of North America. Arrowsmith's map, published in 1795, noted that the Rocky or Stony Mountains were only "3520 Feet High above the Level of their Base." In that case, the western mountains should not be a formidable barrier.

Nor was anyone in 1803 sure of the course of the Missouri River, which drained much of the Louisiana Purchase. But there was a belief, encouraged by contemporary maps, that the river somehow provided a water route to the Pacific. Cook and Vancouver notwithstanding, the idea of the Northwest Passage would not die; the wish was still there, and so the allure. There were two sources of renewed interest in the idea. The Columbia River had recently been discovered and was known to have a latitude virtually the same as that of the Missouri's headwaters. British explorers had come back with reports that there was only a short portage between the Missouri's headwaters and those of the west-flowing river, presumably the Columbia.

And so it was that the Americans felt they might be on the verge of discovering the long-sought Passage to India and of possessing some of the richest lands of the world.

It had long been a part of the American geographical lore that to the west lay riches and the good life. Lewis Evans, the colonial mapmaker, wrote in 1756 of his vision of a continental American empire vested, he said, "with all the Wealth and Power that will naturally arise from the Culture of so great an extent of good Land, in a happy Climate." The reports out of Kentucky reinforced the image, particularly when John Filson wrote in 1784 of Kentucky as "the land of promise, flowing with milk and honey, a land of brooks of water . . . a land of wheat and barley, and all kinds of fruits." In his illuminating book in 1950, *Virgin Land: The American West as Symbol and Myth,* Henry Nash Smith described the promise of the West becoming the "Garden of the World" as "one of the dominant symbols of nineteenth-century American society."

A Passage to India and the Garden of the World—these were the "twin themes" of geographical lore at the time of Lewis and Clark, according to John Logan Allen, whose *Passage Through the Garden* in 1975 examined the Lewis and Clark expedition in the context of the geographical images that spur men to explore. These were the expectations that provided impetus for Thomas Jefferson's plan for a transcontinental expedition.

Before the purchase President Jefferson, whose many talents included surveying, had commissioned Meriwether Lewis and William Clark to explore the Missouri River and locate "the most direct and practicable water communication across this continent for the purpose of commerce." They were instructed also to study the inhabitants, observe the "soil and face of the country," and make astronomical observations so that a map of the region could be constructed.

And so, on May 14, 1804, only a few months after the purchase, Lewis and Clark set out from St. Louis with a party of forty-five in a sturdy keelboat. Lewis was a soldier, a protégé of Jefferson, and a brilliant though melancholy leader. He had been groomed for the assignment by the best minds of the American Philosophical Society in Philadelphia, where he learned natural history, astronomy, and the rudiments of surveying. Clark was much different. A bluff and bold soldier, he was schooled by the frontier and proved to be an instinctive geographer and gifted diplomat in his dealings with the Indians.

The Missouri was their highway west and northwest and west again. With sails and oars the two explorers navigated upstream to the Mandan country of North Dakota, where they wintered and, sitting around the log fires at night, learned much about the country around and ahead of them. The Mandans and other Indians made these things known to the explorers by word, gesture, and yet another demonstration of the ubiquity of the cartographic idea. The Indians understood the utility of maps as a form of human expression. They drew charts on the ground with a stick or on a hide with a piece of charcoal, and their sketches were far superior, in specifics, to any of the maps Lewis and Clark had with them or had seen in Philadelphia or St. Louis. For the Indians made their drawings from a sure knowledge of the land. Clark made copious notes and reproduced the Indian charts in his notebooks.

From the Mandan villages the explorers journeyed into Montana in the spring of 1805 and then through the mountain passes of Idaho. They discovered the headwaters of the Missouri and the west-flowing Snake. Even more important, they discovered that the Rocky Mountains were not a single ridge, as had been portrayed on the maps, but a more formidable series of several towering ridges; and that the Missouri did not connect with any water route to the Pacific. So much for latter-day dreams of an easy passage to Cathay.

Down the Snake and down the Columbia Lewis and Clark made their way in dugout canoes, collecting plant specimens, rock samples, animal hides, and Indian artifacts. They continued to supplement their own observations and chartmaking with information from the Indians. One sketch in Clark's journal bears the inscription: "This Sketch was given to me by a Skaddat [Klickitat] a Chupunnish [Nez Percé] & a Skillute [Dalles] at the Falls of Columbia 18-April 1806."

As often as possible, the two explorers took astronomical observations to plot their positions, though some were more accurate than others. Some of Clark's calculations, based on estimates of longitude, grossly underestimated their distance of travel. But the explorers made no pretense of being refined mappers; theirs was a journey of discovery, the initial step in mapmaking.

Late in 1805, Lewis and Clark reached the Pacific and, in a sense, established a cartographic connection between the eastern United States and the coast that only recently had been charted by George Vancouver. There, where the Columbia emptied into the Pacific, Clark stood in the rain and carved on a tall pine: "William Clark December 3rd 1805. By Land from the U. States in 1804 and 1805."

On their return trip Lewis and Clark explored even more of the Missouri's tributaries, including the Yellowstone. They arrived back in St. Louis on September 26, 1806, completing one of the most fruitful journeys of exploration in history. They returned with an understanding of the broad physiography of much of the American interior, with a detailed description of the Missouri and its tributaries, and with reports of an abundance of wildlife and other resources. They laid to rest the dreams of a navigable water route to the Pacific, but identified at least five passes through the mountains—Lemhi Pass in the Beaverhead Range, Bozeman Pass, Gibbon Pass, Lewis and Clark Pass, and Lolo Pass in the Bitterroot Mountains—by which future travelers could reach the Pacific. They established the basis for the future American claim to the Oregon territory. And they enforced recognition by everyone that the North American continent was much wider than had been generally supposed.

Not until 1814 was there published a map based on the expedition. Lewis had died in 1809, a suicide or murder victim, in a backwoods Tennessee inn. Clark had settled down in St. Louis to the life of a shrewd fur trader, civic leader, official and unofficial commissioner of western Indian affairs, and student of western geography. Clark was so busy, in fact, that he left the drafting of the map to Samuel Lewis (no kin), a Philadelphia cartographer who worked from an "original drawing" by Clark. Though not a scientific cartographic work in the usual sense, the Lewis and Clark map conveyed a wealth of new information and further aroused in a nation the drive that would shape its destiny for the entire century. As Bernard DeVoto, a scholar of western exploration, wrote: "It satisfied desire and it created desire: the desire of the westering nation."

Those who followed Lewis and Clark would be acting as agents of Manifest Destiny, mapping explorers influenced by and perpetuating a sense of national purpose, which was expansionist and optimistic. If there was no Passage to India, there was still the Garden of the World to be mapped in all its fecund promise. Seldom has the relationship between mapmaking and political economy been so profound.

A central institution in the scientific mapping of the American West was the Corps of Topographical Engineers. From 1838 until the Civil War this small elite branch of the Army, with never more than thirty-six officers on its roster at one time, laid out national boundaries, charted wagon trails and railroad routes for the westward migration, and conducted wide-ranging expeditions of exploration. Most of the topographical engineers were sophisticated men equally at home in the company of scholars and in the saddle, West Point graduates trained in engineering and versed in the natural sciences. The most famous of all these men was John Charles Frémont.

Handsome, brilliant, eager, Frémont plunged into his western explorations with a cavalier's dash and an inspired sense of his own and the nation's destiny. He personified the intimate tie between mapping and politics. He became a national hero in his own time. He would later help found the Republican party and be its first presidential candidate. But in the late spring of 1842, then only a twenty-nine-year-old lieutenant, Frémont was in St. Louis preparing for the first of three mapping expeditions that would secure his place in history.

Though one of the few in the Corps of Topographical Engineers who was not a West Point alumnus, Frémont was an experienced explorer and mapper. He had accompanied an expedition of the corps on a reconnaissance of the Cherokee country of Georgia, North Carolina, and Tennessee. In 1838 he came under the tutelage of the immigrant French scientist Joseph N. Nicollet. Over a period of two years Nicollet and Frémont explored Minnesota and the Dakotas, an expedition that, it was said, proved to be Frémont's "Yale College and his Harvard." The expedition, when combined with Nicollet's earlier ones on the lower Mississippi, provided the most mathematically accurate map of the river from Natchez to its headwaters in Minnesota. This was based on some 90,000 readings of latitude, longitude, and altitude above sea level. To determine altitude, Nicollet made the first successful mapmaking use in America of the barometer. Nicollet's measurements along the Mississippi and his subsequent maps served, as was their stated purpose, as points of departure for the scientific mapping of the trans-Mississippi West.

Such an enterprise was already taking shape in the mind of Frémont, who at about the same time had come under the influence of Thomas Hart Benton, senator from Missouri and foremost among the western expansionists. Frémont learned scientific discipline from Nicollet and political fervor from Benton. After an interview with Benton, during which he was promised command of his first major western expedition, Frémont wrote with characteristic enthusiasm: "My mind had been quick to see a larger field and differing and greater results. It would be travel over a part of the world which remained the new—the opening up of unknown lands; the making unknown countries known; and the study without books—the learning at first hand from nature itself; the drinking first at her unknown springs—became a source of never-ending delight to me."

With Frémont in St. Louis in May of 1842 was a red-faced German named Charles Preuss, who if it had not been for the money would rather have left Manifest Destiny to the likes of Frémont and remained at a drafting board back in Washington. Preuss was a curious, reluctant participant in the mapping of the West. He cursed his lot, but was ready to do his job and do it more capably and reliably than any topographer who had gone before him on a western expedition.

Preuss was born in Germany in 1803 and given the full name of Georg Carl Ludwig. He studied geodesy and for a number of years was a surveyor for the Prussian government. In 1834 he immigrated to the United States and found work with the Coast Survey in Washington, where he became known simply as Charles. He resigned in 1839 to take a job with a mining company, but that apparently did not work out, for when he came to Frémont's attention in December 1841 Preuss had been unemployed, except for odd jobs, for nearly two years. Preuss needed work and Frémont needed a topographer, and so the two, as unlike as they were, got together—Frémont the archetypical Western hero-explorer, Preuss the grumbling tenderfoot.

The first Frémont expedition was directed to explore and map the country between the Missouri and the Rocky Mountains. On a steamboat between St. Louis and Chouteau's Station (near present-day Kansas City), Frémont met Kit Carson and hired him as a guide, which was the beginning of a lifelong friendship. After buying rifles, ammunition, mules, and supplies, the party of twenty-five set out from Chouteau's Station and proceeded along the Kansas, Big Blue, and North Platte rivers. This route, the Oregon Trail, was already being used by the vanguard of emigrant caravans. It was Frémont's assignment to chart the trail, determine where to establish outposts, and scout out favorable passes through the Rockies.

This took Frémont as far as the Wind River Mountains of Wyoming. There he decided to scale what he took to be the highest peak in the chain. Once on the icy summit, standing with four others in the party, Frémont made his supreme romantic gesture. He raised a special American flag with a field bearing a triumphant eagle clasping a bundle of arrows and a pipe of peace in its claws. He had dramatized the national resolve: America triumphant and strong, from sea to sea. In his journal Frémont wrote: "We had climbed the loftiest peak of the Rocky mountains and looked down upon the snow a thousand feet below; and standing where never human foot had stood before, felt the exultation of first explorers."

No matter what else Frémont did on the expedition, when word of the flag-raising reached people back East, his fame was assured.

But what of Preuss? Poor Preuss had a morose and melancholy personality. He disliked the hardships of the expedition, was lonesome for his wife and family, complained of Frémont's leadership, and found the western plains anything but romantic. He was just not cut out to be an explorer. Frémont, however, had nothing but praise for Preuss's skill and the "pleasure he felt in the execution of his duties." In his *Memoirs,* Frémont wrote of the "years of faithful and valuable service" by Preuss and of "his even temper and patient endurance of hardship."

Preuss must, then, have kept his unhappiness to himself and to his diaries. The diaries came to light a century later and were translated from German and published in 1958. They add few important details to our knowledge of the Frémont expeditions. But, as they were written by the campfires at night, without thought of publication or personal aggrandizement, the diaries serve as a revealing counterweight to the lofty romanticism of Frémont's reports and, in general, to the prevailing literature of exploration in which drudgery and hardship are submerged or are conditions to be endured, preferably with stoic hardihood, in the interest of the greater purpose. Preuss did not see it that way at all, not in his private thoughts.

Frémont wrote vividly and admiringly of the mountain men on their racing horses, dashing across the prairie and chasing buffalo; Preuss described his painful ineptitude. On June 6, 1842, two days out of Chouteau's Station, he wrote in his diary: "This is my first day of horseback riding. Fortunately I got a gentle horse. Yet the unaccustomed effort, little though it was, made me rather stiff." His performance must not have improved much, for on July 20 he wrote: "Frémont obviously realizes that I cannot do the work so well on horseback. Riding ahead of the party or stopping here and there is now out of the question. Hence he has given orders to prepare a seat for me on one of the carts. From there I can, in all comfort—if that word can be used under the circumstances—'copy' mountains and rivers to my heart's desire."

Preuss found the plains monotonous. "Eternal prairie and grass, with occasional groups of trees," he wrote after a few days on the trail. "Fré-mont prefers this to every other landscape. To me it is as if someone would prefer a book with blank pages to a good story. . . . To the deuce with such a life; I wish I were in Washington with my old girl."

He groused about the "dirty cooking," calling the cook a "rascal." He was "annoyed by that childish Frémont," without giving his reasons, though in other instances Preuss found Frémont guilty of changing his mind on a whim. The chronometer gave him endless trouble, but considering its rough transport by wagon, the trouble was probably inevitable. Nonetheless, this provoked another complaint about Fré-mont: "So far I can't say that I have formed a very high opinion of Fré-mont's astronomical manipulations. . . . I wish I had a drink."

Preuss was not an explorer, but he kept going. Even when the party prepared to enter hostile Indian country, he braced himself, concealed his trepidations, and went along, because, he said, it would be a "dis-grace" not to go. Preuss kept going, and wherever he went, he sketched the landscape, displaying a skill at field topography that was described by contemporaries as unsurpassed in the United States.

His unromantic nature may have been his most singular asset as a topographer. Preuss mapped only what he saw, which had not been the practice of most of his predecessors, and he was careful not to let imag-ination embellish what he saw. His diary provides an example: "I thought I saw three large groves of trees and shrubs at the foot of the low mountain range which runs along the river for a mile. Since this is an important feature for my topography in this treeless region, I pre-pared to record it dutifully in my book as soon as I should be at right angles with the groves. But how easily one can be deceived; my woods, which would have looked nice on the map, turned out to be three immense herds of buffalo."

Even so, as a mapping venture Frémont's first expedition was of only limited value. The party made sixty-eight observations of latitude and longitude, but unaccountably failed to pinpoint the exact location of South Pass, the Oregon Trail's principal route through the Rockies and one of the expedition's primary mapping objectives. However, Frémont was a mere five kilometers off in his calculations of the distance between Chouteau's Station and the pass—which was not bad, consid-ering the chronometer troubles. Some six hundred meteorological read-ings were taken to establish altitudes, but Frémont underestimated the height of the peak he climbed with such flourish and named for himself. (And Frémont was quite wrong in believing that it was the highest peak in the Rockies; at least twenty others are taller.) The one map Preuss prepared for Frémont's report to Congress covered only the trail from Fort Laramie to the Wind River Mountains.

Frémont's report, nonetheless, stirred the public mind. Written with verve and color, with the help of Frémont's wife, Jessie, Benton's daughter, the report launched many a prairie schooner west in search of adventure, wealth, and a new life in the lands Frémont had seen from his lofty peak and Preuss had surveyed from the seat of his rude cart.

In the spring of 1843 Frémont again led an expedition west, determined to go farther and accomplish more. His orders were to "connect the reconnaissance of 1842 with the surveys of Commander [Charles] Wilkes on the coast of the Pacific Ocean, so as to give a connected survey of the interior of our continent." On this expedition, as on the first, his topographer was Charles Preuss, whatever his personal feelings about Frémont and the miseries of outdoor life.

At first, Frémont followed the route of his first expedition. The Oregon Trail was now alive with traffic—caravans of settlers, priests bound for frontier missions, a Scottish sportsman, itinerants of every description, all looking for the West as portrayed by Frémont. Then Frémont veered to the south, scouting out a new trail with his party of thirty-nine and a load of surveying instruments—a refracting telescope, two sextants, two pocket chronometers, two barometers, six thermometers, and a number of small compasses. The expedition crossed the Rockies in Colorado, moved north into Wyoming, and dipped down to explore, in Frémont's words, the "still and solitary grandeur" of the Great Salt Lake.

Frémont and Preuss conducted only a reconnaissance survey of the lake, but did make observations to fix its location on future maps. Sketches of the lake had appeared on maps even before the trapper Jim Bridger tasted its brine in 1824 and thought he had discovered an arm of the Pacific Ocean. But Preuss was to prepare the first reasonably accurate map of the area, and Frémont was to be the first to herald the region's attractions. Though to many an eye the basin surrounding the lake might have appeared barren, Frémont described it as "truly a bucholic place." The description persuaded Brigham Young, then in Illinois, that the country of the Great Salt Lake was the place for his Mormons to settle.

The expedition turned north again and followed the Snake River to the Columbia River and the Oregon territory. Preuss's disposition was no sunnier than before, and the damp and cold of the Northwest was hardly a tonic. "It is certainly terrible," Preuss wrote in his diary at Walla Walla on October 27, 1843, "what a poor devil has to contend with in this country in order to make an honest living."

Less than a month later, relations between Preuss and Frémont became seriously strained. The expedition had reached the Dalles and set up base camp. Frémont decided to leave the rest of the party and push on down the Columbia to Fort Vancouver. He wanted Preuss to accompany him, but Preuss infuriated him by refusing to cut off his beard to make himself "presentable" to the authorities at the fort. Frémont offered to accept Preuss's resignation. In a marginal note added to the Preuss diary, Mrs. Preuss said: "Here Frémont became so mad that he wanted to challenge Carl to a pistol duel because the latter did not want to cut off his beard."

Somehow—no account provides an explanation—the two men settled their differences. Frémont and Preuss proceeded by canoe to Fort Vancouver, where the Willamette joins the Columbia, near present-day Portland. This Frémont considered the conclusion of his primary survey. He had joined his reconnaissance of the interior with Wilkes's coastal survey.

But, instead of returning the way he came, Frémont chose to expand his survey by heading south for California. This took the expedition down through western Nevada, which Frémont recognized as the western edge of the Great Basin, the great depression between the Sierra Nevada and the Wasatch Mountains of Utah. It was a brilliant stroke of geographic conceptualization. He somehow grasped that the basin he had seen at Salt Lake and the one there in Nevada were the same. Across the space on the map Preuss drew later, Frémont wrote: "The Great Basin; diameter 11° of latitude, 10° of longitude; elevation above the sea between 4 and 5,000 feet: surrounded by lofty mountains: contents almost unknown, but believed to be filled with rivers and lakes which have no communication with the sea, deserts and oases which have never been explored, and savage tribes which no traveler has seen or described."

The description of rivers with no communication to the sea was another of Frémont's important discoveries altering the Western map. Some earlier maps kept alive the legend of Rio Buenaventura, which was said to flow westward from the Rockies to the Pacific, another hoped-for water route to the far west. But no one could be sure if there

was any truth to the legend. Frémont's march south to California eliminated the Buenaventura from the map. Pioneering mappers sometimes
erase as much from the map as they add to it.

On the march the expedition slogged through the worst wintry conditions. Frémont's critics often cited this as proof of his foolhardiness
and of his lust for glory at any risk. It got so bad in the Sierras that one
man, suffering hallucinations, wandered off from the camp, never to be
seen again. The Indians in the party began singing their death songs.
Preuss got lost for three days in March. Without food or firearms, the
man who was an incessant critic of expedition food was reduced to placing his hand in a nest of ants and then licking them off his hand.

By the time Preuss found his way back to the party, Frémont had
descended through the foothills to the warmer valley, to the relief of all.
The valley of California in early spring had a salubrious effect even on
Preuss's spirits. In his diary he was, for him, uncommonly cheerful:
"The weather is beautiful now. I am in best of spirits. The mule-meat
bouillon gives some strength after all. Until now, we fools have been
pouring it away and have eaten only the dry, woody meat. As long as salt
and coffee last, this life can be called wonderful."

But the work of Preuss had only begun. The first of Preuss's major
maps was produced in Washington in 1845 to accompany Frémont's
report to Congress on the second expedition. Preuss was, typically, conservative in his work. With few exceptions, he incorporated only the
information that had been gathered by the first two Frémont expeditions.
Positions were carefully computed from astronomical observations. The
profile of the landscape along the routes of exploration was derived from
barometric readings of altitude or, when the barometers had been broken, from the temperatures of boiling water, the same technique for
determining altitude that was used by the pundits of the Himalayas. The
map depicted for the first time the nature and extent of the Great Basin
and the true location of the Great Salt Lake. The map showed the
Columbia to be the only river that led from the Rockies to the Pacific.
The map gave the most comprehensive picture to date of the Sierra
Nevada, showing it to be a more formidable barrier than had been indicated on earlier maps. For the Pacific coastal regions Preuss drew upon
the reliable surveys of Wilkes, but otherwise he confined himself to what
he and Frémont had seen and measured, from Missouri to Fort Vancouver, from the Northwest down into California and back to Missouri.

According to Carl I. Wheat, a leading authority on cartography of the
American West, the 1845 Preuss map "changed the entire picture of the

West and made a lasting contribution to cartography." According to William H. Goetzmann, in his comprehensive history of the Corps of Topographical Engineers, the map by Preuss, based on the Frémont expeditions, was "the first of a series of scientific mappings of the western country [and] a landmark in the progress of geographical knowledge."

The next map by Preuss was, in effect, the first road map of the American West. It was published in 1846, in seven sections, with the title *Topographical Map of the Road from Missouri to Oregon Commencing at the Mouth of the Kansas in the Missouri River and Ending at the Mouth of the Wallah-Wallah in the Columbia.* This was the map of the Oregon Trail, at a scale of ten miles to an inch, and it was drawn with the emigrant in mind. Working from Frémont's field notes and journals, Preuss included many useful details—distances from the trail's starting point, locations of Indian war grounds, weather conditions, game and fuel and water. If Frémont's words inspired people to head west, Preuss's accompanying maps pointed the way.

While Preuss was at work on the Oregon Trail map, Frémont returned to the West on his third expedition. It was probably just as well that Preuss stayed behind, for the expedition turned out to be more military than scientific or cartographic, and the source of bitter controversy. In the spring of 1845, when Frémont set out, war with Mexico over annexation of Texas seemed imminent and, if it broke out, was expected to spread to California, which was owned by Mexico. Frémont had secret instructions for military action in California in case of war. Indeed, when news of the declaration of war reached California in 1846, Frémont led a battalion of American volunteers in the "conquest" of the future state. He became the first civil governor of the territory, which put him in immediate conflict with Stephen W. Kearny, the general sent by Washington to establish a government and an army officer senior to Frémont. Kearny had Frémont arrested for insubordination and returned to Washington, where he was convicted by court-martial and sentenced to be dismissed from the army. Though bitter, Frémont was not one to retreat from public life. He completed a memoir on his third expedition, and began seeking private support for one more western venture.

Edward M. Kern, Frémont's topographer on the third expedition, did manage to gather some mapping data, which Preuss used in producing his third important map. Though entitled *Map of Oregon and Upper California,* it was actually a map of all the territory west of the 105th meridian, based on all three Frémont expeditions as well as other available sources. Since Preuss completed the map in the summer of 1848,

shortly after news of the discovery of gold had reached Washington, he was able to indicate for the first time the location of the California gold fields, with the inscription "El Dorado or Gold Region." Preuss's 1848 map was also apparently the first to apply the term "Chrysopylae or Golden Gate" to the entrance of San Francisco Bay.

Preuss must have contracted the California fever himself. Why else would he have agreed to join Frémont's fourth expedition, a rash privately financed scheme to cross the southern Rockies in midwinter in search of a railroad route? Preuss never got a chance on the expedition to use his topographic skills. He was fortunate just to get out alive.

In December 1848 Frémont led the party into the San Juan Mountains of southern Colorado. It was bitter cold, and the snow became deeper and deeper. They lost their way. Game was scarce, provisions ran out, and the pack animals died from exhaustion and the cold. A detail sent off for help, Preuss said in his diary, had to resort to cannibalism. Those in the main party who were too weak to go on were left to die by the campfire. As Preuss wrote, "They would lie down, wrapped in a blanket, and shortly yield up the ghost." The survivors turned back and finally made it to Taos in mid-February. Preuss the tenderfoot had endured while more hardened outdoorsmen had faltered and died.

Preuss then proceeded with Frémont to California. He worked there for more than a year as a surveyor, and may have had an idea of settling. California was the only place in his travels with Frémont where his normally gloomy spirits seemed to brighten. But in 1850 Preuss suffered a sunstroke while out surveying and was forced to abandon California. He went on one more survey, a railroad survey to Arizona in 1853, but his constitution was no longer equal to the demands of an arduous outdoor life. In September 1854, after his return to Washington, Preuss hanged himself from a tree limb.

Frémont's days as pathfinder and exploring mapmaker, his days of greatest glory and achievement, had also come to an end. He became a senator from California, the Republican presidential nominee in 1856, a Civil War general, a multimillionaire in mining, impoverished through unfortunate railroad investments, and territorial governor of Arizona.

At the height of Frémont's fame his critics contended that the significance of his exploits had been exaggerated. They called him an adventurer, not a true explorer. What real discoveries had he made? The

*Map of the Pacific Coast, from the
surveys of John C. Frémont, 1848*

critics pointed out that most of the lands Frémont mapped had already been visited by earlier explorers or trappers, and they were correct. In defense of Frémont, however, Goetzmann contended that his contributions to geography rested not on any primacy of discovery but on the comprehensiveness, presentation (the reports and maps), and trustworthiness of his explorations.

This too was the value of the work of the entire Corps of Topographical Engineers. As Goetzmann wrote: "They brought the element of trustworthiness inherent in the scientific method to the making of western maps where only the uncertain perpetuation of myth, resting on dubious authority, existed before. Though they made relatively few major geographical discoveries, their account of what they did see had a much greater factor of dependability."

By 1854, half a century after Lewis and Clark launched their expedition, a decade after Frémont and Preuss charted the Oregon Trail, a wealth of western cartographic material had accumulated, and it seemed an appropriate time to compile it all in a "general map," a master map. Not only appropriate, but necessary. Each topographer had mapped the route of his party's reconnaissance, each supplying a piece but never fitting the pieces into a whole. The pieces came primarily from the Topographical Engineers. Other than Frémont, they were William H. Emory, who surveyed the far Southwest and produced the first reasonably accurate map of that region in 1853; Emory and Amiel Weeks Whipple, along with the freelance surveyor A. B. Gray, who surveyed the new Mexico–United States boundary and surrounding lands; Randolph B. Marcy and James H. Simpson, who explored the Navajo country; Howard Stansbury and John W. Gunnison, who mapped the Mormon country; Robert S. Williamson, who crossed the Sierra Nevada in search of railroad routes; and many others, many of whom were increasingly involved in the Pacific Railroad Survey, their marches of reconnaissance often extending 3,000 kilometers and adding more lines of survey across the continent.

Strictly speaking, cartography is the drawing or compiling of maps. The explorers and surveyors go out and make the measurements and gather the information from which the cartographers draw their maps. Sometimes, as in the case of Preuss, the field work and drafting-table work are done by the same person. But, where the scope is broad and the sources of information many, maps are more often the product of a compilation. They represent the accumulated work of many people, brought together under the supervision of one person, the compiler. The value of the map depends of course on the integrity of the compiler; he sifts available information, discards the fantasy, selects the most accurate data, and comes up with a judicious synthesis of the geographical knowledge of a region.

The man assigned to compile the master map of the West was Gouverneur Kemble Warren. A New Yorker who seemed destined for great things, Warren had graduated second in the 1850 class at West Point and, as a young lieutenant in the Topographical Engineers, practiced civil engineering in the Mississippi delta. He had the enthusiasm and ambition of a Frémont. He had the good looks, too, with his black mustaches and flowing hair. This was the young lieutenant who began work in 1854, at the age of twenty-four, on the first comprehensive, scientific map of the American West.

Warren had to pore over reports and maps and data going back to 1800. He found himself, he said, faced with "two perplexing difficulties." These were the determination of what is reliable information and the reconciliation of discrepancies found to exist even in maps of reliable explorers.

Later, in the memoir accompanying the map, Warren explained some of his methods of analyzing and selecting material for inclusion in his map:

Comparatively few points in this large area have their latitudes and longitudes determined with precision, and the surveys and explorations vary in accuracy, by almost insensible degrees, from the determinations of a boundary line to the crude information of the Indian, or the still more vague representations of the imaginative adventurer.

In some large sections we possess no information, except from uncertain sources. In these parts the rule was adopted to leave the map blank, or to faintly indicate such information as is probably correct.

Where discrepancies are found on comparing the maps of reliable explorations, especially in relation to geographical positions, the principle has been carried out of considering that explorer's maps the most accurate whose experience and means of observation were the most perfect. Where these advantages appeared equal, a mean of the results was adopted. In other cases, a less proportionate value was given to the inferior, and in some cases it was even rejected. It is evident that the combination of the materials of different maps in one has necessarily required some distortion of the originals; but, in all cases, much caution was observed to make this distortion as little as possible.

Warren thus conceded that the map was of uneven accuracy. Having accepted this as inevitable, he began his compilation at the Mississippi and worked west. Because of the earlier work by Nicollet and the more recent Land Office surveys, this part of his task was relatively easy and swift. The eastern portion of the map was compiled and engraved in 1854. The actual drawing of the map was done by Edwin Freyhold and F. W. Von Egloffstein, experienced artists and draftsmen.

In the summers of 1855 and 1856, Warren interrupted his work to go on expeditions for the Topographical Engineers, first up the Missouri to explore the Dakota country, then to the Yellowstone and Powder rivers country. This produced even more material for the master map. It also gave Warren practical field experience that made him appreciate all the more the problems of mappers and the possible shortcomings of their measurements and maps.

Seeing discrepancies, he often had to put himself in the surveyor's position and then try to figure out what went wrong. When he examined the output of the Frémont expeditions, Warren came away with sincere admiration for Preuss's "skill in sketching topography in the field and in representing it on the map." This skill, he concluded, "has probably never been surpassed in this country."

Warren completed the map in 1857, and it was published in 1859 as part of volume II of the *Pacific Railroad Reports,* the thirteen-volume compendium of knowledge gained during the many railroad-route surveys of the previous fifteen years. Warren's map was drawn to a scale of 1:3,000,000. He conceded that "this is too small to adequately represent the topography and character of the country, except in a very general way." Such a small scale was necessary, Warren said, to encompass the entire trans-Mississippi West on a single sheet and to portray its broad outlines and important features at a glance. The map located forts and Indian villages, lakes and rivers, mountains and basins, existing and projected wagon roads, and mountain passes. In his assessment of Warren's *General Map,* Carl Wheat declared that "subsequent efforts in the way of maps may properly be deemed merely filling in the detail."

The filling-in would have to wait. The outbreak of the Civil War disrupted western mapping and led to the demise of the Corps of Topographical Engineers. Many of the officers of the corps distinguished themselves in combat, sometimes on opposite sides. Emory became known as "Bold Emory." Andrew A. Humphreys, who had supervised Warren's map compilation, earned a reputation as "the fighting fool of Gettysburg." It was at Gettysburg that Warren, chief engineer of the

Army of the Potomac, cast his topographer's eye across the battlefield and recognized the strategic importance of Little Round Top. He led troops to seize the hill before the Confederates could, an action that proved decisive to the Union victory.

Yet, though an authentic hero of Gettysburg, Warren spent the last part of his military career under a cloud. During one of the final battles of the war, he was charged with failing to act decisively and summarily relieved of his command. A court of inquiry exonerated him seventeen years later—months after his death, in 1883, at the age of fifty-two. G. K. Warren is memorialized in bronze at Gettysburg and by the first reliable map of the American West. The map was, as Warren himself said, "the culmination of more than three centuries of effort to come to grips with reality in this vast and complex area."

"We are now ready to start on our way down the Great Unknown," the indomitable leader of the little expedition wrote of the morning of August 13, 1869. The party had been on the river almost three months. They had camped the last three days at the mouth of the Little Colorado, resting, repairing their boats, and determining their latitude and longitude as part of their mapping work. Now they were set to resume their journey into the unknown.

The leader, John Wesley Powell, described the moment:

Our boats, tied to a common stake, chafe each other as they are tossed by the fretful river. They ride high and buoyant, for their loads are lighter than we could desire. We have but a month's rations remaining. The flour has been resifted through the mosquito-net sieve; the spoiled bacon has been dried and the worst of it boiled; the few pounds of dried apples have been spread in the sun and reshrunken to their normal bulk. The sugar has all melted and gone on its way down the river. But we have a large sack of coffee. The lightening of the boats has this advantage: they will ride the waves better and we shall have but little to carry when we make a portage.

We are three quarters of a mile in the depths of the Earth, and the great river shrinks into insignificance as it dashes its angry waves against the walls and cliffs that rise to the world above; the waves are but puny ripples, and we but pigmies, running up and down the sands or lost among the boulders.

We have an unknown distance yet to run, an unknown river to explore. What falls there are, we know not; what rocks beset the channel, we know not; what walls rise over the river, we know not. Ah, well! we may conjecture many things. The men talk as cheerfully as ever; jests are bandied about freely this morning; but to me the cheer is somber and the jests ghastly.

With some eagerness and some anxiety and some misgiving we enter the canyon below and are carried along by the swift water through walls which rise from its very edge.

The second day, August 14, they found themselves where the river entered a granite gorge narrower than any they had seen on this trip. This was the Grand Canyon of the Colorado, a place of mystery, beauty, and danger.

They became aware of "a great roar ahead" and moved forward more cautiously until they were above a long broken cataract with ledges and pinnacles of rock obstructing the water and "a mad, white foam." The walls of the canyon rose perpendicular from either side of the river, leaving no foothold for a portage. The nine men in the three wooden boats had to run the rapids or turn back.

"There is no hesitation," wrote Powell, and down the river they went, "down again into the trough, up again on a higher wave, and down and up on waves higher and still higher until we strike one just as it curls back, and a breaker rolls over our little boat." Miraculously, the boats survived the whirlpools, waves, and jutting rocks.

Several days later, they approached another narrow gorge with threatening cataracts. It had been a long, harrowing journey from Green River Station in the Wyoming territory. Their rations had dwindled to little more than unleavened bread and coffee. And there, on August 27, as one member of the party wrote in his journal, they faced "the worst rapid yet seen."

It was a time of discontent and decision. Three of the men, Oramel and Seneca Howland and William Dunn, announced that night their intention to leave the expedition and go overland to the Mormon settlements about 120 kilometers to the north. Powell spent the night pacing up and down on the narrow beach along the river. By his calculations they should be only about 70 kilometers in a direct line from the mouth of the Virgin River. And there it should be an easy journey upstream to a Mormon settlement—and the successful conclusion of the journey.

"To leave the exploration unfinished, to say that there is a part of the canyon which I cannot explore, having already nearly accomplished it," Powell declared, "is more than I am willing to acknowledge and I determine to go on."

At daybreak on August 28, after a breakfast "as solemn as a funeral," Powell wrote that the Howlands and Dunn still thought it "best to leave us." They were given a share of the scanty rations and some guns and ammunition. "It is a rather solemn parting," Powell wrote. "Each thinks the other is taking the dangerous course."

On the morning of August 29, the remaining six men found themselves downstream, emerging from the canyon into open country. After rounding a bend early the next afternoon, they came upon three white men and an Indian hauling a seine. They had reached the mouth of the Virgin River. Their journey of 2,400 kilometers down the Green and Colorado rivers and through the uncharted Grand Canyon was over—one of the great adventures of exploration.

Unfortunately, though Powell and his men did not know it then, Oramel and Seneca Howland and Bill Dunn had been murdered on the plateau, perhaps by Shivwits arrows.

But for Powell the expedition was only the beginning of his investigations in one of the most extensive unexplored areas that had been left blank on Warren's map—the Grand Canyon country and the surrounding "Plateau Provinces," which extended across southern Utah, northern Arizona, and northwestern New Mexico. Much has been made of the drama of the first expedition: John Wesley Powell, the bearded son of a frontier preacher, one-armed Civil War veteran, and determined self-made scientist, conquering the Colorado. And Powell, the conqueror, going on to Washington and becoming a mover-and-shaker of western development and a leader of American science. Too little, however, has been made of the second expedition, which was initiated in 1871. It turned out to be scientifically and topographically more important than the first expedition, owing in large part to the efforts of Almon Harris Thompson.

Powell himself recognized the shortcomings of the first expedition. The few biological and geological specimens collected had been lost. The mapping measurements were not as complete or as reliable as Powell wished. Astronomic measurements had been made about every eighty kilometers; the observations were taken for latitude with the sextant, and for longitude by the method of lunar distances. The meandering course of the river was determined by compass observations from point to point, with the intervening distances estimated. Barometric observations three times a day were the source of elevations along the river's edge. But many of the records of these observations were lost with the death of Oramel Howland, who had been in charge of the mapping.

Consequently, Powell planned the second expedition more as a topographic survey than an adventure in discovery. He wanted to survey the river and its most important side canyons and to map "as broad a belt of country as it was possible" on either side of the Green and Colorado rivers. An appropriation of $12,000 from Congress enabled Powell to embark on the survey.

The crew for the second expedition was, as for the first, made up of relatively inexperienced men. Powell again ignored professional scientists or experienced explorers in favor of friends and relatives. Thompson, the man who would be second-in-command and the chief topographer, was Powell's brother-in-law. But, as Wallace Stegner, one of Powell's biographers, observed, this was "a far better choice, actually, than most of the brother-in-law appointments" so rife in the time of the Grant Administration.

The thirty-one-year-old Thompson was a Civil War veteran, a school superintendent in Illinois, and the husband of Powell's sister. Though he loved mathematics and science, he had never worked in the field as a surveyor or mapmaker and would have to learn on the job. That he did, and in time became an expert topographer.

On May 22, 1871, the party pushed their three boats out into the stream at Green River Station and headed south. They had already established with a fair degree of accuracy the altitude, latitude, and longitude of their point of departure. The altitude above sea level of Green River Station had been fixed by the railroad surveys. The latitude was determined by observations with the zenith telescope. The longitude was fixed by telegraphic time signals between there and an astronomic station at Salt Lake City.

In descending Green River, the expedition established astronomic stations at distances averaging 70 kilometers by river, or about 40 kilometers by direct line. At each station the men took latitude observations by sextant. Thompson and Stephen V. Jones, the assistant topographer and another schoolteacher, charted the river's course by sighting ahead at each bend with a prismatic compass and making an estimate of the distance to each sighting point. Along the way they continually estimated the width of the river and the height of the canyon walls.

Describing the survey, Powell wrote:

The lines between stations on the river were used as a series of base-lines, the lengths, of course, only approximately determined, and an intricate network of triangles was projected to salient points on either side of the river. From a vast number of points thus fixed, the surface contour of the country was sketched so as to include a belt from twenty to fifty miles wide, the parties making frequent trips from the river into the interior of the country. At each of these astronomic stations barometric readings were recorded in hourly series, and as we proceeded down the river tri-daily barometric readings were made, all referred to the water's edge. With the river as a baseline for hypsometric work, altitudes were determined by triangulation and by barometric methods, using both mercurial and aneroid instruments. Thus all of our altitudes in this region are related to the river.

The routine of river mapping had Powell champing at the bit, and in July he left the party for the first of many times. Powell went to Salt Lake City, where his wife was pregnant and ill. He also made several excursions to Indian villages to observe their languages and customs. All the while Thompson led the expedition slowly and methodically downstream until, in late October, they ran Glen Canyon and reached the mouth of the Paria.

From there they proceeded overland west to the Mormon village of Kanab, which would be their winter headquarters. Kanab, Thompson wrote in his diary, was "a stockaded square of log houses with some few neat adobe houses outside." Irrigated land produced substantial crops of potatoes and corn. Fruit trees had been planted. It was Eden to the river-weary men, for they arrived there in a serious state of depletion. One suffered a badly wrenched leg, another a scorpion bite. Old war wounds were acting up, and the effects of scurvy took a toll. Poor diet and exhaustion brought Jones and John F. Steward into Kanab on stretchers.

The expedition set up a camp of tents outside Kanab. They enjoyed the social life of the community, traded for supplies, and spent long hours transcribing and plotting the survey measurements onto a map. On December 12, soon after Powell rejoined the party, he and Thompson rode their horses out of the camp for a talk. Powell was restless; he felt that he should be in Washington appealing for an additional appropriation to expand the survey. He asked Thompson what he thought of the idea. Thompson said, "Go ahead." After all, it had been Thompson's expedition almost from the start, in fact if not name.

Powell left for Washington in February, and Thompson and the others continued their surveying by laying out a 14½-kilometer base line from which the primary triangulation could be conducted. The line was measured on a flat stretch of land southeast of Kanab. The alignments were made with a theodolite. The latitude of the northern extremity of the line was determined by an extended series of observations with the zenith telescope, and the longitude by telegraphic time signals from Salt Lake City. The line itself was measured with wooden rods, leveled on trestles and trussed to prevent sagging and warping. Every 30 meters of distance was marked off by a stake.

The progress was exceedingly slow, as reflected by the cryptic entries in Thompson's diary:

December 14th. Took the transit to "Gap" to locate the meridian line. . . . Friday, the 15th. Tried to locate meridian, but so cloudy could not. Saturday, December 16th. Cloudy. Waiting. . . . December 31st. 1871. Kept the Sabbath by riding out. Got an observation that will enable us to locate our line tonight. . . . Tuesday, January 2nd. Ranged out line below camp and placed a stone at north end. . . . Thursday, the 4th. Went over the line and took angles. . . . Tuesday, January 9th. Worked on line, went to Kanab and to Indian camp. Jones and Mac went to flag on "Cedar Ridge" to build monument. . . . Friday, January 12th. Went to end of line. Located stone on south end of line. Went to dance at Kanab this evening. . . . Saturday, February 17th. Went to Navajo well. Climbed to Point "B." Triangulated. Could see Points D and E and Signal Station 4 base to point "C.". . . . Wednesday, the 21st. Finished "Base Line."

And so the methodical Thompson worked through the early months of 1872:

Sunday, March 10th. Got rations and came to Pipe Spring. Willie Johnson commenced to work on map. Fred [Frederick S. Dellenbaugh, a distant cousin of Thompson] copied bearings. . . . Wednesday, March 13th. Clem [W. Clement Powell, a cousin of the Major] went to Kanab for more chemicals [for photographic development]. Fred at work on bearings. Traded in old buckskin for a little mare. . . . Saturday, March 16th. Johnson, Fred and myself worked, plotting bearings. Jones on observations. Captain [Pardyn] Dodds shoeing.

From the base line the surveyors moved out across the plateau with a network of triangles. At the major survey points mounds were built and flagstaffs erected, and in the clear desert air it was found possible to make the sides of the triangles from 40 to 50 kilometers long. Much of southern Utah, from the Colorado on the east to the Nevada line, was thus measured and mapped.

In the spring of 1872, Thompson led the party out of Kanab on an exploratory traverse of the region between the Colorado and the high plateaus, during which they discovered the last unmapped river and mountains in the conterminous United States.

Thompson set out on the journey because he wanted to recover one of the expedition's boats that had been abandoned the previous year at the mouth of the Frémont River and because he wanted to learn something about the unexplored plateau country. In a region that Thompson inked on the map as the Kaiparowits Plateau, the party were puzzled when they came upon a large stream that they could not ford. Could this be the headwaters of the Dirty Devil River? From a high ridge the next day Thompson scanned the vast panorama (the so-called Wild Horse Mesa of Zane Grey fiction) and realized that the stream could not be the Dirty Devil. It was another tributary of the Colorado, one that had gone undetected by Powell in 1869 and by Powell and Thompson in 1871. In his diary that night Thompson wrote and underlined the words: "Is not Dirty Devil."

Since it flowed out of what the Mormons called Potato Valley, Thompson tentatively named the stream Potato Creek. Later he entered it on his map as the Escalante River, after the first white man known to have crossed the Utah wilderness. According to Stegner and other Powell scholars, it was the last river added to the map of the United States, aside from Alaska and Hawaii. And from the same ridge Thompson sighted the last mountain range to go on the map, the Henry Mountains of Utah. They were named for Joseph Henry, secretary of the Smithsonian Institution and a friend of Powell's.

Meanwhile, after months of plotting measurements, the expedition had completed the map based on their river journey the previous year. It covered the course from Green River Station down the Green to the Colorado and on to the mouth of the Paria. When Powell returned from Washington in the summer, he regrouped the expedition for another river trip—from the Paria down through Marble and Grand canyons.

It was, if anything, a more perilous voyage than Powell's first canyon adventure. The water of the Colorado was higher, swifter, and more "fretful." Sometimes the boats moved so fast they could not be controlled. But for his life jacket Powell would have drowned in a whirlpool. As the boats plunged through one boiling cataract, Dellenbaugh heard Powell exclaim: "By God, boys, we're gone!"

When the party reached safety at the mouth of Kanab Creek, in the depths of the Grand Canyon, Powell and Thompson decided against continuing the rest of the way. The boats were battered and the men shaky. Packers who met them with food at Kanab Creek had brought news of an Indian uprising that posed a grave threat if the expedition should proceed much farther downstream. All things considered, Powell and Thompson decided that they would have to rely on the 1869 notes and measurements in mapping the rest of the Colorado's course through the Grand Canyon. According to Goetzmann, it was "the only time in his life that Powell backed off from a challenge, but in this instance it probably saved the lives of all of his men and with them the future of the Powell Survey."

Thompson headed back to Kanab. With Fred Dellenbaugh, John H. Renshawe, and Stephen Jones, he settled down to the task of map drafting. They lived in tents with wooden floors as protection against the dampness. In one of the tents they set up a large table for drafting. There they worked through the fall and well into winter.

At times it was a frustrating labor. The plotting of bearings and measurements revealed obvious inaccuracies in their surveying work, particularly in the beginning when the amateurs were learning on the job. "Swear on the outside about it—Old Boy—but thank God in your heart that I have not chucked the whole business in the fire," Thompson wrote Powell early in 1873. A few weeks later, he said the map was almost completed, adding that "it does not satisfy me in many respects, but it is perhaps passable."

In February 1873 Thompson wrote in his diary: "Got map finished. Fred and Jack started [with it] for Panguitch or farther."

Fred Dellenbaugh, who had signed on with the expedition as a boat-man, had turned out to be Thompson's favorite associate. He was young, cheerful, eager to learn, and, above all, enthusiastic. He would go on to become a world traveler and writer, but for the moment he was entrusted with getting Thompson's map, encased in a homemade tin tube, to an eastbound train in Salt Lake City. Dellenbaugh made it through the snow and freezing cold and sent the map on its way to Powell in Washington.

Thompson's map, at a scale of 2 miles to an inch, encompassed much of southern Utah and the part of northern Arizona north of the Grand Canyon, including a long stretch of the Colorado River and its tributaries—nearly all of which had been blank on Warren's map. For some of the remote parts of the area Thompson's survey was still the basis for maps produced nearly a hundred years later.

Afterward, he turned his attention to new mapping surveys in which he ranged northward across the high plateau country, eastward beyond

Paria, and westward to Pine Valley Mountain. He was assisted by Jack Miller, the expedition's photographer, and Renshawe. The so-called Powell Survey, with Thompson remaining in charge of most of the topographic duties, conducted its mapping operations until 1879.

The influence of Thompson's maps and Powell's geologic observations was profound and almost immediate. In his reports, Powell used Thompson's maps to illustrate some bold generalizations about rivers, canyons, mountains, and the processes of erosion. This was a new approach to American geology. Unlike most of the other scientific explorers of the West, who spent their time collecting fossils and contemplating the geological past or had a narrow eye out for mining prospects, Powell was concerned with structural geology and the ongoing processes shaping the Earth. Why, Powell asked, did the Green River that he had navigated and Thompson had mapped flow right through the Uinta Mountains? Why did it not find an easier course around the mountains?

"The answer," Powell wrote, "is that the river had the right of way; in other words, it was running ere the mountains were formed; not before the rocks of which the mountains are composed, were deposited, but before the formations were folded, so as to make a mountain range. The river preserved its level, but the mountains were lifted up; as the saw revolved on a fixed pivot, while the log through which it cuts is moved along. The river was the saw which cut the mountains in two."

The idea was not wholly new, but it had remained for Powell to expound it so graphically and convincingly. Powell had looked at the contours of mountains and the courses of rivers, as mapped by his expeditions, and recognized that there was more to geology than fossils and rocks, more than a search for resources to mine. His insights, elaborated upon and refined by others, ushered in the new science of physical geology, or geomorphology. Geology, more than ever, became the broad science of the Earth.

Powell also forced people back East to think more realistically about the West. The concept of the West as the Garden of the World lived on in the national psyche. To those many explorers who spoke of much of the West as the American Sahara, the irrepressible expansionists answered that settlement of the country would cause an increase in the timber through artificial planting and protection against prairie fire. One promoter, an amateur scientist named Charles Dana Wilber, had people believing him when he confidently declared: "Rain follows the plough."

But Powell warned that the old methods of agricultural settlement would be disastrous in much of the West. Most of the land, he said, was arid and could not "be redeemed for agriculture, except by irrigation." Powell recommended, therefore, that the public lands of the West be mapped not only topographically but also by "scientific classification"— that is, classified as mineral lands, coal lands, pasturage lands, timber lands, and irrigable lands. He also proposed reform of the homestead laws to reflect the differences in land potential between the humid East and the arid West. Powell met defeat at the hands of the developers, but the dust-bowl experiences of the 1930s were a sad form of posthumous vindication. Today, the land-classification map is one of the most valuable applications of cartography.

While Powell and Thompson were mapping southern Utah and the Grand Canyon region, three other major western surveys were out in the field, and soon their overlapping interests and competition for government funding provoked bickering rivalries. These were the surveys of Clarence King, Ferdinand V. Hayden, and George M. Wheeler. King, a Yale-trained geologist whom Henry Adams called "the best and brightest man of his generation," was instructed by the War Department to conduct a survey, both geological and topographical, from California eastward to the 105th meridian, which runs through Colorado, along the 40th parallel. Hayden, a geologist and physician working for the Department of the Interior, concentrated his surveys in the area of Nebraska, Colorado, Wyoming, and Montana. His work led to the creation of Yellowstone National Park. Wheeler, an Army officer carrying on the tradition of the old Corps of Topographical Engineers, spent eight seasons surveying more than 500,000 square kilometers in southern California, New Mexico, Utah, and Nevada. In 1873, Wheeler and Hayden met in southern Colorado and proceeded to map each other's territory, a clash of interests that reverberated back to Washington and set in motion a congressional investigation of the competing surveys.

After years of maneuvering in political and scientific circles, Powell and King persuaded Congress and the President to merge the surveys and establish a single civilian agency for the geological and geographical exploration of the United States. The result was the United States Geological Survey, founded in 1879. King was appointed the first director, but in 1881 Powell took over and shaped the Survey as one of the nation's principal mapping institutions. Thompson became the Survey's chief geographer, and continued in that position until his death in 1906.

From the outset Powell recognized the importance of topographic mapping, which he considered the prerequisite to accurate geological work, and obtained authorization to map the whole country. Powell's plan was to divide the map into quadrangles bounded by parallels and meridians. The smallest-scale quadrangle map would cover a space of one degree of longitude by one degree of latitude; this would be used to portray desert regions, where it was deemed that a scale of four miles to the inch would be sufficient. The largest-scale quadrangle, one mile to the inch, would cover an area 15' by 15'; the nation's more densely populated regions would be so mapped, representing the shape of the land by contours, the classification of the land (desert, swamp, arable, etc.) by different colors, and important cultural features (towns, canals, railroads, roads) by appropriate symbols and lines. Upon these base maps could be overprinted the surface geology, land-use information, or just about any other scientific data. These were to be, in Powell's mind, all-purpose maps.

Powell, looking ahead, made sure that the maps could be easily revised from time to time. In testifying to Congress in 1884, he explained: "The culture will always have to be kept up. So far as the natural features are concerned it will never have to be changed. . . . On the present plan, those sheets go on copper. Now, copper is of such a nature, you can make changes at any time. It is a soft metal, malleable and mobile, so that you may erase a line or put a new line upon it, as the case may be, so that a practical mechanician can cut that town out or put a new town in, or if a road has been changed, he can change that road on the map."

The maps of the earlier surveys had been printed from stone. When an edition was printed, the stone was cleaned off and smoothed down for reengraving of the next map. To reprint or revise a map required an entirely new engraving.

But Powell underestimated the mapping task. He told Congress the entire United States, except for Alaska, could be thus mapped in twenty-four years. It has yet to be fully accomplished. By 1894, when Powell retired as director of the Survey, he reported that one-fifth of the United States had been surveyed and mapped according to his quadrangle plan. Today, the Survey continues to add to the topographic quadrangle series and to make the maps even more detailed and larger-scaled. The emphasis now is on the 7.5' quadrangle. At this scale, 1 inch on the map represents 24,000 inches, or 2,000 feet, on the ground. The map covers a 4-sided, almost rectangular area bounded by 7.5′ of longitude and 7.5′ of latitude. Because of the converging longitude lines, the actual area covered ranges from about 70 square miles in the south to 50 square miles along the Canadian border. On such maps, as well as the 15′ quadrangles, it is possible to depict all the streets and houses of a town in addition to the other physical and cultural features shown on topographic maps. The Survey reported in the 1970s that topographic mapping in the 7.5′ or 15′ series is available for more than three-quarters of the United States.

This, then, is the legacy of John Wesley Powell—the exploration and mapping of much of the Colorado River and the surrounding plateau country, a new understanding of geological forces, the Geological Survey, and its familiar quadrangle maps. Powell died in 1902 at the age of sixty-eight. But every year hundreds of hardy souls descend the Colorado through the Grand Canyon in neoprene rafts or wooden dories, reliving Powell's adventure in the "Great Unknown" and knowing where they have been and are going by means of one of the quadrangle maps of Powell's Geological Survey.

Meters, Meridians, and
a New World Map

During the last third of the nineteenth century, cartography began to reflect the greater internationalism in commerce and the sciences. Mapmakers still had much to do in their own countries, particularly outside Europe, for it was estimated in 1885 that less than one-ninth of the land surface of the globe had been surveyed or was at the time being surveyed. But they had progressed far enough to appreciate that they could not continue to operate without some international standards. For too long maps had been prepared with little or no consideration of uniform standards of measurement, meridians, scale, or symbols. Each country seemed to go its own way, developing its own traditions of cartography and publishing maps of the world from a rather chauvinistic perspective. Unless the trend could be reversed, mapmaking could never hope to achieve status as an unambiguous form of communication.

With improvements in travel and communications, the mapmakers of Europe and America began to meet regularly at international conferences to discuss these shortcomings of their profession and the many inconsistencies of their maps. National pride and the traditional conservatism of the cartographic profession often acted as a drag on the proceedings. Even so, important steps were taken during the period to establish a standard for linear measurement, to agree upon a prime meridian, and to embark upon a cooperative endeavor to produce a standard world map.

There was nothing particularly complicated about any of it—simply an application of common sense.

Although it was 1875 when representatives of twenty nations met in Paris to adopt the Convention of the Meter, this standard of measurement had its practical origins in Revolutionary France in the last years of the eighteenth century. Few countries needed weights and measures reform more than France did at that time.

England had at least some semblance of standards, based on a number of royal decrees. The yard as a measure of length can be traced back to the early Saxon kings. They wore around their waists sashes or girdles that could be removed and used as convenient measuring devices. Tradition holds that Henry I decreed that the yard should be the distance from the tip of his nose to the end of his thumb. The length of a furlong (furrow-long) was established by early Tudor rulers as 220 yards. This led Elizabeth I to declare in the sixteenth century that henceforth the traditional Roman mile of 5,000 feet would be replaced by one of 5,280 feet, making the mile exactly 8 furlongs. These units, though derived in unscientific ways, became the basis of the English system that was adopted throughout the English-speaking world.

In France the standard often differed from province to province, creating a veritable babel in the communication of measurements. This babel and the arbitrary basis for the various units of measurement had upset French scientists for years. The toise, for example, which equaled six French feet (slightly longer than English feet) and was used most often in cartography, was supposedly derived from half the width of the main gate of the Louvre.

Finally, in 1790, the National Assembly of France requested the French Academy of Sciences to "deduce an invariable standard for all the measures and all the weights." The Assembly said that the objective should be a decimal system of linear measure whose fundamental unit would be based on a "natural" standard.

Two principal methods for arriving at the fundamental unit were considered. One was to be related to the length of the movement of a pendulum set for one second. The other was to be based on the size of the Earth—in particular on the distance from the North Pole to the equator.

The pendulum method had been suggested more than a century earlier by Gabriel Mouton, vicar of St. Paul's church in Lyons. Charles Maurice de Talleyrand, the leader in the National Assembly, recommended its adoption. The method did seem attractive because it would certainly be easier to measure the movement of a pendulum than—as the French knew from the travails of their geodetic predecessors in Peru and Lapland—to survey the arc of a meridian. But the scientists of the Academy raised serious objections. Such a method could not be uniform and international for the well-known reason that the length of a pendulum set for a one-second swing differs from place to place on Earth because of gravitational variations.

And so the Academy recommended the alternative method, which had also been an idea of Mouton's. The standard unit of linear measurement would be one ten-millionth of the meridian distance between the pole and the equator as determined by a new geodetic survey. Measures for capacity (volume) and mass (weight) were to be derived from the unit of length, thus relating the basic units of the system to each other and to nature. Furthermore, the larger and smaller versions of each unit were to be created by multiplying or dividing the basic units by ten and its multiples. The Academy assigned the name *mètre*, or "meter," to the unit of length. This name came from the Greek word *metron*, meaning "a measure." The metric unit of mass, called the gram, was defined as the mass of one cubic centimeter (a cube that is one one-hundredth of a meter on each side) of water at its temperature of maximum density. The cubic decimeter (a cube one-tenth of a meter on each side) was chosen as the unit of fluid capacity and given the name liter.

The National Assembly approved the system in 1791, and tentative values of the metric units were established from the results of earlier geodetic surveys of the arc of the meridian. These values were subject to adjustment upon completion of the new survey undertaken by Jean Delambre and Pierre Méchain, who were astronomers and geodesists. They set out to remeasure the difference in latitude between Dunkerque and Barcelona; both cities were on the same meridian and both were at sea level.

Working in the midst of a political and social revolution proved difficult and frustrating. More than once Delambre and Méchain, these strangers out in the countryside carrying unusual instruments, were accused of being foreign spies, or royalist agents, for they were always flying white flags (white was the royal color) to mark survey stations. The surveyors were often detained and sometimes thrown in jail.

A letter from one member of the survey party described their troubles in the suburbs of Paris: "When I arrived at Saint-Denis I had to show my passports and I obtained a permit to remain, but the magistrate warned me that even with that aid I would not travel a quarter of a league. And, indeed, a half-hour after, in passing through Épinay, we were arrested. They found that our instruments had not been designated with sufficient clarity in our passports; they wished to seize them; I was required to spread them on the ground and explain their use. No one understood the explanation I made. . . . After three hours of debate we were forced to remount our vehicles with an armed guard and were taken to Saint-Denis."

As far as is known, no surveyor wound up in a tumbrel, and by 1799, after seven years, the Dunkerque-Barcelona survey was completed. From the measurements scientists in Paris prepared a platinum bar the length of a standard meter. This became the prototype meter—the *Mètre des Archives*. The measured prototype meter changed the estimated meter by less than 0.3 millimeter—less than 3 parts in 10,000, indicating that the earlier surveys had been fairly accurate. Furthermore, the meter of 1799 was itself inaccurate by a similar factor. It is now known that the length of a standard quadrant of meridian is 10,002,001.23 meters, a little more than 2 parts in 10,000 greater than the quadrant length established by the survey of Delambre and Méchain.

French scientists immediately converted to the metric system. The French chemist Antoine Lavoisier, who helped to devise the system, boasted: "Never has anything more grand and simple, more coherent in all its parts, issued from the hand of man."

The most immediate effect of the metric system on cartography was the introduction of a simpler and more universal language for expressing map scale. This new language was the representative fraction, an outgrowth of the new decimal thinking of the metric system. The fraction states the relative size of the map and the area of the Earth it depicts: one unit on the map being equal to X units on the ground. This can be expressed as a fraction or a ratio. For instance, $\frac{1}{50,000}$ or 1:50,000, $\frac{1}{100,000}$ or 1:100,000, and so forth. By this method of scale expression, any unit, such as 1 inch or 1 centimeter on the map, represents 50,000 or 100,000 of the same units on the ground.

In his history of pre-twentieth-century cartography, *The Story of Maps*, Lloyd A. Brown said that the fractional scale appeared as early as 1806 in a revised edition of the *Atlas National de la France,* edited by P. G. Chanlaire. Brown also noted that the maps carried a supplemental explanation that 1 ligne was equal to 300 toises. In other words, the cartographers of France were slow to convert to metric measurement. And so was the rest of France. Not until 1840 did France make its use compulsory.

Other countries began to follow suit, but by no means all of them. During the world exposition of Paris in 1867, the show of tremendous industrial development in the world impressed many scientists with the need for unification of standards of weights and measures. At the exposition, the scientists said, "man found himself in the presence of the immense variety of products . . . whose quantities were described in all kinds of measurement standards." The scientists formed an international committee, and the French government invited other countries to send delegates to Paris for an International Commission for the Meter. This action eventually produced the Convention of the Meter in 1875. The signatory nations, including the United States, agreed to set up and maintain at common expense a permanent International Bureau of Weights and Measures, which is located at Sèvres, just outside Paris. Even though the United States was a non-metric country, it participated in the effort because the metric system had become the standard for scientific research and, being the most precise, was used as the basis for defining non-metric units like the yard and pound.

A century after the treaty of 1875 the United States, one of the last non-metric holdouts, was in the process of converting. And so were the nation's mapmakers. The basis for the length of the meter, however, had since changed. In 1960 it was decided that the definition of the meter would be the length of a certain number of wavelengths of light emitted by krypton atoms. Such a definition would be more invariable throughout the world than one based on arc measurements. Fortunately, the new definition left the length of the meter essentially the same.

Cartography has been one of the beneficiaries of the more precise and uniform system of measurements. But it was the science of geodesy that, in a large sense, made it possible—one more example of geodesy's importance in mapping the world. For the metric system of measurements was perhaps the most useful scientific, cartographic, and practical result of the long years of inquiry into the exact shape of the Earth by geodesists like the Cassinis, Maupertuis, La Condamine, and Bouguer.

T he equator is where it is because that position is halfway between the poles. It is the line of zero latitude, the prime parallel, and it could not run anywhere else. But the line of zero longitude, the prime meridian, is another matter. It is an arbitrarily designated line of longitude, and could be almost anywhere—and at one time or another, on one map or another, it has been. This, of course, made for confusion when pinpointing the coordinate positions of places.

Ptolemy, for example, chose the Fortunate Islands for the prime meridian because they were considered the westernmost extremity of the habitable world. This was a better reason in his day than others had in their day. A French king simply decreed the island of Ferro in the Canaries as the prime meridian for all French maps, but around 1800 it was shifted to the meridian of Paris, primarily for patriotic reasons. Other nations tended to follow the same practice. Depending on the national origin of the map, the world's prime meridian ran through Toledo or Cádiz or Madrid; Cracow; Copenhagen; Pisa or Naples or Rome; Augsburg or Ulm or Tübingen; Peking; St. Petersburg; Washington or Philadelphia; and the Royal Observatory at Greenwich.

At about the time the metric system was devised, in 1800, Pierre Simon Laplace, the French mathematician, urged the adoption of a single prime meridian. "It is desirable," he wrote, "that all the nations of Europe, in place of arranging geographical longitude from their own observatories, should agree to compute it from the same meridian, one indicated by nature itself, in order to determine it for all time to come. Such an arrangement would introduce into the science of geography the same uniformity which is already enjoyed in the calendar and the arithmetic."

But nature gave no clear direction, as it had in the case of the meter. During the nineteenth century, however, many nations began to accept the Greenwich meridian, primarily because of the ubiquity of British Admiralty charts in navigation. But France held out for Paris; Spain, for Cádiz. In 1881 there were still fourteen different prime meridians being used on topographic survey maps alone.

At the invitation of the United States government, twenty-five nations sent representatives to the first International Meridian Conference, which opened in Washington in October 1884. The purpose was to agree upon a single prime meridian. In addition, and of particular interest to a country as wide as the United States, spanning 57 degrees of longitude, the conference was asked to consider establishing a system of standard time related to zones of longitude.

The Washington conference approved, with few dissenting votes, a resolution to adopt a prime meridian "passing through the center of the transit instrument at the Observatory of Greenwich." From this meridian longitude would be reckoned in two directions up to 180°, east longitude being plus and west longitude being minus. One disadvantage was that such a prime meridian divides both Europe and Africa into east and west longitude. But the opposite of the prime meridian, the 180° meridian, was thus more fortunately located, for it is a line running down through the Pacific and intersecting few land areas. As such, it provides a convenient international date line. When it is noon at Greenwich, it is midnight along the 180° meridian, the beginning of the next calendar day.

As people in Harrison's day recognized, there is a direct correlation between longitude and time, but the idea of dividing the world into 24 time zones (of 15° longitude each) did not and could not win acceptance until the matter of the prime meridian was resolved. Thus, the meridian conference in Washington in 1884 simplified and standardized timekeeping as well as navigation and mapmaking.

Rare is the map today that does not recognize the Greenwich longitude as the prime meridian, the fundamental line of the Earth's coordinate system.

I n the same spirit of budding scientific internationalism that produced treaties on the meter and the prime meridian, geographers had begun to gather every four years to exchange ideas and discuss common frustrations and aspirations. It was at the fifth such International Geographical Congress, at Berne in 1891, that Albrecht Penck, a young professor of geography at the University of Vienna, stood and read a proposal for an International Map of the World.

This would be the most ambitious attempt to date toward achieving cartographic uniformity. All sheets of the map would be drawn on a scale of 1:1,000,000 (1 centimeter = 10 kilometers, 1 inch = 15.78 miles), and thus it was referred to as the 1/M or Millionth Map. Penck recommended that sheets be drawn on the same projection so that they could be readily put together with a minimum of distortion. The plan also was to use standard symbols and colors.

The idea was reasonable and desirable, but for years there were studies and debate, and no action on the Millionth Map. In 1904, at the eighth International Geographical Congress, Penck made an even more effective case for the map by exhibiting three trial maps compiled according to his proposed standards. These were a French series of sheets covering the Antilles, Persia, and China; a German series comprising part of eastern China; and a British series of eighteen sheets toward a projected map of Africa. At last, after more debate and foot dragging, the International Conference on the International Map of 1913, meeting in Paris, reached agreement on a set of standards for the Millionth Map.

One obstacle had had to be settled by arbitration. The French finally agreed to the Greenwich prime meridian, instead of Paris, and the British accepted the metric system as the official system for expressing distances and elevations above sea level.

Other stipulations included the following: Political boundaries were to be ignored in dividing the world into sheets, which was foresighted, considering the many changes since 1913. Each sheet was to cover 4° latitude and 6° longitude. The meridians were to be straight lines and the parallels, arcs of circles; the projection should be modified polyconic. It would be a hypsometric map on which successive altitudes would be indicated by a system of color tints. Though the Latin alphabet was to be used exclusively, place names would be spelled according to local usage—*Firenze* instead of Florence and *Wien* instead of Vienna. Each sheet would be inscribed *Carte Internationale du Monde au* 1 000 000.

Progress on the Millionth Map has been slow and uncertain. First there was the interruption of World War I and the reluctance of many governments to support the work on the sheet covering their part of the world. Even though most of the sheets were to be compilations not requiring new surveys, the undertaking was expensive and time-consuming.

For example, it took $570,000 and 25 years for the American Geographical Society, a private organization, to complete the first edition of its *Map of Hispanic America*. It was a compiled map, with the exception of data derived from a 1928 expedition to the sources of the Amazon. The Society made a point of stating that the map was not "officially" part of the International Map of the World, "lest the Hispanic-American countries should feel that the Society was presuming to take upon itself an enterprise which was the prerogative of their governments." Nevertheless, the Society said that the 107 sheets were "compiled and reproduced in conformity with the International Map of the World on the scale of 1:1,000,000."

Since World War II many nations have resumed work on their share of the Millionth Map. The United Nations acts as the clearinghouse for the national cartographic agencies publishing sheets for the map. In a 1973 report on the state of the map, the UN noted that nearly 1,000 sheets had been published but that about half of them were based on "inadequate surveys and outdated information." Still not covered at all are large areas in North America (northern Canada, the Arctic islands, and Greenland), the Pacific Ocean, and a few areas in Africa. "A rough estimate," the UN report concluded, "indicates that about half of the land area of the globe requires further IMM work."

When Albrecht Penck first proposed the Millionth Map in 1891, mapmakers had reached across all the continents except Antarctica and across the oceans. They had measured the Earth to a more or less fine degree of accuracy and developed the skills of topographic mapping. They had probed the hinterlands and charted the coastlines. They had blazed the trails and staked out the boundaries of empire.

By the turn of the twentieth century the mappers of the world had done all that, and yet it was insufficient. There was so much still to be mapped, more than they knew, more than they could imagine being mappable.

PART *Three*

Bright Angel Point

Across the great chasm, filled to the rim with silence two kilometers and two billion years deep, a strange new sun appeared in a flash on the horizon. The bright orange ball flickered, then shot a column of light to a target 16 kilometers away on the north rim.

For those who had come to the Grand Canyon to look and to meditate, it must have been a puzzling signal, another visual stimulus to be processed by minds already overloaded to numbness by the grandeur that was everywhere. For those of us who had come to measure and to map the Canyon, it was a reassuring signal. The laser was working.

A beam of helium-neon light was carrying from rim to rim and, reflected by quartz prisms, was returning to its source in such a way as to provide the most precise measurements ever made of the Canyon. These measurements would give Brad Washburn the distances between all prominent features on his map, as his theodolite angles had provided the relative elevations and directions.

"The whole thing we're doing now is totally different from fifteen years ago," Washburn remarked.

A few days earlier, B. Q. "Buddy" Cutshaw had arrived from Tennessee with the Rangemaster. The instrument came packed snugly in a two-meter-long mahogany chest lined with deep-pile carpeting. As we looked on expectantly, Cutshaw unpacked and erected the instrument, which when mounted on a tripod resembled a large television camera. This new model, Cutshaw explained, cost $18,000 and was being used in surveying by a number of highway departments and civil engineering firms. It operated on the same principle as the lasers used in the Apollo Project to determine more precisely the distance between the Earth and the Moon. For our purposes it would be used to measure accurately and swiftly the distances between the many control points in our survey of the Grand Canyon.

A measurement by laser beam is accomplished by what is called "phase comparison." Cutshaw peered through a telescope built into the laser instrument, focusing roughly on the distant point to be measured. He then directed the laser's intense, coherent beam of light (laser is the acronym for "light amplification by stimulated emissions of radiation") on the target of reflecting prisms. The prisms were configured so as to reflect light pouring in from angles up to 20°. When the reflected beam returned, the Rangemaster converted it to an electrical signal. A small digital computer inside the instrument corrected for distortions in light reflectivity, due to temperature and barometric pressure, and to take into account the curvature of the Earth. With its timer and its programmed "knowledge" of the speed of light, the instrument determined the distance between two points—the laser and the reflector—by measuring the time it took the light beam to make the round trip.

A dull click, a moment's wait, and then several amber numbers appeared on a panel at the back of the Rangemaster.

"Surveyors a hundred years ago used to be happy with an error factor of 1 part in 5,000, about a foot to every mile," Cutshaw commented after a few test measurements. "Our accuracy with this thing is 1 part in a million."

Our party split into two teams for the first of many laser measurements. Cutshaw, Washburn, and Washburn's wife, Barbara, remained on the south rim, keeping the Rangemaster with them. Wendell Mason, a surveyor, and I flew by helicopter across the Canyon to Bright Angel Point. We carried a two-way radio, a barometer and thermometer, and a set of seven reflectors.

The Bright Angel Point station was situated on a boulder resting at the edge of a 600-meter precipice. As in the case of nearly all our mapping stations, Washburn had been there months before, had drilled a hole in the rock, and had implanted a pipe. Into the pipe was screwed a target, a large square piece of metal painted luminescent orange, on which the theodolite had already sighted for measuring angles.

We extended the radio antenna and sought to establish contact with Washburn, who was across at Yavapai Point. We heard Washburn tell us that Cutshaw had already focused the Rangemaster's telescope on our orange target. So we quickly unscrewed the target and replaced it with the cluster of quartz prisms. Mason radioed barometric and temperature readings for the computer's calculations. These had to be taken at both ends of the line, then averaged, to determine the medium through which the laser beam traveled.

Then we saw it. The strange light, the tiny man-made sun. We were told to stand by. Cutshaw wanted to take at least eight readings, just to be sure.

While we waited, Mason spoke of surveying he had done all over the world. He nodded in the direction of the laser light. "To do that straight line with a steel tape, and of course it couldn't be done, but if it could, it would take three days," he said. "And here we are doing it eight times in a matter of minutes."

The thought was to Mason the surveyor almost as incredible as the Canyon itself.

When our southern sun had disappeared, Mason and I packed up the prisms, switched off the radio, and clambered down off the boulder at Bright Angel Point. We hiked back to the helicopter for the ride to our next mapping station.

The Winged Mappers

CHAPTER FIFTEEN

Two biologists once went on an expedition to study the atoll of Ifalik, in the Caroline Islands, from the standpoint of its human ecology: the interrelations of man, his culture, and the environment. First, they counted, weighed, and measured the 260 people living on the one square kilometer of coral and sand. This was the standard first step in an anthropological study. Next, when they were ready to get to work on the environment—the land, the reef, the lagoon, the sea, the plants and animals—the two biologists, Marston Bates and Donald P. Abbott, decided that the "logical first move here was a map, a topographic survey of the situation."

Their description of "mapping the boondocks," contained in their book *Coral Island: Portrait of an Atoll* (1958), suggests a number of fundamental points about the utility and practice of modern mapping. Now, as before, maps are considered an essential part of science, and not only the physical sciences, but more and more the social sciences as well; there is still much to be learned about the world and ourselves through mapping. Now, as before, amateurs sometimes find themselves perforce cast in the role of cartographers. Now, more than ever before, new technologies are expanding the range, precision, and versatility of mapmaking by amateurs and professionals alike.

Of these new technologies the first and most broadly significant in the twentieth century was aerial photography. It certainly gave Bates and Abbott a leg up in their mapping of Ifalik, as they acknowledged.

The map problem [the biologists wrote] was greatly simplified by the fact that a little previously a Navy reconnaissance plane had made a series of vertical and oblique aerial photographs of the whole atoll. The Ifalukians were fascinated by these photographs. They seemed to have no trouble at all in grasping the significance of either the vertical or oblique shots, and most people easily recognized the major landmarks, even locating their particular family dwellings. They knew the atoll thoroughly, and the aerial photographs gave them a chance to stand off and look at the whole thing at once, for the first time.

To stand off and look at the whole thing at once. This was indeed a major advance in mapmaking, like always having a mountaintop view of the terrain—only better, because more terrain could be surveyed more quickly and the images faithfully recorded for future use. But there is more to making a map than taking a few aerial photographs.

The two biologists spent one morning drawing a base map for the island of Falarik from a mosaic of the vertical photographs. Then they set out to fill in details of trails, households, and vegetation. Since they were amateurs at mapping and had none of the modern surveying instruments, their operation was reminiscent of early reconnaissance mapping.

Bates and Abbott gave the following description of mapping a trail on the island:

> We had a Brompton compass . . . and a triangular ruler marked with conventional scales, but no measuring tape more than six feet long and no plane table.
>
> We first tried improvising a tape by tying knots at two-yard intervals in 100 yards of cod-line. When we had finished this tedious job, we discovered the darn thing stretched and tangled too easily to be of any use. I [Bates] then dedicated myself to trying to learn to pace accurately, and soon managed a pace that fitted nicely when used on the base map with the 1:2500 scale of the triangular rule. This "Bates pace" was a purely arbitrary unit, but it worked well enough for our purpose. I learned to make nice adjustments in scrambling over coral boulders and around breadfruit trees and through fern thickets so that I could come out with the proper count. In crossing the island, plotting the course on the outline map made from the air photos, we generally found an error of five or ten feet (it was about 900 feet across Falarik). This would probably have been disgraceful for an engineer but it pleased us immensely.

In this way Bates and Abbott went about the atoll, mapping trails and vegetation zones, doing it rather crudely by twentieth-century standards, but accurately enough for their scientific purposes and much better than they could have hoped to do in the days before aviation. Theirs is a modest example of how the merger of aviation and the camera has revolutionized cartography. This merger led to aerial photogrammetry.

The idea of converting aerial photographs into some kind of map goes back many years, long before the introduction of the airplane at Kitty Hawk in 1903. In describing the new invention of photography to the French Academy of Sciences in 1839, Dominique François Jean Arago, a prominent geodesist, spoke of the possible practical uses of the new art, including "the rapid method which topography might borrow from the photographic process." Within a decade the French Army was using photographs from terrestrial stations as an aid in mapping. One officer, Aimé Laussedat, seeking a higher perspective, attempted to loft a camera suspended from a kite, but with little success.

Aerial photography may be said to have begun in 1858 with a feat by Gaspard Félix Tournachon, a Paris photographer who called himself Nadar. He loaded his cumbersome wet-plate apparatus into the basket of a balloon and cast off to an altitude of some seventy-five meters. It was awkward trying to prepare the sensitized photographic plates in his makeshift darkroom, and the gas with which the balloon was inflated had a deleterious effect on the plates. Notwithstanding all this, Nadar succeeded in taking a photograph of the village of Petit Bicêtre. Despite spots and small defects in the negative, the houses of the village could be clearly recognized. Honoré Daumier, the artist, drew a caricature of a balloonist with the caption saying that Nadar had elevated photography to the "highest" art.

Meanwhile, Laussedat had continued experimenting with this new technology and, in 1859, developed the first camera designed specifically for mapping. It was really a camera and theodolite combined, and with it Laussedat made several surveys in the Paris environs, sometimes working on rooftops and church steeples. He would photograph an easily identifiable object twice, once each from different points a predetermined distance apart, and thereby fix the exact position of the object. Once a number of such objects were plotted, the topography of the surrounding area could be drawn and filled in with a fair degree of accuracy. The camera, in effect, replaced the plane table and transit. Laussedat's technique marked the beginning of photogrammetry, which is the science of obtaining reliable measurements by means of photography.

Much later, in 1906, Theodore Scheimpflug of the Austrian Army conceived of a camera system that would make aerial photogrammetry possible in at least rudimentary form. He designed a large camera with eight lenses. One pointed directly down for vertical coverage, while the other seven were arranged in a circle to take separate oblique views. All of the exposures were made simultaneously on separate pieces of film. The eight photographs were combined into a single panoramic photograph reaching from horizon to horizon. But Scheimpflug found that the balloon was too unstable for any practical application of his new camera.

W orld War I brought together the airplane and photography. Although most of the aerial photography in the war was for the purpose of military reconnaissance, not mapping, the experience left no doubt that the two technologies had a joint future.

A few of the people who saw that future and helped to bring it about are still living. Virgil Kauffman is one of them, and he relishes the memory of his long and vigorous life as a pioneer in aerial mapping. Kauffman was five years old when Wilbur and Orville Wright made their first flight at Kitty Hawk. He started flying himself in the roistering, open-cockpit days during and after World War I. Who could have imagined that the airplane, this frail and rather frivolous machine, would be of much use in precision mapmaking? Not Kauffman at first, or the mapmakers. But flying got into the blood of many people, ingenious people who wanted to see what could be done in the air.

There was Talbert Abrams, for example, who had begun taking photographs in 1916 from the top of the Dime Savings Bank Building in Detroit. As a Marine Corps pilot during the war, he flew aerial photographic missions along the Gulf Coast of Texas, over Florida and the Bahama Islands, and over Haiti. Later, he flew on the first airmail route in the United States, between New York and Washington. In 1922, Abrams, like a number of pilot-photographers, went into business for himself, forming the Abrams Aerial Survey Corporation. One of his first contracts was to help the state of Michigan locate all of its roads more accurately.

In the early 1920s, Sherman M. Fairchild established a corporation to produce cameras specifically designed for aerial photography and to conduct aerial surveys. He was the son of a wealthy New York congressman and had worked with the Army Signal Corps science and research division during World War I. Fairchild's earliest inventions included cameras designed to be electrically driven, instead of hand-cranked, and preset to take photographs at any interval required. The Fairchild Aerial Camera Corporation became a leader in the development of aerial mapping cameras, photogrammetric equipment, and the lunar mapping camera used by the Apollo astronauts. The company also built the first enclosed-cockpit airplane in the United States, the FC-1, which was used for many years in aerial mapping.

Another talented innovator in the science of aerial photogrammetry was W. Sidney Park. In Louisville, Kentucky, after World War I, Park installed the first automatic pilot in a mapping aircraft, built his own aerial camera, and designed a trainer for instructing aerial mapping photographers. In presenting Park an award of the American Society of Photogrammetry in 1969, Russell K. Bean of the Geological Survey called him "the complete photogrammetrist—he builds his own aerial camera, constructs the mount and installs it in his own airplane, flies the plane himself and takes the pictures, makes his own plates, operates his own stereoplotting facility, builds his own towers for field surveys, gets the control [ground measurements] with his own men."

George W. Goddard stayed in the Army after the war, taking the lead in shaping military photo-reconnaissance technology. He organized the first Army aerial photographic mapping unit, which got in considerable practice photographing the Mississippi River. "Many people think of aerial reconnaissance as a principal service to the ground forces in time of war," Goddard said. "They do not realize that peacetime aerial mapping use of reconnaissance is of great importance."

Virgil Kauffman's story, as much as any other, illustrates how aerial photography gradually came to be an accepted and invaluable part of peacetime cartography.

After the war, in which he learned both photography and flying as a soldier, Kauffman joined a struggling Philadelphia company, Aero Service Corporation. All it had was an old Curtiss JN4, the Jenny, a World War I training plane, which the company flew as an air taxi between Philadelphia and the Jersey shore and sometimes as far as New York. Kauffman soon took control of the company and, to stave off bankruptcy, redirected the business to aerial photography.

It was a primitive operation. Kauffman would handle the stick while a photographer leaned out of the open Jenny. When the photographer was ready, Kauffman would slow the craft from its cruising speed of 100 kilometers an hour and then shut off all power, trusting that when he pressed the throttle ahead the engine would catch again. ("Once or twice, it didn't," Kauffman said. "I made a fast landing.") With the aircraft dead and drifting smoothly, the photographer aimed his hand-held camera and snapped a picture at an oblique angle. The shutter speed was slow, but so was the airplane, and consequently blurring was no problem.

For taking vertical pictures, which achieved a more maplike view of the terrain, Kauffman cut a hole in the floor of the fuselage and mounted a custom-built camera over it. The pilot would sight on a target ahead and with a stopwatch would time the interval until the plane should be directly over it. The photographer would crank the camera, try to level it, and then take the picture. This would be repeated every few minutes, each picture overlapping part of the previous one to insure complete coverage. The pilot had to fly back and forth along parallel lines, trying his best to keep the plane from drifting off course.

In this fashion Kauffman produced photomaps of New Jersey, the first such maps of an entire state, and sold them to municipalities, utilities, and real-estate developers. His first major contract was to photograph the entire Tennessee River basin and assemble the pictures into mosaics for study of drainage patterns, cultivated lands, forests, roads, and population centers.

By this time, Kauffman and the Aero Service crews were equipped with bigger planes, multi-lens cameras with faster shutter speeds, and easier-to-handle film. "We called our pictures 'maps,'" Kauffman recalled. "It didn't take us long to find out they weren't maps, but they were useful in the preparation of maps."

Another Philadelphia company had made considerable progress in the 1920s toward understanding how to make precision maps from aerial photographs. The company, Brock and Weymouth, Inc., had developed instruments with which high-resolution vertical aerial photographs could be translated—by a method involving stereoptics—into contour maps of great detail and accuracy. Differences in elevation as small as half a meter could be plotted with the aid of some ground control. The result was maps based on the science of photogrammetry, not patchwork mosaics.

Despite several successes, including a survey of the Hoover Dam area, Brock and Weymouth had trouble attracting customers. Most people still could not accept the idea that a map produced from aerial photographs could be as good as or better than one made by the time-honored, foot-slogging ways.

So when Norman Brock, a partner, heard that the United States Geological Survey was mapping a section of the Pocono Mountains known as the Bushkill Quadrangle, he had an idea as to how he might finally convince traditionalists of aerial mapping's value. He flew an airplane over the area, photographed the heavily forested quadrangle, sent in ground-control surveyors, and then produced a map of one-third of the quadrangle. Brock took his map to Washington and asked that it be compared with the Survey's version.

In April 1927, Claude H. Birdseye, the Survey's chief topographic engineer, wrote Brock that he was impressed by the "marked superiority of the photographic method." Birdseye spent two days on the ground, examining certain points of difference between the two maps, and told Brock: "I would be very glad to use your map in preference to the one made on the ground in the publication of the entire quadrangle."

The Survey published only a limited edition of the map it had prepared by ground methods. Not until 1943 did the Survey finally issue a map of the Bushkill Quadrangle based on aerial photogrammetric work. By then the Survey had developed its own multi-lens camera system, which proved to be more practical than the Brock method, and had taken the lead in converting to photogrammetric mapping.

Brock and Weymouth had the misfortune of being ahead of its time and was forced out of business during the Depression. Kauffman bought the company's aerial cameras and contour-plotting instruments. With these instruments and a growing squadron of airplanes, Kauffman's crews completed thousands of kilometers of flood-control surveys, highway-location studies, city mapping, and industrial-plant-site mapping. On the eve of World War II, he obtained government contracts to photograph and map such potentially strategic areas as the Caribbean, the Canal Zone, and Newfoundland. When Kauffman's crew arrived in Newfoundland in a single-engine plane, they found that an entire Air Force photographic squadron had spent three months trying without much success to get good pictures. The cloud cover had made the task seemingly impossible. Kauffman's crew was up at daybreak the next day, and they flew before the clouds had formed.

After World War II, Aero Service and other aerial survey companies photographed and mapped several million square kilometers of the Earth's surface. They made many of the first extensive surveys of the Middle East, Africa, and South America. In some of the more remote places, the aerial surveyors' intentions were not always understood: frightened Africans once threw spears at a low-flying Aero Service aircraft; suspicious moonshiners in Appalachia took a few rifle shots; and Saudi Arabian police fired on and arrested a pilot for "attacking" Riyadh. Such incidents became part of the legend of the spirited profession of aerial mapping.

Another part of the legend grew from the increasingly impressive achievements. When Aero Service mapped the 230-kilometer eastern extension of the Pennsylvania Turnpike in 1948, the company delivered finished topographic maps of the route in 145 days. According to the turnpike engineers, Aero's air-ground teamwork helped get the job done five times as fast and five times as cheaply as it would have been done if conventional ground-survey methods had been used.

Through the eye of the aerial camera the mapmaker's vantage point had been moved from the hilltop to a point hundreds and thousands of meters above the ground. He could now see both sides of the distant mountain range instead of only the near side. If he got high enough, he could even photograph the curvature of the Earth. The first such photograph was taken in 1935 by Albert W. Stevens, an Army captain, who rode a balloon to an altitude of 21,720 meters over the Black Hills of South Dakota.

There is more to aerial photogrammetry than flying and taking pictures. For every $5 spent on aerial photography, $95 has to be spent on converting the photos to published maps. An aerial photograph does not, in itself, constitute a map. Because of variations in ground elevations, tilting of the aerial camera, lens and film aberrations, and other phenomena, the aerial photograph usually gives a distorted picture of the terrain. Furthermore, a single aerial photograph affords no means of measuring variations in the height of the terrain.

For converting the information contained in the aerial photograph into accurate topographic maps, various stereoscopic plotting instruments have been developed. The basic principles were first worked out by Karl Pulfrich, a German inventor, and by Erich von Orel, an Austrian, early in the century. Pulfrich invented the stereo-comparator. This was an instrument that projected overlapping pairs of pictures into a three-dimensional optical image and had a measuring index—the so-called floating mark—by which it is possible to determine the exact position of any point in the common area of both pictures. Von Orel simplified the operation with the stereo-autograph. This connected the floating mark to a pencil or pen on a special plotting table so that a road, for example, could be followed by the floating mark and plotted simultaneously by the pencil. In a similar way, it later became possible to plot contour lines by causing the floating mark to trace the outline of a feature at a fixed elevation.

Some of the early plotting instruments, with their awkward assemblies of viewing optics, hand wheels, and mechanical arms, looked like Rube Goldberg contraptions. But improvements continued to be made between the world wars. A few years after the Bushkill episode, Birdseye led the Geological Survey into extensive utilization of aerial photogrammetry, first in the topographic mapping of the Tennessee River valley, then in projects throughout the country.

Now, though there are wide differences in the mechanical details and in the degree of accuracy of photogrammetric equipment, practically all the stereoscopic instruments operate on the same principle. The following is a description of a typical operation by the Geological Survey, from flight to map manuscript to map production:

The task of photographing the area to be mapped is contracted to one of approximately 75 private flying companies in the United States. The aircraft flies a number of parallel runs, taking strips of successive vertical photographs. Each photograph overlaps the next one by 60 percent, and each strip overlaps adjacent strips taken on parallel flight runs by 10 to 25 percent. For the Survey's standard 1:24,000 quadrangle maps, a flight altitude of 3,600 meters is usually prescribed, as it affords an acceptable compromise between the area covered and the amount of detail captured on a photograph, or the resolution. The flights are usually restricted to a season when there is no snow, no leaves, no flooding, and few obscuring atmospheric conditions.

The camera is generally a single-lens instrument with a 6-inch (152-mm) focal length. This too is a compromise choice. The longer the focal length (the distance between the plane of the film plate and the lens), the more detail can be recorded on the photograph. But the greater detail is thus achieved at the expense of coverage. The camera is shock-mounted, with the lens projecting through an opening in the belly of the aircraft. Some cameras are operated by automatic timing controls, but often the photographer prefers to look through a view finder and manually control the exposures.

When the film has been exposed and delivered to the Survey, the negatives are processed and from them are produced transparent positive prints, or diapositives, on glass or film. Diapositives of a pair of photographs taken at successive exposure stations are then inserted in the projection system of a stereoscopic instrument. The operator of the instrument adjusts the projectors to reconstruct the different angles from which the photographs were shot. Then he observes the overlapping photographs through a stereoscopic viewing system and sees what is, in effect, a miniature model of the photographed terrain. The left eye sees one photograph and the right eye sees the other, and the result is a three-dimensional view.

After the model has been thus relatively oriented, the operator must search for control points—identifiable points common to both photographs for which latitude, longitude, and elevation have been established. Such points used to be fixed by ground survey or by instrumental aerotriangulation—a process of extending control through a strip of photographs from a few known control points. Through a recent application of automatic data processing to photogrammetry, it is now possible to perform analytical aerotriangulation of large blocks of photographs, meaning that it is possible to extend both horizontal and vertical control from points of known position, which can be many miles distant, to points whose positions are needed for map control. The key to this operation is the development of highly complex computer programs for the mathematical solutions.

By a few judicious measurements in the field a surveyor tests the accuracy of the office compilation and verifies the photogrammetrist's interpretation of map detail. Was that break in the woods a trail or a mountain stream? Is that building a barn or a small factory? He also adds detail that was obscured by dense foliage or by heavy shadow. He obtains information on political boundaries, place names, and similar data not obtainable from the photographs.

Preparation of the map for publication begins by the making of a film negative of the map manuscript at reproduction scale. The image on this film negative is photochemically reproduced on several coated plastic sheets. These sheets will serve as guide copy for separate engravings required for each color of the printed map. A standard Survey map has five colors—black for "culture" (buildings, roads, place names); brown for contour lines; blue for bodies of water; red for major highways and concentrations of houses too dense to be shown individually; and green for woodland.

Each "color separate" is prepared by a technique known as scribing. Features to be printed in the same color are scribed on a coated plastic sheet. That is, the scriber, working over a light table, uses sharp etching tools to cut the required map lines and symbols through the thin coating paint of the plastic sheet. If the particular sheet is to be the "separate" for brown, for example, the scriber etches out thin lines where contours are to be shown and leaves the rest of the sheet's photographically opaque paint untouched. Mistakes can be painted over and corrections rescribed.

By multiple exposure of the scribed sheets and negatives of type for place names and through a series of sensitizing and developing steps, a photochemical color proof is prepared on plastic. This proof is reviewed and edited. A lithographic plate is made for each map color by exposing the appropriate scribed sheet.

Then the map is ready to go to press.

The prefix "ortho," from the Greek *orthos,* means straight, right, direct, or correct. An orthophotomap thus is a cor- rect map made directly from aerial photographs. But is such a map possible?

Until the 1950s, the prospect seemed remote at best. Scale in an aerial photograph varies in different parts of the picture because of camera tilt and ground relief, which means that higher terrain will be closer to the camera and thus at a different scale. If only some way could be found to produce a uniform-scale photograph, the result would be the much-desired combination of the scale reliability of a map with the wealth of detail afforded by a photograph.

Russell Bean, of the Geological Survey, showed the way with his invention of the orthophotoscope, the first prototype of which was built in 1956. The instrument combined conventional stereoscopic mapping equipment with a means of exposing sensitized film, bit by bit, through a narrow slit in a movable screen on which the terrain model was pro- jected. Tilt displacements were eliminated by proper orientation of the projectors. The elevation of the film was varied according to the terrain as the scanning proceeded. The film was thus differentially exposed, always at the correct elevation to eliminate relief distortion. When the scanning was complete, the film was developed as a negative and the orthophotographs were printed from the negative.

By the 1970s, improved orthophoto instruments, including computer- assisted devices, provided a quick, inexpensive method of providing maps where none existed before or of making swift revisions of exist- ing maps. A single orthophotograph or a precisely controlled assembly of several could be readily converted to an orthophotographic map. It was generally published in standard map format and related to a geodetic reference system. Cartographic symbols, including contours, boundaries, and labels, could be added to suit the area and intended use of the map. In a black-and-white quadrangle format, these maps are called orthophotographs. When color-separation techniques are used to show different kinds of detail in different colors, the product is called an orthophotomap.

Such mapping has found increasing application. According to Rupert B. Southard, Jr., a chief of the Geological Survey's topographic division, orthophotomaps are an effective way to represent flat swamp and marshland, flat desert terrain, or highly eroded areas where contours do not adequately portray the physiographic details. The first published orthophotomaps were 1:24,000 quadrangles of the Okefenokee Swamp in the southeastern United States. They were ideal, Southard said, "because there are few contours, there is a low density of cultural features, and standard maps show almost nothing but the marsh symbol pattern."

The Survey has since published orthophotomaps of the Everglades in southern Florida, the flat lake country of northern Minnesota, the coastal plains of Virginia and North Carolina, the Great Salt Lake area of Utah, and the oil-rich Prudhoe Bay region of Alaska. In this last case, two important requirements were satisfied better by orthophotomaps than by regular line maps: ground detail was much more complete, and the maps were available promptly to meet the needs of oilmen.

In an experiment in Fort Wayne, Indiana, large-scale (1:2,400) orthophotographic maps were put to use in urban planning. An area of 360 square kilometers was covered in 440 sheets. Fort Wayne officials reported using the maps effectively for studying highway relocation and street widening, locating sewer lines, laying out bicycle trails in city parks, and planning schoolbus routes.

Aerial photogrammetry has advanced far beyond Virgil Kauffman's days in the open cockpit, far beyond those strips of pictures he used to assemble and call maps. In only a few decades the airplane and the camera have become conventional mapping tools. The newer technologies are electronic computers, automation techniques, electronic distance-measuring instruments, inertial navigation systems, remote sensing, and applications of space sciences. Although the full impact of the new technologies has yet to be realized, at the close of the twentieth century the results were already bringing about a revolution in cartography.

Radar Over the Amazon

For centuries much of the Amazon Basin, nearly 60 percent of Brazil, had been Brazilian in name only. It was remote and largely inaccessible, much of it unexplored, and all of it underdeveloped. It was a place of dark waters and dense rain forests, crocodiles and carnivorous fish, sleepy river towns and isolated Stone Age people. Exotic and forbidding, it was commonly known as Brazil's Green Hell.

The French geodesist La Condamine was the first foreigner to travel the length of the river, the world's largest. In 1745, on his return from Peru, where he had helped prove that the Earth bulged at the equator, La Condamine measured the width and depth of the Amazon and observed the angles and positions of many of its tributaries. From this, and from some earlier but rough Jesuit charts, he produced the first map of the river with any basis in scientific observation.

In his great reconnaissance of South America in 1799–1800, Alexander von Humboldt made some seven hundred sextant observations that became the framework for the first good base maps of the continent. Humboldt, the son of a Prussian baron, was a scientist of commanding intellect, devouring curiosity, and incredible endurance. He was acclaimed in his day as "the greatest man since Aristotle." In 1799, with a French botanist, Aimé Bonpland, Humboldt sailed on a Spanish mail boat to Venezuela, eluding a British naval blockade. The two scientists explored the Orinoco River, verified the existence of a stream connecting the Orinoco with the Amazon system, and charted the upper reaches of the Rio Negro, a major tributary of the Amazon. In the area where the Orinoco and Amazon systems join, they barely endured the unremitting onslaughts of mosquitoes. As Humboldt wrote, "We could neither speak nor uncover the face, without the mouth and nose being filled with insects."

During their five years in the New World, Bonpland collected more than 2,000 plant specimens. Humboldt experimented with electric eels, deliberately sustaining countless shocks, and dissected an alligator to examine its respiratory system. Upon his return to Europe, Humboldt left almost no observation unreported. He produced 33 volumes, illustrated with 1,425 maps and plates, to tell the rest of the world about that mysterious region. The expedition and the report established Humboldt as one of the founders of modern scientific geography.

Except for an occasional scientific expedition and the explorations of rubber companies, little progress was made in mapping the Amazon in the century and a half after Humboldt. The area was considered as unmappable as it was inpenetrable.

Then, in 1970, in a spirit of nationalistic fervor, Brazil launched a crash program to construct a highway through the Amazon basin, from the Atlantic coast 5,000 kilometers to the border of Peru. Construction began before the route into the interior could be adequately mapped. Only the hastiest aerial surveys were conducted, and their coverage was limited by the clouds. Surveyors, living in lean-tos and working out of construction camps, did not always keep much ahead of the bulldozers. They would run into rivers that were not always where they were assumed to be. Consequently, the trans-Amazon highway frequently had to recross the same stream several times. The additional cost in bridge building might have been saved if the maps had been better.

Not long after the road was started, the Brazilians learned about how side-looking radar could be used in cartography. While it was too late to undo construction mistakes already committed, they quickly seized upon this new kind of radar as the mapmaking tool they had long sought and had almost given up hope of ever finding.

Side-looking radar is only one of several ways of viewing Earth by what has come to be known as remote sensing. Mapmaking perspective in the twentieth century shifted from the crow's nest and the mountaintop to the airplane and then the spacecraft. But it was not only this lofty perspective that transformed cartography;

there was also the increasing capacity to see beyond the limited sensitivity of the human eye. Through various forms of remote-sensing technology, mapmakers could observe the hidden contours and nature of a place without seeing the area in the conventional sense. An insight by the exquisitely observant Marcel Proust (although he had other things in mind) applies to the critical role of remote sensing in extending cartography's reach. "The only true voyage of discovery," he wrote in *Remembrance of Things Past,* "is not to go to new places, but to have other eyes."

The telescope could be considered the first remote-sensing instrument, although it lacked the important ability to record what it saw. The camera solved that, and provides what is still the most widely used means of remote sensing. In standard photography, the scene is recorded as humans see it, on film sensitive to electromagnetic energy in the form of visible light. New digital cameras incorporate electronic sensors even more efficient in collecting light. An added advantage is that picture data in digital form can be readily processed by computers to accentuate certain details and alter perspective or scale. Each new image from spacecraft exploring the planets and from the Hubble Space Telescope focusing on distant galaxies is a stunning example of the increasing remoteness of photography's subjects. In many of its new forms, remote sensing moved beyond ordinary telescopes and visual photography to record a wide range of phenomena beyond human vision.

The principle behind remote sensing is not new; only the methodology is. Every kind of feature on the face of the Earth tends to reflect or emit radiant energy in distinctive amounts at certain specific wavelengths. Electromagnetic energy travels as a wave, and each part of the energy spectrum, each wavelength, is especially suitable for providing information about some natural or man-made object. Each object, for instance, has its own unique distribution of reflected, emitted, and absorbed radiation. This applies to rocks and trees as well as plants and animals, to materials in buildings and roads as well as in ships and aircraft. A catalog of the "spectral signatures" of many natural objects has been compiled through years of testing.

The most familiar part of the spectrum is, of course, visible light. This includes all the radiations at wavelengths detectable by the human eye or the camera—red, orange, yellow, green, blue, indigo, and violet, the colors varying by wavelength. Red, for instance, has the longest wavelength, and violet has the shortest. A rose is red or yellow and a leaf green because of their particular absorption characteristics.

Wavelength energies shorter than the visible are, in descending order of length, ultraviolet, X-ray, gamma ray, and cosmic ray. With the development of new films, photography has been extended to the sensing of ultraviolet. But its application is limited on Earth because of the strong atmospheric absorption of ultraviolet and shorter-wavelength radiation. Astronomers, however, have made good use of ultraviolet sensing devices in spacecraft to analyze the gaseous constituents of planetary atmospheres and stars. Airborne gamma-ray spectrometers determine composition and conditions (such as moisture content) of terrain materials.

Wavelength energies longer than the visible are infrared, microwave, and radio. The infrared region of the electromagnetic spectrum is one of the most widely used in remote sensing. It opens up a world of unseen thermal radiation.

Photographic film sensitive to a limited range of IR radiation was developed in World War II to detect camouflage, but a more common method of obtaining aerial IR imagery is use of mechanical devices equipped with photo-electric sensors. Such devices, called radiometers, can perceive subtle differences in surface thermal radiation, the difference between healthy and diseased or dead plants, between a heated and unheated building, between warm and cold ocean currents. The measurement of temperature differences and the mapping of their distribution can reveal impending volcanic activity, geothermal energy, moisture conditions in soil (variations in moisture affect surface temperatures), and smoldering forest fires. In an aerial infrared survey, the United States Forest Service spotted in the course of a year two hundred forest fires that might otherwise have escaped early detection.

Within the microwave region, radar offered the most immediate potential for extensive remote-sensing surveys.* Unlike the other sensing methods, which are passive detectors and recorders of radiation, radar is an active sensor. It transmits a narrowly beamed radio signal and records the signal as it is reflected from some distant surface. Since the signals are affected by the roughness and altitude of the reflecting ground surface, the returned signals can reveal with varying degrees of resolution the contours and relief of the landscape several miles beneath the radar-beaming aircraft, or of the landscape of a faraway planet. What is more important in surveying, radar can operate effectively day or night and through all but extremely dense clouds and rain.

---

*A non-electromagnetic method of remote sensing, using the airborne magnetometer, is also widely used, and is described in the next chapter.

Early radar systems were developed independently in Britain, France, Germany, and the United States during the years prior to World War II. At first, radar attracted some attention as a possible weapon, the "death ray" of many a science-fiction tale. But Robert Watson-Watt, a descendant of the famous James Watt, suggested the initial use to which radar was eventually put during the war—the detection and location of aircraft. (The name "radar" is an acronym for "radio detection and ranging.")

In the late 1930s Britain constructed a network of radar antennas along its coast and managed for many months to conceal their true purpose from the Germans. A rotating radar beam swept the sky, and if it was blocked and reflected by some obstacle, a moving blip of light appeared on a cathode-ray tube. The display screen was another idea of Watson-Watt's. By the size of the blips and their position on a grid of lines superimposed on the screen, a trained observer could judge the type of object detected and its location. The first of these radar installations went into operation in 1940, in time to prove decisive against German aircraft in the Battle of Britain. This same technique, though considerably improved, is now employed by meteorologists to plot rainstorms, by air controllers to direct traffic in the skies, and by the military in its early-warning system against enemy aerial attack.

For a while radar seemed an unpromising tool for making maps. Ranging measurements made with radar had, to be sure, been employed to discover errors in existing maps. In World War II, bombers used radar to find their positions accurately by measuring the distances, or ranges, to two fixed ground stations at which radar beacons were located. One such station was on the island of Corsica, in the presumably well-charted Mediterranean. On the first mission carried out in Italy with this system, the Allied bombers missed their targets by about 1,000 meters, nearly 50 times the expected error. After an investigation ruled out all other possible sources of the error, it was determined that the position of Corsica in relation to Italy was misplaced on the map— by about 1,000 meters! But as for using radar to produce images of the terrain that could be used in cartography, the World War II systems simply lacked the necessary resolution.

Improvements were soon to come, the result once again of research for warfare. In the 1960s, after years of experimentation, side-looking

Radar imagery converted into a map
of the Amazon River in Brazil, 1972

radar was installed on military aircraft for all-weather targeting and reconnaissance work.

Just as the telescope's resolving power increases with the size of its mirror, radar's resolution increases with the width of its antenna. Fine radar focusing, revealing great detail at long distances as well as in the foreground, would require an antenna of such width that it could not be flown on an aircraft. In the earlier systems, the antenna rotated and had to be short enough to fit inside a streamlined cover, or radome, on the aircraft. This seriously restricted the system's beam resolution. Side-looking radar, on the other hand, employed an antenna that pointed its beam steadily toward one side of the aircraft path during a straight flight. The width of the antenna could thus extend along much of the length of the aircraft. It provided, in addition to sharper resolution, wider coverage.

The returned images from side-looking radar are like photographs taken in early morning or late afternoon. The oblique angle of the radar beam, like the angles of early or late sunlight, produces more sharply etched images of mountains, ridges, and other relief. Moreover, such radar images are less distorted at greater angles than aerial photographs.

But even if cartographers had been allowed to use this classified technology, the early forms of side-looking radar would have had a serious drawback. Since a focused beam will be in focus only at one range, the system could not produce images of uniform resolution over a wide area. Overcoming this problem, and at the same time further increasing the effective antenna width, required a bit of electronic trickery. The trick is called synthetic-aperture radar, the most modern version of side-looking radar. A small antenna, as little as one meter long, can be mounted on an aircraft and operated with the same effectiveness, in terms of resolution, as a hundred-meter antenna. The instrument sends out a stream of precisely phased pulses of microwave energy and then records and times the return signals which have been reflected by a distant surface. Ten, twenty, or more pulses can be digitally recorded and processed as a group. In this way, the effective width of the radar antenna can be many times the length of the aircraft, with comparable improvements in the observations used for making maps.

As the technology existed at the time of the Brazilian survey, the return from each pulse illuminated a scan of a cathode-ray tube, which was recorded onto a moving photographic film, strip after strip. The succession of strips added up to a scaled-down hologram of the radar illumination, just as the reflections from coherent visible light add their energies to produce holograms—which can show objects in three dimensions. The holograms of apparently formless radar return signals were then illuminated with laser light and projected to produce recognizable images appearing almost like photographs of the Earth.

A specialized computer measured and sorted the thousands of overlapping return signals, some from near and some from far, and factored in the aircraft's velocity. This resulted in resolutions of a quality and uniformity unattainable by conventional airborne radar.

Side-looking airborne radar, or SLAR, was first deployed as a mapping tool in the areas where it seemed indispensable, such as the cloud-covered tropics. In 1967, before the system was released for commercial use, the U.S. Army Corps of Engineers tested SLAR (not the synthetic-aperture model, however) over the Darien Province of Panama and part of northwest Colombia. As in most jungle areas, the almost perpetual cloud cover had frustrated aerial photographers for several decades, and land surveys were only approximations of reality. Maps prepared from SLAR results, declassified a year later, provided the first complete overview of that region and first accurate look at a possible site of a new Panama Canal. The course and direction of flow of one of the rivers were found to be incorrect on existing maps. A short mountain range was found to be misaligned by about 90°. Other mountains had gone altogether unmapped.

After SLAR was given over to general use in 1970, surveys in Ecuador proved the imagery superior to infrared photography for preparing drainage-system maps. A survey through the clouds over New Guinea demonstrated that SLAR clearly revealed geologically interesting cracks and faults that had not been seen in aerial photography. It took only seven days to obtain SLAR imagery for the entire country of Nicaragua, and this was then translated into comprehensive maps within four months.

B razil's Ministry of Mines and Industries, which had responsibility for developing Amazonia, established the RADAM (Radar Amazon) Commission to put side-looking radar to use in mapping the country's great frontier.

The commission approached Aero Service of Philadelphia, now a division of Litton Industries, to take on this assignment. Homer Jensen, the company's vice president and a geophysicist with long experience in aerial surveys, knew that few people outside the military understood how to operate SLAR equipment, and even fewer had the experience needed to apply it to a cartographic project of such scope. Jensen went to Goodyear Aerospace Corporation, which had built SLAR systems for the military, and found out what was required. He needed an airplane big enough to carry all the necessary electronic equipment, and powerful enough to fly high above the more turbulent weather. Aero Service bought a used Caravelle, and Goodyear engineers outfitted it with the SLAR antenna and related control electronics.

Jensen also learned that the navigation system was an absolutely essential part of a SLAR mapping mission. "The system," Jensen said, "had to guide the plane down a straight path, without any curving or drifting. No human being could navigate that well. I'm all for human beings, but they are too slow, their reaction time too slow for this kind of work. People just can't react fast enough or accurately enough to prevent slight deviations in a flight course, what the engineers call aircraft phugoid. You really have to do it electronically."

The aircraft was, therefore, equipped with an inertial guidance system, which is a complex of gyroscopes, accelerometers, and computers. It was the same kind of unit that guides airliners across ocean routes and not unlike the sophisticated guidance systems for the Apollo spacecraft. The gyroscopes sense the aircraft's attitude and flight direction. The accelerometers monitor velocity changes. They generate a constant stream of electrical signals to the auto-pilot, precise instructions for maintaining a predetermined course and speed, and to the SLAR, information for fine-tuning the orientation of its antennas. The system was capable of providing navigation accuracies of better than 1 kilometer, without any updating, after a flight of 3,400 kilometers. Only in this way could the Caravelle fly its daily parallel pattern and serve as a stable platform for precise, relatively distortion-free radar mapping.

But, even with all its electronics, the Caravelle could not have succeeded without that hardy frontier breed, the shoran station operators.

Shoran, the abbreviated name for short-range navigation, is a precision position-fixing radio system that had been employed in most post-World War II aerial mapping. A network of four to six shoran stations was established throughout the area being surveyed at any given time. Each station was positioned from fixes on a navigation satellite, with the longitude and latitude being established to within 10 meters of geodetic accuracy. When the survey aircraft would come within range of the station's radio beacon, the aircraft's transmitter-receiver would broadcast a pulsed radio signal to the station. The ground transponder picked up the signals and rebroadcast them back. By measuring the round-trip times of the signals, an airborne computer could calculate distances to the station. By simultaneously receiving the broadcasts from two such stations, the aircraft's longitude and latitude at the time of each radar observation were determined to an accuracy of about 75 meters. This enabled mappers to place objects in all radar images close to their true location.

In Brazil shoran stations had to be set up before the Caravelle could begin flying. Most were placed close to jungle airstrips or river landings so that the operators could go in by light aircraft or by boat. But a few of the operators had to pack their gear in by mule. They chose whatever high ground they could find and then erected 30-meter towers to clear the trees. At most stations, one operator was from Aero Service, the other Brazilian. A husband-and-wife team occupied one station.

In the summer of 1971 the Caravelle and its crew arrived in Brazil and made their base at Belém, a city on the Atlantic coast near the mouth of the Amazon.

The Caravelle made two flights a day almost every day for a year. Only mechanical or electrical trouble or high-altitude turbulence occasionally broke the routine.

The crew would arrive at the airfield at 6:30 A.M. to check out the aircraft and the weather and to warm up the electronics. Several mechanics and the SLAR engineers from Goodyear would go through their check lists and early-morning mission chores. An electronic engineer would carefully align the inertial navigator. All this took nearly an hour, and then the Caravelle would take off and climb through the clouds to its regular cruising altitude of 12,000 meters—the plane's operational limit, and a height that afforded the greatest range for both the aircraft and the SLAR. All the flights followed a north-south course. Each 32-kilometer-wide swath overlapped the previous swath by at least 20 percent and sometimes, to get stereo effects, by as much as 60 percent.

Later, Aero Service and Brazilian cartographers processed the film and produced mosaic base maps at a scale of 1:250,000. Because the radar images covered vast areas and were illuminated uniformly by the SLAR's multiple pulses, the maps were virtually free of the patch-quilt effects of many aerial mosaics produced by conventional photography. Seen through a stereoscope, the terrain in the radar maps appeared in striking three-dimensional relief—and the terrain turned out to be more rugged than had been assumed. The radar mappers thus exploded the myth of the Amazon as a basically flat rain forest. In fact, lowlands extend over only about 20 percent of the area, mostly along the Amazon

River and its tributaries. The rest is hilly terrain, often giving way to mountain ranges. "People always thought the Amazon was flat because the old-time explorers stuck to the rivers," said Otto Bittencourt Netto, technical director of Project RADAM. "They rarely ventured much further than the river banks because the jungles are so impenetrable."

The radar mapping led to other discoveries. The courses of many tributaries were charted correctly for the first time. A previously unknown stream, several hundred kilometers long, was discovered in the jungle of northwestern Brazil between the Tefé and Coari rivers; it was christened Rio Radam. Mountain ranges indicated on old maps turned out to be far from their supposed locations, and a region marked out as a national forest reserve was found to be savannah.

The results of the survey were incorporated into an eighteen-volume book on the geology, mineralogy, soil and vegetation, and land-use potential for the whole Amazon basin. In addition to the radar images themselves, six interpretive maps were published: on geology, geomorphology, soils, vegetation, agricultural potential, and proposed land-use categories, all at a scale of 1:1,000,000. The Amazon basin, which had defied cartographers for centuries, was finally mapped.

Here, at last, was cartography's opening into many of Earth's remaining inaccessible regions. The Brazilians were so pleased with the results that they decided to have their entire country mapped by radar.

Years later, Homer Jensen came to have mixed feelings about the achievement. "It was a phenomenal success technically," he said. "We completed the maps and they're useful." But events had tempered his initial pride. Much of the rain forest surveyed by radar was being cut, burned, and bulldozed by ranchers and developers. "The impact was not brought about by those maps," he added, "but it was facilitated by them—made easier and quicker."

The acquisition of skills and knowledge animates human history for good and ill, whether it is knapping a stone ax or forging steel, making gunpowder or mixing pesticides, splitting the atom or drawing a new map. It would have been contrary to human nature if the skills of mapmakers and the knowledge conveyed by maps had not been applied time and again by mankind against nature, the strong over the weak, the standard-bearers of one value system over those of another. The exercise of power, inevitably, is a dark side to the history of cartography.

In the fifteenth century, the Florentine Goro Dati embodied the ambivalence of human nature. He eagerly gathered wealth and dutifully gave to the poor. He swore vows of chastity on Fridays and then broke them. He attended lectures by scholars and thought of ways to turn their knowledge to profit. He was familiar with the schematic T-O maps of Christian geographic orthodoxy, but as an ambitious merchant-adventurer, he knew and found more useful the practical sailing charts of the Mediterranean. In a book of verse, Dati wrote:

> With maps that lay out like a frieze
> The winds, the ports, the seven seas,
> Pirates and merchants sail the main,
> Hunting for plunder or for gain.

Maps have long been instruments and representations of dominion. For countless merchant-adventurers like Dati, they revealed new trade routes and sources of natural resources. For missionaries, they opened eyes to lands where there were souls to convert to the true religion. For explorers, they inspired dreams of penetrating the unknown for knowledge, fame, and fortune. For rulers, a better knowledge of their lands through maps strengthened their power to control them. And on maps beyond their lands, rulers beheld the possibilities of empire and colonization. When President Jefferson instructed Lewis and Clark to go and see what lay beyond, the explorers came back with the discouraging news that there was no navigable way to the Pacific, but their knowledge and maps invited further explorations and the eventual displacement of the native populations from their homelands. "Every map," Stephen Hall has written, "presages some kind of exploitation."

Maps, for example, played an important role in the European conquest and colonization of Africa in the nineteenth century. They traced the routes by which explorers pushed inland, marked the locations of mines and trading centers for commerce, guided soldiers in their seizure of land, and then provided documentary evidence in the claiming of conquered territory. In a study of cartography and empire-building in West Africa, Thomas J. Bassett, a University of Illinois geographer, cited examples of European maps as instruments to legitimize the colonizer's dominion over the colonized. One not-so-subtle case in point was the art on the borders of many maps. An 1845 French map portrays a French soldier standing over and showing a seated Arab a map of Africa. "It is a moment of instruction, of imparting knowledge to noble but inferior subjects," Bassett observed. "The superior foreigner is shown as possessing greater knowledge of the continent than do its inhabitants. The map, held by the officer, also symbolizes French power to civilize through conquest. The power of the map is synonymous with the power to delimit, to control, and to develop territory."

Complicity between mapmakers and the powerful is unavoidable, given the usual relationship between them, whereby the former are employed by the latter. In any event, the politically, economically, or militarily powerful are the ones with the means, and most likely the will, to use maps to their own ends. A mapmaker like Jensen, once he had contracted to conduct the Amazon radar survey, had little or no control over how the subsequent maps were used. He could do no more than acknowledge in maps the Manichaean conflict inherent in all technology and knowledge. "The fact that you have complete information," he said, "permits intelligent people to make decisions which will protect, as well as aggressive people to make decisions which will destroy."

Since the Amazon survey, radar mapping has advanced rapidly in sophistication and a multiplicity of uses. Airborne radar systems have penetrated other usually cloud-covered terrains to gather mapmaking data over millions of square kilometers of the world. These data have contributed to the discovery of mineral and energy resources and to the identification of pollution and other environmental hazards. Improved versions of synthetic-aperture radar, collecting data at three different wavelengths, now can discern subtle variations in surface texture, vegetation, and ground moisture. Their microwave pulses can also penetrate as deeply as five meters into dry ground.

At the same time, imaging radar developed for spacecraft by military intelligence services became available to nonmilitary users. From orbital heights, these instruments can cover a broad area of Earth's surface in a single view and return data for mosaic maps of entire countries and regions. Spaceborne radar is capable of producing three-dimensional measurements for generating computer versions of topographic maps, called digital elevation models, that can be used for a large number of scientific, cartographic, and military applications. Such high-resolution maps, embracing nearly 80 percent of Earth's land surface (all except the polar regions), are at hand.

Radar mapping has shown promise in archeology and exploration. In 1977–78, a side-looking radar system, developed for eventual use by spacecraft mapping cloud-shrouded Venus, was tested in an airborne survey of the rain forest of Guatemala. When archeologists examined

This three-dimensional perspective image, looking up the Tigil River, and the western side of the volcanically active Kamchatka Peninsula, Russia, shows how data collected by the Shuttle Radar Topography Mission (SRTM) can be used to enhance other satellite images. Natural shading provided by a Landsat 7 image acquired on January 31, 2000.

the results, they discovered the long-hidden remnants of an elaborate network of canals dug by the Maya. The canals were the first strong evidence that the Mayan civilization had a relatively advanced agriculture.

A similar radar system, used on American space shuttles, has given archeologists even more to think about—revealing views of the canals, reservoirs, and buried remains at the ancient Cambodian city of Angkor, capital of the Khmer empire that ruled much of Southeast Asia from the ninth to the thirteenth century. Long after a forest had overgrown the old capital's ruins, archeologists were able to see in radar images traces of agricultural fields, tracks and paths, and hitherto undetected structures.

In 1981, the first imaging radar system carried by a space shuttle led to a surprise discovery in the Egyptian desert. The radar saw through the sand, down several meters, and revealed a hidden landscape of dry riverbeds and a possible campsite. Digging there, geologists found a spot where two rivers had once converged and Stone Age people had left some of their tools. Later shuttle flights returned similar radar data locating ruins along the fabled Silk Road under the desert sands of western China.

A Los Angeles filmmaker and adventurer, Nicholas Clapp, liked what he heard about radar. His dream was to search the Arabian peninsula for the "lost city" that T. E. Lawrence, Lawrence of Arabia, called "the Atlantis of the sands." From the start Clapp had appreciated the value of maps, old and new. Ptolemy-based maps drew his attention to a promising search area in southern Oman, at the edge of the Empty Quarter. The space shuttle's imaging radar gave him a detailed picture of topography there, as well as the location of a "hot spot" inviting closer inspection. American and French satellite photography (see Chapter 22 for more on this mapping technology) proved even more illuminating. Images made in invisible near-infrared wavelengths revealed the tracks of ancient caravan routes, long buried in the sand. The tracks led to and from one particular site, near the remote waterhole of Shisur.

On an expedition to Shisur in 1991, explorers and archeologists came upon ruins in a limestone sinkhole. Another type of radar, operated on the ground, showed archeologists where to dig for buried stone columns, foundations, and pieces of pottery. Some of the artifacts were 4,000 years old. Guided by ancient maps and sharp-eyed surveys from space, Clapp's expedition had discovered what appeared to be the lost city of Ubar, the fabled entrepôt of the frankincense trade thousands of years ago. If not Ubar, it was some equally important place in ancient times.

Radar mapping's surpassing achievement so far has been to expose the contours of mountains, plateaus, lava flows, and broad plains of another world: Venus.

The demure Venus, Earth's nearest planetary neighbor and remarkably Earthlike in size, mass, and density, had always concealed itself in a thick veil of enveloping clouds. The Pioneer Venus spacecraft, in 1979 and 1980, was the first to part that veil to any significant degree. It returned radar images of more than 90 percent of the Venusian surface. From these scientists prepared a small-scale map; the smallest objects that could be resolved in the images measured 50 to 140 kilometers. In 1985, two Russian spacecraft used radar to map the northern polar region at much finer resolution. Finally, the *pièce de résistance* of Venus mapping was served up by the aptly named Magellan spacecraft, which began circumnavigating the planet in 1990.

Outwardly, Magellan seemed to be nothing more than a four-meter-diameter radio antenna. On each orbit, the antenna was pointed toward the planet, transmitting radar pulses at the surface and catching them on the rebound. The technology, developed by NASA's Jet Propulsion Laboratory in Pasadena, California, had been amply tested looking for those Mayan ruins and on several space shuttle flights. Now, over the course of more than 6,000 orbits, until its demise in October 1994, Magellan managed to map a whole world.

Magellan's radar, at its best, resolved the Venusian surface down to anything 200 meters in width or length, a view nearly ten times better than that of any previous spacecraft. The images, processed and plotted methodically, filled in the map of continent-size highlands, broad plains, large volcano-like mountains, and plateaus with strangely cracked and wrinkled surfaces extending thousands of kilometers. The planet that was supposed to be Earth's twin bore little or no resemblance. Venus, it appeared from the images and maps, had gone through a tremendous upheaval some 500 million years ago, rather recent in geologic time. The cataclysm, of mysterious origin, had drastically altered the planet's surface and replaced it with a fresh one.

Seeing the picture mosaics and resulting maps, one of the scientists, David Harry Grinspoon, felt a "certain psychological and visual threshold was crossed." Mapping a place made it seem real for the first time. "Thanks to Magellan," Grinspoon wrote, "we now have global maps of Venus that are in many respects superior to those we have of Earth, since most of Earth's surface is under the oceans."

Deep Horizons

Some still believe that it is possible to locate minerals and fluids with a divining rod—a forked stick which, if held in the right way by the right person, supposedly twists downward toward the concealed resource. The idea goes back at least to Biblical times, but is viewed with skepticism by nearly all scientists. Instead, the search for knowledge of the Earth's interior and for natural resources therein is now the pursuit of geologists and geophysicists, not dowsers, and subterranean mapping is one of their indispensable tools.

Geologic mapping, like other mapping, is a way of conveying complex information in a graphic form. This information involves the distribution of rocks and many kinds of geologic features—the sequence and thickness of rock formations and other subsurface structures—over the Earth's surface and to considerable depths below it. But, unlike topographers, who gather their information primarily by direct observation and measurement, geologists have to go about their mapping in much less straightforward ways.

Their methods are more subjective and interpretive. Rarely can geologists see their subject whole. Although the walls of the Grand Canyon, for example, are like a geologic cross-section map in full scale, in most places geologists cannot see the interior and must make educated inferences to arrive at such a map. Their inferences are based on the character of the topsoil, the distribution of ridges and valleys, and the occasional exposures of subsurface formations at widely scattered points, in cliffs, ledges, and streambeds, and in road cuts and other excavations. Only after the many bits of information are put together on a map, each in its proper position, does the geologic whole begin to emerge in recognizable form. In other words, while the topographer can usually see the panorama first and then reconstruct it in map form, the geologist never sees the whole until he has mapped the many parts.

New instruments and techniques are now taking subterranean surveying to greater depths, but the fundamentals of geologic mapping remain much the same as those practiced by William Smith, a geologist of the late eighteenth and early nineteenth centuries.

T he first person to study geology by making a map appears to have been the French naturalist Jean Étienne Guettard, who prepared a mineralogical map to accompany a paper arguing that England and France were part of the same geologic region. The map was published in the proceedings of the Royal Academy of France for 1751. In the following years Guettard traveled extensively and managed to gather enough information to complete sixteen sheets of a proposed geologic map of France.

But most historians of geology credit William Smith with the conception and execution of the first geologic map in the modern form. His classic map, published in 1815, was entitled *A Delineation of the Strata of England and Wales, with part of Scotland; exhibiting the Collieries and Mines, the Marshes and Fen Lands originally overflowed by the Sea, and the Varieties of Soil, according to the Variations in the Substrata; illustrated by the most descriptive names.*

The hand-colored map was the product of a quarter of a century of work by an extraordinarily keen and industrious observer. Born in 1769, Smith was the eldest son of an Oxfordshire mechanic, and became a surveyor's assistant at the age of eighteen. While holding the rod and dragging the chain, he observed the different kinds of soil and rock and wondered if there might be some order where there appeared to be only chaos. This was, and is, the instinct of a geologic mapper.

When his work took him down into coal mines, Smith took note of the way that certain rocks containing certain fossils and other distinctive material always lay in a definite order. He remarked that the strata lay one above another like "superimposed slices of bread and butter." He thought there must be some order to the structure of the Earth beneath his feet.

For several years Smith traveled about England laying out routes for canals, logging more than a thousand kilometers in a year and always making copious notes on the geology he encountered, particularly the patterns of strata. He observed the strata of rocks exposed on hillsides and in the cuts for the canals. He came to know rocks so well that by looking at the contours of a hill and the character of the countryside he could accurately predict the subsurface stratification. This earned him the nickname "Strata Smith."

The fossils that attracted Smith's attention led him to an important insight. Smith was the first to realize the significance of the fact that different strata had different fossils. This, he realized, provided a means of establishing the ages of the strata. And, once he had proved to his satisfaction that "the same species of fossils are found in the same strata, even at a wide distance," Smith knew that he had a way of extrapolating from isolated observations to produce maps of the substrata over extensive areas.

Smith's ideas about fossils and strata were not readily accepted. He was a self-taught geologist who worked for a living, and worked hard, at a time when science in Europe was still a gentleman's pursuit. But similar discoveries were being made in France, by Georges Cuvier, and so the doubts about Smith were eventually dispelled.

Between his many surveying and engineering jobs, and despite frequent diversions to collect and study fossils, Smith recorded his remarkable fund of geologic knowledge in the form of a map. Twenty colors were employed to depict different geologic formations. The lowest part of each separate formation was more deeply shaded than its upper reaches, a device which made the map more easily intelligible.

Prices of the Smith maps ranged from an equivalent of $25 to nearly $60, at the time a princely sum and indicative of the considerable value placed on maps of the Earth's subsurface as early as 1815. Unfortunately, Smith had spent heavily on his great map and had made several bad investments in quarries and mines. In 1819 his debts caught up with him, and he was confined for ten weeks in the King's Bench Prison.

Not until 1831 were Smith's services to geology and geologic mapping publicly acknowledged. The Geological Society of London awarded him the first Wollaston Medal. A small pension made his remaining years more comfortable. He died in 1839 from influenza, which he contracted while on one final outing to collect fossils and observe strata.

S mith would probably fit right in on a typical modern geologic mapping trip, at least on one that continues the practice of charting the depths by conventional surficial observations.

The standard geologic maps show the distribution of geologic units at or near the surface of the earth. The map can be of loose materials such as river sediment, glacial debris, beach deposits, or sand dunes— surficial geology. Or it can be an attempt to show the distribution of consolidated igneous, metamorphic, or sedimentary rock units that underlie the topsoil—bedrock geology. How good the map is depends on the ability of the mapper to interpret what he sees at the surface. As Norman L. Hatch, Jr., one of the U.S. Geological Survey's more experienced surveyors, remarked: "Geology is an enormously interpretative thing, not like math, where two and two are definitely four."

In 1965, Hatch and two other geologists undertook a mapping project in the Berkshires of western Massachusetts. Their task was to examine a rugged, sparsely populated 80-square-kilometer region and then prepare a map whose different colors would portray the kinds and ages of rocks that lie beneath the trees, grass, and topsoil. Their methods were those of conventional field geologists.

They discovered that an Amherst College professor had produced the last geologic map of the area in the 1880s. But he had worked with a poor topographic base map, and the then-current understanding of geology was by modern standards inadequate. Hatch came across an 1831 publication whose author, with incredible aplomb, had entitled it *Final Report on the Geology of Massachusetts.*

The geologists also took advantage of available geophysical data on the region. Aeromagnetic surveys established the magnetic intensity of the underlying rocks, a clue to their composition. A gravity survey gave them the density of the rocks. A seismic survey would have been of doubtful value, since the region's subsurface strata are too folded and jumbled, not generally horizontal.

The survey took about seven man-months of fieldwork. In drawing the map, the geologists had to assume that each rock type they saw represented a layer that was once continuous over the whole area. They also assumed that these layers occurred in the same order, from top to bottom, over the whole area. If the final geologic map was consistent with everything they knew about this and surrounding areas, and did not, for example, imply that one layer was older than another when fossils showed the opposite, the map would be inferred to be the best possible representation of the surface and near-surface distribution of rock formations.

At the bottom of the finished map Hatch prepared a cross-section of what the subsurface layering of the Berkshire region would look like if one dug a trench 600 meters deep along a line. The underground projection, Hatch said, was based on what he saw on the surface. In this case, the depth cross-section extended down to sea level; but it could go as deep as the geologists wished to go with the data they had in hand.

Even though the geologic map is interpretive, it has proved to be a useful method of recording and presenting data in compact and systematic form. Many kinds of scientists and engineers use such maps. Because certain types of rock or geologic structures are associated with certain kinds of mineral deposits, geologic maps can help find new places to explore for natural resources. Highway engineers use them to locate sources of construction materials and to predict foundation and excavation conditions. Hydrologists use them to locate underground water. And, because soil is commonly a product of the disintegration of the bedrock underneath, geologic maps are helpful to soil scientists in classifying soils for agriculture.

The increasing pressure to find more minerals and petroleum, along with normal human curiosity, has led to the development of sophisticated new techniques for probing deeper into the Earth's interior and mapping its unseen strata.

A direct method of subsurface exploration is known as well logging, the recording of various physical or chemical properties of the strata penetrated by a well. In a logging operation, measuring probes, or sondes, are lowered on cable into the drill hole. Insulated conductors in the cable pass power to the sonde and transmit signals to the surface. As the sonde is pulled up the hole, the measurements made are recorded, with depths noted, as a well log.

Some of the logging devices take advantage of the fact that different rocks have different electrical properties. Electrical currents are thus passed through the ground adjacent to the well, measuring for differences in the conductivity or resistivity of surrounding rocks; the presence of oil or gas contributes to a higher resistivity reading, since oil and gas are nonconductive. Gamma-ray probes are used to identify features by their emission of radioactive elements; clays and shales, for example, seem to contain more radioactive material than clean sands and limes. Another nuclear-type probe bombards the surrounding ground with neutrons and detects reflected radiations, which can reveal the presence of hydrogen in the form of either water or hydrocarbons. Other types of sondes are employed for determining the densities and angle and direction of strata penetrated by a borehole. Usually by identifying characteristic wiggles, breaks, or other features that persist on logs in different wells over an area, geologists can produce subsurface cross-section maps rich in detail.

But well logging can be slow and expensive. Much of the new emphasis in subsurface mapping involves geophysical methods of seeing what is down there without having to dig or drill. The three principal methods used by geophysicists in subterranean mapping are gravity, magnetic, and seismic surveys.

Gravity.   By measuring slight variations of the Earth's gravitational attraction at given points, geophysicists can make some deductions about the nature and structure of subsurface geology. Variations in gravity arise from the topography of an area and the different densities of materials beneath the surface. The effect of topography—latitude, elevation, and nearby mountains—can be eliminated by compensating corrections. The corrected data are then plotted on a map, and the lines, called *isogals,* are drawn to connect points of equal gravitational attraction. Isogals are thus the contour lines of a gravity map.

With sensitive gravimeters it is possible to relate the changes in gravitational pull to changes in rock types in underlying structures. These provide clues to geophysicists. Thick, dense subsurface structures increase gravity, causing positive gravity anomalies; these may be associated with heavy ore deposits or structures containing petroleum. Lighter rock and salt domes diminish gravity, causing negative gravity anomalies; salt domes often act as traps for petroleum.

The effectiveness of the gravity technique in locating geological anomalies was first satisfactorily demonstrated in Texas in the 1930s. At the time the pendulum and the torsion balance were the favored measuring devices. The torsion balance acts on the principle that a suspended weight may be deflected by a large force of attraction and so may not hang absolutely vertically to the surface.

Geophysicists making gravity surveys now rely primarily on the gravimeter. It is a highly refined version of an ordinary spring scale, to which is attached a small object as a weight. As the meter is moved from place to place, changes in the weight of the object are caused by differences in the gravitational attraction. The gravimeter is designed to magnify the slightest weight change so that it can be detected and recorded.

But gravimeters are so sensitive to accelerations that they cannot be used aboard aircraft and are seldom used aboard ships. Consequently, magnetic and seismic surveys are more widely employed by geophysicists in mapping.

Magnetic. The magnet was used as early as 1640 to discover ore bodies in Sweden, and the dip needle, an adaptation of the compass, was used to locate similar deposits in Wisconsin as late as 1915. But subsurface mapping with the compass proved to be a slow and tedious task, for the instrument had to be firmly mounted and leveled at each measurement point to get an accurate reading. This drawback has been overcome with the development of the magnetometer.

An airborne version of the magnetometer was developed in the 1930s by the Gulf Research and Development Company and was used with moderate success in World War II as a submarine detector. Tests over known ore deposits proved so successful that as soon as the war was over, aerial survey companies added magnetic surveying to their repertory.

A more modern type of magnetometer depends on an application of atomic physics. The sensing elements in the device are spinning hydrogen nuclei, or protons, which precess, or wobble, around lines of force in the Earth's magnetic field. The frequency of precession depends on the strength of the field, which can thus be measured.

Flight crews call the magnetometer "the maggie" or simply "the bird." In its early form, it weighed about thirty kilograms and was encased in a plastic container that looked like a small bomb—so much so, in fact, that authorities in a South American country once held a crew under house arrest for five days while they made sure that the bomb-shaped object cradled under the belly of the aircraft was indeed harmless.

For many years the survey planes had to follow the Earth's contours roller-coaster fashion, always about 150 meters above the ground. This made for a bumpy ride, and even veteran fliers often got airsick. In flight the bird was lowered by cable so that it trailed about 25 meters below, away from any magnetic interference from the plane itself. Once, flying a low-altitude survey over rugged terrain in Peru, a pilot came to a mountain that rose more steeply than he had estimated. The suspended bird clunked into the mountain and was demolished.

Now, however, the maggies have been improved so that aircraft can fly with the device attached to a wing or fuselage and maintain higher and smoother altitudes.

In one of the most ambitious applications of magnetic mapping, Canada's Department of Energy, Mines and Resources surveyed the whole of that country's vast Precambrian Shield, the repository of a wealth of mineral deposits. Some 7 million kilometers were flown by survey aircraft, and more than 7,000 contour maps have been published. Since the magnetometer measurements were recorded in digital form, rather than in the more classical way on a roll of chart paper, the map-making process could be carried out by computer, with automated plotting machines accurately drawing the magnetic contours.

The contour lines of a magnetic map are called isogams, a gamma being the standard unit of magnetic force. An isogam connects points of equal magnetic value. The contoured values may be either positive (magnetic "highs") or negative (magnetic "lows").

A strong magnetic high is a good indication of ore deposits, particularly iron. Only indirectly do magnetic surveys suggest subterranean structure. This occurs when geophysicists are able to use magnetic data to determine the location and disposition of the basement rocks, because they are igneous and contain iron, and hence, by inference, can determine the lay of the sedimentary rocks that might contain petroleum.

Seismic.   When John Milne, an English geologist, went to Japan in 1875, an earthquake shook the land soon after his arrival and changed his career. In an effort to trace the sources of the frequent Japanese earthquakes, Milne borrowed the ideas of other scientists and produced the first seismograph, which is to the earth scientist what the telescope is to the astronomer—a tool for peering into inaccessible regions.

Milne used seismographs to make the first world-wide map of the zones of earthquake origin. Other scientists, using Milne's instruments and techniques, identified the sources and measured the velocities of seismic waves and, in so doing, developed an understanding of the Earth's deepest interior.

In 1909, for example, a Yugoslav physicist, Andrija Mohorovičić, noticed sharp changes in the velocity of seismic waves traveling deep through the Earth. This was how he discovered the discontinuity that forms the boundary between the Earth's crust and its mantle. Four years later, scientists were able to measure the radius of the Earth's core as being about 3,200 kilometers.

By then seismology had in a sense mapped the gross features of the Earth's interior: the crust, which is a thin surface skin (from about 6½ kilometers thick under the oceans to about 50 kilometers under high mountains); the mantle, which extends from the crust about halfway toward the center; and the dense, partly molten core.

Economically, it was the crust that counted. If only seismic waves could be interpreted to detect more subtle layering in the Earth's crust, this might point the way to hidden resources. But the occurrence of earthquakes was too unpredictable to be of any real help in detailed subsurface mapping.

During World War I, a German scientist, Ludwig Mintrop, conceived the idea of creating artificial earthquakes to produce seismic waves. He induced shock waves in the ground as a possible means of locating enemy artillery emplacements. Though of little consequence in the war, the system was patented and applied in 1924 to search for possible oil-bearing rock formations in Texas. It was a success. From the differences in the return time of refracted seismic waves it was possible to locate shallow salt domes, the low velocities of the denser surrounding rocks contrasting sharply with the high velocities from the salt. This early seismic mapping method is known as refraction seismology, utilizing horizontally traveling seismic waves. It required large amounts of energy to generate the shock waves and resulted in an ability to map one or two levels of subsurface strata.

Today's routine method of seismic mapping utilizes the seismic reflection method. Sound waves are generated at the surface by small explosions, pneumatic devices, or vibrators. The sound penetrates deep into the crust. Velocity and density differences at the boundaries between different types of rock reflect part of the wave. Each reflecting boundary is called a horizon. The returning echoes are detected and timed by a sensitive listening device on the surface, a type of seismograph.

A miniaturized version of the seismograph, a rugged but sensitive device called a geophone, has been developed for easy handling in seismic surveys. The geophone consists of a magnetic circuit fastened firmly to a cylindrical case less than a centimeter in diameter and a coil suspended within the magnetic field. The case is coupled to the ground with a spike. As the surface shakes from the seismic waves, the magnetic structure moves with it but the coil stays still. The effect is that of a magnetic field moving past a conductor. This generates a current that is amplified and recorded as tracings on strips of paper or on magnetic tape.

The round-trip time of the signal is clocked to give an indication of the depth of the reflecting layer. From long experience geophysicists have learned the travel times of signals through topsoil, various kinds of rock, and deep sediments. The strength of the return signal is analyzed through computer processing to learn something of the nature of the reflecting layer; porous rock in which natural gas is trapped, for example, reflects a much stronger seismic echo than does rock filled with water. The result, after the survey has been repeated along a line, is a map showing a cross-section of the Earth's crust to depths of several thousand meters. The sequence and shape of the various layers stand out in profile, like a sliced layer cake.

These profiles are seismic cross-sections used to construct seismic maps, which are now a routine part of the exploration for oil and gas.

In the earlier days of seismic mapping, "seis" parties went around drilling holes to plant dynamite charges, which were detonated to create shock waves. Everywhere they drilled they made mounds of dirt that reminded people of the many little dirt piles dug up by a bug, the doodlebug. So seismic surveys became known, in the colorful language of exploration geophysics, as doodlebugging.

But today's doodlebuggers usually leave not a trace, with their faster, nonexplosive methods. An operation by Conoco, Inc., in Oklahoma is typical of seismic surveying.

Men known as jug hustlers, or jerbs, unwound long black cables over a straight line of more than 5 kilometers. Attached to the cables were clusters of geophones, often called jugs, about 6 clusters of 56 jugs each for every kilometer. Having so many geophones helped distinguish the real signals from random noise. Then five heavy trucks, the thumpers or vibrators, lined up nose to tail like a train of circus elephants. Each truck was equipped with a mechanism, called a Vibroseis, that presses a steel plate to the ground, vibrates it, and sends tremors of seismic waves deep below the surface.

Ahead of the caravan was the recording van, in which two men took turns directing the operation and monitoring the seismic signals. When the head jerb reported by radio that all the jugs were planted, and when each thumper driver called in ready, the observer in the recording van sent a radio signal that simultaneously activated each of the Vibroseis plates. The rear wheels of the trucks rose in the air as the steel vibrating plates shook the ground. The vibrations started fast, 40 cycles a second, and then slowed down, to 7 cycles. The sound waves were reflected off the substrata, in ways suggestive of different rock densities and porosities, and the reflected signals were picked up by the geophones and fed by cable to amplifiers inside the recording van. A computer inside the van converted the data from electronic impulses into numbers and summed up all the different sound reflections into traces on a seismic graph. After each sweep of vibrations, which lasted less than a minute, the trucks moved forward about one meter, waited for another radio signal, and repeated the vibrations. And so it went all day.

Surveyors had been there earlier with plane tables and levels to map elevations and lay out the grid for the seismic survey. During reconnaissance seismic mapping, the lines are many kilometers long and spaced several kilometers apart. For more detailed work, the lines are usually less than two kilometers apart and are crossed with still more lines so that small grids are developed.

At the end of a survey the raw data, recorded on magnetic tape, are turned over to computer experts for special processing. Computers sort out the millions of seismic signals, measure the spurious noise, and come up with charts of how the Earth's rock layers would look if the crust could be sliced open. From these maps geophysicists can make shrewd guesses as to where oil or gas might have accumulated.

In particular, geophysicists examine the seismic maps for evidence of faults, salt domes, or anticlines, which are upward bulges of an underground rock layer. These structures can be identified and located by a careful analysis of the arrival times of reflected seismic signals.

Ignored in the processing for many years were variations in the strength of the reflected signals. This began to trouble Carl H. Savit, a scientist at Western Geophysical Company, which does seismic exploration under contract to the oil companies.

In 1960 Savit published a scientific paper suggesting that, in removing variations in signal strengths by filters and volume controls, geophysicists were discarding valuable information relating to the nature of the reflecting layers, data that might help distinguish between water, trapped natural gas, or oil.

By the late 1960s Savit had persuaded his company to introduce recording systems that would preserve all amplitude data and to install more computers. Other companies did likewise. They kept their techniques and results a secret, in the hope of maintaining an edge over competitors. In 1974 their secret methods became known to the public as the "bright spot" technique, after the stand-out pattern that appears on seismic charts where a rock layer is more likely to contain gas or oil.

While the unusual strength of a returning reflection is highly indicative of gas or oil, bright spot relies on other, confirming indicators. One is the ability to monitor the polarity of a reflection; the direction of movement (the polarity) of a gas-reservoir reflection is opposite to that from a water-filled sandstone. The other is the ability to make such fine interpretations of reflection signals as to detect a perfectly horizontal layer; since very few geologic surfaces are perfectly horizontal, such a surface must be an interface between two substances: either gas over water, gas over oil, or oil over water.

With more discriminating data to work with, geophysicists have begun to analyze seismic maps from a new and perhaps more rewarding angle, called stratigraphic analysis. The emphasis is on the composition of the rocks and sediments.

Charles Payton, manager of exploration research for Conoco, explained this new direction, saying: "In places like Oklahoma and Louisiana and Texas, if you are going to find any more oil and gas, you've got to look for either deeper structures or some different kind of trap than an anticline or a fault. That's where the business of stratigraphic traps comes in. There are many flat planes or layers beneath the surface. Almost every layer changes rock type from one place to another. Some of these may present natural barriers within layers. So we're trying to identify rocks in the layers and look for those barriers that could contain oil or gas."

Many oil companies are experimenting with shear-wave seismology. Sound waves directed vertically down in the ground, compressional or primary (P) waves, are the standard energy used in seismic mapping. Shear (S) waves, directed more horizontally through the ground, cause rocks to rub against each other sideways to the direction of the energy source. One of their most interesting properties is that they do not travel in fluids, which means that shear waves can be used to distinguish trapped fluids, possibly oil, from changes in rock densities.

In few areas of cartography are the economic rewards for innovation so swift and abundant. Bright spot led to important discoveries in the Gulf of Mexico, offshore in Nigeria and Indonesia, and in a few places in the continental United States. Stratigraphic analysis was responsible for new petroleum discoveries near Mobile, Alabama.

As petroleum exploration has moved more and more off-shore, so has seismic mapping. The principles are no different than on land, only the equipment—ships instead of trucks; propane and compressed air guns instead of vibrating steel plates; hydrophones instead of geophones; and navigation systems tied to signals from orbiting spacecraft.

The Baltimore Canyon, an offshore basin extending along the coast of Maryland to Long Island, was the focus of extensive geophysical surveying during the 1970s. Aeromagnetic surveys detected undulations in the canyon's bedrock suggesting the presence of vast and promising traps in which crude oil and natural gas might have collected. Then the seismic ships embarked to make more detailed surveys.

In the summer of 1975 the Shell Oil Company's *Phaedra,* a 55-meter ship with a crew of 30 and millions of dollars worth of electronic gear, was typical of the dozens of seismic ships plying the Atlantic waters and "sharpshooting" the Baltimore Canyon. She covered about 80 kilometers a day, following a precision-navigated criss-cross of courses over the continental shelf off New Jersey.

Behind the *Phaedra* trailed two arrays of compressed-air guns. Once every 50 meters, on a signal from the ship's computer, the guns sent harmless, nonexplosive shock waves pulsing down through the water to the ocean floor. As the sound impulses reach the boundaries between various rock layers, they are reflected back to the surface and detected by hydrophones. An array of 2,900 hydrophones was strung out on a three-kilometer-long plastic-covered cable, which also trailed the ship. The hydrophones measure in milliseconds the time it takes for the air guns' sharp sound pulses to bounce upward.

In the *Phaedra*'s instrument room, the data from each shot were printed out as a pen and paper profile of the interior of the sea bottom. The sharpest echo might be from granite basement rock 600 meters beneath the sand and other rock strata. The echoes were simultaneously translated to digital form and stored on computer tape for further analysis back on land.

Working with measurements of the time taken by the shock waves to travel from the air guns to the reflecting layer and back to the hydrophones, computer processors sort and sift and study the data from every angle and produce maps of the strata several thousands of meters below the ocean floor. One of the maps produced from *Phaedra*'s mission revealed subsurface features 175 kilometers off the New Jersey coast that might hold as much crude oil as the rich East Texas field.

The technology of seismic reflection profiling developed by the petroleum industry is now being used to probe even deeper. In the early 1970s, it became apparent to a number of university scientists that seismic mapping techniques might, with slight modification, be applied to the geologic exploration of the hard rock "basement" of the continent. Whereas petroleum prospectors were satisfied mapping the depths of 10 to 15 kilometers, the scientists wanted to look at levels at least 40 to 50 kilometers down, and eventually all the way to the bottom of the lithosphere itself, which is presumed to be in most places about 100 kilometers deep. This includes not only the crust but a portion of the upper mantle of Earth.

Accordingly, in 1974, scientists organized the Consortium for Continental Reflection Profiling (COCORP) to conduct experiments in mapping the continental basement. The National Science Foundation provided the financial support.

The first experiment took place in March 1975 in Hardeman County, Texas, midway between Amarillo and Wichita Falls. "We selected the site primarily because of its good reflection properties," said Jack Oliver, a Cornell University geologist who was chairman of the consortium. "During exploration for oil in that vicinity, echoes were occasionally recorded for extended periods. These indicated that there were strong reflecting horizons at considerable depths. We designed our own profile experiment to record echoes for 15 seconds, compared with the industry practice of 6 seconds or less, so we would have a theoretical depth capability of 45 kilometers."

The Hardeman results were encouraging, dispelling any initial doubts as to the capability of seismic profiling techniques in hard rock (as opposed to the sediment basins probed in petroleum exploration) and at greater depths. The Vibroseis system of continuous seismic profiling was employed by the consortium scientists. They obtained data along three lines, for a total of 37 kilometers. The data disclosed the presence of a number of homogeneous, irregular-shaped bodies in the rock basement at depths of 12 to 33 kilometers. The scientists suggested that these were solidified igneous "plutons" that had intruded in molten form into the already formed rocks that make up most of the crystalline basement in that area. Strong reflections showed up at depths of 33, 37, and 43 kilometers, the deepest of which may have been a layer associated with the Mohorovičić discontinuity, the boundary between the crust and upper mantle.

The following December, after the Hardeman success, the consortium and its contractor for this phase of the work, the Petty-Ray Geophysical Company, turned their attention to a site of greater geologic interest: the Rio Grande Rift in central New Mexico. This is a long, wide trough that extends from Mexico to Colorado, its principal surface feature being the river by the same name. Like other rift systems, it is a region of instability—micro-earthquakes, numerous thermal springs, and a gradual swelling of the surface.

The Rio Grande experiment produced data from some 25 million individual reflection points. Deep reflections were recorded and mapped at depths of 36, 44, and 51 kilometers. At the extreme western edge of the profile, at depths below 21 kilometers, the scientists detected what may be a lava, or magma, chamber that could be responsible for the area's hot springs and uplifting terrain. The mapping of magma chambers is important because they are a potential source of geothermal energy, could evolve into full-fledged volcanoes with surface eruptions, and may be a major factor in concentrating minerals and other resources in zones immediately above them.

By probing the continental basement, mapmakers could draw with greater certainty the texture of the bottom of Earth's crust—the flip side, as it were, of a surface relief map of continents. Subterranean maps were becoming important interpretive tools of geology.

"We think the big plus for our project is that it yields the highest resolution of any geophysical technique," Oliver explained. "There are other ways of getting information about the deep crust, like measurements of magnetic and gravitational anomalies, and studying the behavior of seismic waves from earthquakes and underground nuclear tests. But none of these other techniques yields information that *looks* like the geological things we see at the Earth's surface.

"Our game is to get detailed information on the deep rocks that has the resolution and structural detail of the shallow basement rocks and outcrops we know about. What seismic reflection profiling really does is extend the eyeball of the geologist from the surface to the subsurface, by providing the highest possible resolution of the fine structure of the deep crust."

This is only one example of an incredible new dimension of modern mapping. Cartographers are now able, with increasing confidence and precision, to map the unseen and thereby claim new worlds for human contemplation.

A Continent
Beneath the Ice

On the evening of February 24, 1898, John Murray, one of the scientific lions of the day, rose to speak in the gaslit hall of the Royal Society in London. A naturalist on the global expedition of the *Challenger* in 1872–75, Murray exerted strong leadership during the voyage and was personally responsible for the writing and publication of the fifty-volume report of the expedition's scientific and geographic discoveries. He was a fellow of the Royal Society, a man of rough geniality and dominating manner, and when he spoke he commanded attention.

Murray's speech to the Royal Society dealt with Antarctica, a subject that increasingly fascinated the minds of the 1890s. Only three years earlier, the Sixth International Geographical Congress had met in London and concluded that an investigation of the Antarctic region was "the greatest piece of geographical exploration still to be undertaken."

The delegates to the Congress recognized that no one had more than a glimmering of what lay down there at the bottom of the Earth, and the maps were hardly edifying. All they showed was a frosting of white spreading out in all directions from the South Pole, then some vague tracings of coastline. But the coastline of what? An ice sheet covering an ocean, as at the North Pole? Only some islands in the polar ocean? Or the coastline of the Earth's seventh continent?

About all the delegates could know, by way of answering the questions, was that James Cook, penetrating the Antarctic Circle for the first time, had disproved the classical and Renaissance notions of a fertile supercontinent in the south. They were also aware that Cook had encountered a sea of mighty icebergs, which, it was assumed, could only come from a polar land of some size. But when some land was finally sighted, nearly a half century later, in 1820, no one knew what to make of the discovery. In fact, no one could be sure to whom the credit for discovery should go.

Many American scholars believe that Nathaniel B. Palmer, a New England sealing captain, was the first to sight the continent, in November 1820. He saw, thrusting north toward South America, a part of the panhandle, which became known as Palmer Peninsula. But not to the British, who named it Graham Land, for the First Lord of the Admiralty (James R. G. Graham), when it was discovered, according to British accounts, by Edward Bransfield and William Smith in January 1820. The evidence for this claim is uncertain. Some scholars believe that the two Britons actually saw an island, not any part of the mainland. Russian historians contend that Fabian G. von Bellingshausen, an explorer for the czar, deserves the credit for discovery. It is unlikely that the matter will ever be settled to the satisfaction of all concerned. In 1964 an international convention agreed to give the panhandle a neutral name, the Antarctic Peninsula, with the northern half called Graham Land and the southern half, Palmer Land.

In any case, between 1820 and the 1890s, little more was learned about the Antarctic. Whaling and sealing ships and a few scientific expeditions skirted the towering cliffs of ice, probed inlets in search of an opening to the pole, observed a smoking volcano, and suspected that they were in the presence of a continent. But they could not be sure, and they were unable to establish a foothold on what appeared to be the mainland.

When Murray began to speak about Antarctica, therefore, his distinguished audience was eager to learn what this man who had sailed into Antarctic waters had to say about that fog-shrouded, mystifying region. He did not disappoint them. For Murray laid out an impressive body of evidence as to the true nature of the Antarctic.

First, he told of the many flat-topped icebergs he had seen when the *Challenger* scouted the region in 1874. "Their form and structure," he said, "seem clearly to indicate that they were formed over an extended land surface."

Second, Murray pointed out, the *Challenger* had dredged up from the sea floor around Antarctica many telltale rock fragments. They were gneisses, granites, sandstones, limestones, and shales. "There can be no doubt of their having been transported from land situated near the south pole," Murray declared.

This was not unlike a report to lunar scientists in Houston in the 1970s. From educated observations and a few rock samples, derived from an expedition of heroic proportions, Murray had reached a hypothesis about the unearthly world of the Antarctic.

He told the Royal Society: "We are thus in possession of abundant indications that there is a wide extent of continental land within the ice-bound regions of the southern hemisphere."

Most scientists of the day were favorably disposed to Murray's "hypothetical continent," and the explorers leaped into action. During the next decade Belgium, Britain, France, Germany, and Sweden sent expeditions to Antarctica. In December 1911 Roald Amundsen, the Norwegian who had broken through the Northwest Passage for the first time, succeeded in reaching the South Pole.

Surveying and mapping were integral parts of the early explorations, first along the coasts, then in the penetrations inland to the pole. This was a tremendous undertaking. The climate was harsh, the terrain rugged. The area to be covered was comparable in size to the United States and Europe combined, and 98 percent of it was covered with ice. Moreover, the ice was often one, two, or three kilometers deep and extended out over the sea in the form of ice shelves, obscuring most of the true coastline.

Could this vast ice cap ever be mapped? This was the first question that Antarctica posed for cartography.

Although most expeditions have engaged in some effort to determine the area of the Antarctic continental ice sheet and then fill in the map with mountain ranges, nunataks (mountain peaks sticking through the ice), high plateaus, and ice shelves, the quality of the results has varied. Aerial photography, begun over Antarctica in 1928 by Hubert Wilkins, played an ever-increasing role in exploration and mapping, but for several decades the accuracy was not very high because of the lack of ground control. Mountain peaks charted by one expedition could not be found when the sites were visited by another. Traverses on the surface produced limited results because of the lack of modern surveying equipment, communications and timekeeping problems, and the atmospheric conditions peculiar to the Antarctic—such as the bewildering mirages that played tricks on surveyors' eyes. Not until years after

World War II was it possible to begin a more systematic topographic mapping effort, aided by the introduction of new mapping tools.

But could the land beneath the ice ever be mapped at all? This posed an even tougher challenge for cartography, one that could not have been met without the benefit of several other important innovations in mapping technology.

Within a decade or so after Murray's speech, geologists began to suspect that reality was hardly as simple as hypothesis. One part of Antarctica seemed to bear little relation to the other. From rock samples found on the scene geologists recognized two distinct geologic provinces. East Antarctica (in the eastern hemisphere, facing mainly the Atlantic and Indian oceans) rests on a foundation of older igneous and metamorphic rocks, a typical continental platform. West Antarctica (in the western hemisphere, facing the Pacific Ocean) is smaller and composed of generally younger rocks with considerable evidence of recent volcanic activity.

When expeditions returned with surveys of the two great embayments, marked by the low-lying ice of the Ross and Filchner ice shelves, pinching Antarctica into its bell-shaped figure, this suggested that it might even be two continents, not one. Scientists theorized that East and West Antarctica might be divided by an under-ice trough linking the Ross and Weddell seas. There was, however, no direct evidence. The pre–World War II surveyors had no means of measuring the thickness of the ice cover and mapping the surface of the hidden land.

In 1935 Richard E. Byrd, the American polar explorer, lamented: "It's a curious but readily accredited fact that long after most astronomers had settled to their satisfaction that there were no canals on Mars no geographer nor geologist could have told you whether Antarctica only 10,000 miles away was one continent or two."

Thus it became necessary in mapping Antarctica, particularly beginning in the mid-twentieth century, to proceed on two levels—the surface of the ice and the surface of the land beneath the ice.

In 1946 the United States Navy's Operation Highjump, involving 13 ships and 25 planes, initiated the most extensive aerial reconnaissance of Antarctica up to that time. Some 70,000 photographs were taken during 64 mapping flights, covering an area of some 550,000 square kilometers. This included about 60 percent of the continent's coastline, nearly half of which was found to have been incorrectly charted by earlier expeditions.

But the Navy, in its report on the expedition, conceded that the photography was mainly useful as "reconnaissance" rather than for mapping. Two essential ingredients had been missing—precision navigation by the planes and sufficient ground control for the areas photographed.

The plane crews had no aids to navigation except brief glimpses of the Sun and compasses, which behaved erratically so near the pole. Navigators were usually unable to detect aircraft drift over the featureless terrain and in the uncertain winds. In sorting out the photographs afterward, technicians found, for example, that two aircraft that were supposed to have flown widely separated courses actually followed almost identical routes. Two other flights, which were to have flown the same track, produced photographs of completely different areas.

An attempt to correct one of the deficiencies, the lack of ground control, was made in the austral summer of 1947–48 during the Navy's Operation Windmill. Icebreakers pushed their way to islands off Wilkes Land and dropped off survey parties, which took theodolite measurements and laid out base lines for triangulation. Other triangulation parties were shuttled from the icebreakers to inaccessible islands or to the mainland by helicopter. This was the first systematic use of photogrammetric techniques in Antarctic mapping, though the results were hardly sufficient to the need.

By 1961 photographic coverage of Antarctica totaled more than 1,000,000 square kilometers; but this still consisted mostly of exploratory photography, and ground control remained meager. Too often ground-control surveys had been only a secondary function of multifaceted traverses. A surveyor who accompanied the party as navigator obtained positions and elevations for prominent topographic features whenever he had time. Theodolites were used to intersect peaks from base lines laid out carefully by steel tape and to measure solar altitudes for computing astronomic positions. Elevations were determined by altimeter readings made at the instrument station and then transferred to the peaks by means of measured vertical angles and computed distances. All of this was good as far as it went. But seldom were the control points sufficiently accurate or numerous for the purposes of reliable aerial photogrammetry.

A turning point in the topographic mapping of Antarctica came with the Topo North-South project of the austral summer of 1961–62. In a single season "on the ice," three topographic engineers of the United States Geological Survey produced mapping control for about 160,000 square kilometers of the continent, extending from Cape Adare along

the rugged western edge of Ross Ice Shelf inland to the head of Beard-
more Glacier, 510 kilometers from the South Pole. Their 2,400-kilometer
traverse ran along the backbone of the Transantarctic Mountains.

Several innovations contributed to their success. It was the first
expedition in which the establishment of mapping control was the only
objective, not a catch-as-catch-can sideline. Helicopters were used to
transport the surveyors to mountain peaks, which represented an
invaluable gain in mobility. Newly developed electronic distance-
measuring devices were deployed by the helicopters to measure distances between traverse stations with greater speed and accuracy.

Over a period of two months, the three surveyors—William H.
Chapman, William C. Elder, and Ezekiel R. Soza—occupied 63 traverse
stations, an average of 30 kilometers apart. Soza would go by helicopter
to the forward peak while Chapman and Elder occupied the rear peak.
At each forward peak Soza erected a wooden target for the other surveyors to sight on with their theodolite. On most of the peaks, the ones
bare of snow and ice, Soza also implanted brass geodetic tablets for
future identification.

Soza's other task was to operate the remote unit of the tellurometer.
This was one of surveying's newest and most promising types of instruments. Instead of direct measurement of distance with a steel or invar
tape, an extremely slow procedure, it was now possible to exploit the
modern techniques of electromagnetic wave transmission and electronics in making base-line measurements. The tellurometer can operate in
daylight or darkness; it is a two-way microwave system in which a mod-
ulated radio signal travels from a master transmitter to a remote unit,
where it is retransmitted back. The distance between the two points is
derived from the measured transit time of the radio signal and the
known velocity of radio waves. At that time tellurometer measurements
were practical for distances of 15 to about 40 kilometers.

As soon as Soza radioed that he was ready with the remote unit,
Elder would switch on the master tellurometer and transmit the modu-
lated radio signal. Distances were thus determined to an accuracy of
about 5 centimeters in 35 kilometers. Meanwhile, Chapman handled
the theodolite. He took several angle measurements off Soza's wooden
target and, in addition, swung around to measure the angles of clearly
visible peaks on both sides, bringing many inaccessible points into the
triangulation network and thus greatly increasing the control for future
mapmakers.

In succeeding years other parties conducted the same kind of control surveys over more and more of Antarctica, concentrating on the mountainous regions within range of resupply aircraft from McMurdo Station, the primary United States base. They provided enough control to produce 1:250,000-scale topographic maps covering about 820,000 square kilometers.

This method of electronic traverses did not at first provide satisfactory results over flat snow surfaces, partly because of optical refraction close to the ground and accompanying difficulties with communications. Portable survey towers, from which the instruments could be operated high above the ground, might have been the solution anywhere but in Antarctica; the transportation of such towers proved too difficult. The Geological Survey, therefore, decided to put the helicopter to yet another use. Using some recent experience in Alaska, the Survey had developed its Air-Borne Control (ABC) system, in which a tellurometer-type instrument is elevated in a helicopter that hovers over a marked point on the ground. The helicopter-borne instrument can then be observed by two ground stations equipped with compatible electronic equipment and theodolites. The helicopter thus marks one point in the surveyors' triangle.

Accordingly, during the 1960s, through the perseverence of dozens of Survey parties, Antarctica served as an important proving ground for two new means of advancing the mobility and skill of precision surveying—the helicopter and electronic distance measuring.*

D uring the same period, while the ground parties fixed the locations of identifiable landmarks, a concerted effort was made to get as much mapping-quality aerial photography as possible. The photography would provide the content for the maps, as the control provided the mathematical framework.

The task of aerial photography fell mainly to the Navy's Air Development Squadron Six (VX-6, or the "Very Exceptional Six") and a photogrammetrist from the Geological Survey, William R. MacDonald. No aerial mappers of more ordinary terrain ever had to contend with so many difficulties: vagaries of an unforgiving climate, disorienting mirages, tenuous logistics.

The season for flying and mapping in Antarctica, as for nearly all other activities, runs from October until late January. This is the austral summer, when the weather moderates somewhat, the sun shines

---

*See Chapter 22 for discussion of some space-age innovations applied to Antarctic mapping.

twenty-four hours a day, and planes fly into McMurdo with supplies and the continent's seasonal population of scientists, technicians, and support crews.

A typical mapping mission went something like the following:

The alarm clock rang at 5 A.M. in MacDonald's hut at McMurdo Station. His first waking thoughts were of the weather, and so, dressed in his parka, he trudged over to the weather hut. Any storms in sight? What about cloud cover? He needed to know if the outlook was favorable in any of his preplanned mapping areas. And he had to be relatively certain that if he took off from McMurdo he would have a runway somewhere where he could make a landing.

MacDonald then conferred with the squadron skipper, and they decided to fly. The runway, Williams Field, was eight kilometers from "the Hill," the base camp, over a stretch of solid sea ice. MacDonald and the flight crew rode snow tractors to the field, the drivers ever watchful for cracks or potholes in the ice. Tractors have been known to break the ice and plunge out of sight and beyond rescue.

At the field, the cameras were loaded with film, the camera windows cleaned, and the C-130 (with skis for landing gear) checked out for takeoff. Using whatever maps were available and the reconnaissance photography from Operation Highjump, MacDonald had prepared a plan for the mapping segment of the flight, complete with known landmarks as checkpoints. Each flight had a primary target and, in case of a change in the weather, several secondary targets.

For the mapping runs the plane maintained an altitude of 8,000 meters above sea level. It flew parallel courses 25 kilometers apart, usually running 150 to 350 kilometers long in a north-south direction; the lay of the continent's terrain generally radiates from the pole. After about six hours of flying the mapping tracks, back and forth, photographing continuously, the plane turned and headed back to Byrd Station, an interior outpost. If there was no message to return to McMurdo, the C-130 would be refueled and, with a new flight crew, take off again for a new mapping area.

The aerial photography of the 1960s by the United States and other nations made possible the compilation of 1:250,000-scale maps with 200-meter contours of all the mountainous and coastal areas of West Antarctica. This was done by the Geological Survey with the support of the National Science Foundation. But much of the more featureless interior, particularly the plateaus and plains of East Antarctica, remains to be mapped in accurate detail.

Before the International Geophysical Year of 1957–58 little more had been learned about the landmass beneath the ice than was known at the time of Murray's hypothetical continent and the subsequent debate over whether it was one continent or two. A Soviet glaciologist had gone so far as to suggest that Antarctica might not be a continent at all, but rather a string of islands with only a massive overlay of ice in common. It became, therefore, a major objective of the IGY to explore and map the sub-ice contours of Antarctica.

Toward this end ground traverse parties from several nations set out in snow-going caterpillar tractors (Sno-cats), ranging some 10,000 kilometers over the Filchner and Ross ice shelves, Marie Byrd Land, the Ellsworth Highlands, and the Victoria Land plateau. Each party was made up of five or six men equipped with two or three Sno-cats. The lead vehicle, in a typical convoy, was outfitted with a crevasse detector and navigational gear. Aboard the second vehicle were the seismic and gravity instruments, and in the third were the mess facilities and the party's radio. Cargo sleds in tow behind each Sno-cat carried the rest of the party's supplies. On an ordinary traveling day these parties would travel about 50 kilometers, but on alternate days they would stay in one place to make seismic probes.

Their sub-ice probing techniques were essentially the same as those used in geophysical mapping elsewhere in the world.

The simple spring gravimeter gave them gravity readings in three or four minutes. Since gravity is a manifestation of mass, and rock is three times as dense as ice, local variations in gravity indicated differences in ice thickness. This proved adequate for charting the gross features of buried mountains and valleys. But no two of these sensitive instruments are precisely the same, and broad variations in gravity can reduce the reliability of such readings over an extensive survey. It was the usual practice of the traverse parties to carry out gravity measurements at regular intervals—on an average, at six places—during a 50-kilometer trek. These measurements would serve to trace the configuration of the sub-ice rock surface between two seismic measurement points.

The seismic probes were usually much more reliable indicators of the sub-ice terrain. But it could take virtually a whole day to make a single probe. In this technique, borrowed from the experience of oil-field geophysicists, an explosive charge is set off in a hole a few tens of meters deep. Seismic waves traveling outward from the explosion are reflected from any surface where the density or elasticity increases abruptly—as where the ice ends and land beneath the ice begins. Small seismometers placed at various distances from the explosion record the time required for the reflected waves to reach them. As in echo sounding at sea, the depth of the ice can thus be computed from the round-trip time of the signal.

The results produced several surprises. The Antarctic glacier was thicker than had been estimated. Some gravity and seismic readings in Marie Byrd Land showed ice depths of nearly 5 kilometers. Indeed, most of the sub-ice landmass there lies well below sea level, sometimes by more than a kilometer, and land, instead of rising as the parties advanced into the interior, becomes deeper and deeper. Near the coast of Marie Byrd Land the bedrock was found to be less than 300 meters below sea level, but in the heart of the region, at Byrd Station, it was a kilometer below sea level. This seemed to lend some credibility to the Russian idea that Antarctica was merely an archipelago bound by an oceanic ice sheet. If the ice were to disappear, the range of mountain peaks along the coast of Marie Byrd Land would stand alone as an archipelago comparable in extent to the Philippines.

The sub-ice surveys disproved the notion that the ice always follows the land contours. In the vicinity of 88° west longitude an American traverse party, searching for the supposed trough linking Weddell and Ross seas, was standing on a broad, level ice surface. But only 300 meters below, the seismic readings showed, lay a buried mountain range, presumably the sub-ice southward extension of the Sentinel Mountains.

In summarizing the results of IGY ice soundings, four American scientists—Charles R. Bentley, Albert P. Crary, Ned A. Ostenso, and E. C. Thiel—concluded that there could be no broad channel below sea level to connect Weddell and Ross seas. This, along with the soundings in Marie Byrd Land, seemed to end fifty years of speculation about Antarctica's being two continents.

It began to seem more likely that Antarctica was one vast continent—but by no means shaped like the ice cap shown on the maps. Without the ice cap much of West Antarctica would be submerged, except for the ring of coastal mountains, possibly some "inland" islands, and the Antarctic Peninsula. The peninsula may turn out to be a long mountainous island, separated from the main continent by a narrow channel much as the British Isles are separated from Europe. Or the two may be joined.

The shoreline for the bulk of the continent would probably be somewhere along the Transantarctic Mountains, the Queen Maud Range, and the Horlick Mountains—in some places only 500 kilometers from the South Pole and 650 kilometers south of what now appears to be the north coast of Marie Byrd Land. In other words, East Antarctica turns out to be the true continent.

Beyond the chain of mountains bordering East Antarctica, the traverse parties discovered, lies a rocky plain at or above sea level. The Commonwealth Trans-Antarctic Expedition, led by Vivian Fuchs, crossed much of East Antarctica in eight snow tractors, stopping to do seismic shots every 50 kilometers and making gravity measurements at those points and midway between them. They found the ice depth at the South Pole to be about 3,000 meters, but on either side of the pole the British detected submerged mountains rising to within 500 to 750 meters of the ice surface. Nearly all the soundings along the traverse showed the buried land to be above sea level.

Murray's hypothetical continent had been discovered. East Antarctica, the IGY seismologists all demonstrated, was a continent larger than Australia.

As revealing as the seismic soundings had been, their limitations were all too apparent to those who wanted a broader survey of Antarctica's sub-ice terrain. All seismic measurements in Antarctica were plagued by noise; the continent's raging winds set up vibrations in the sensitive seismometers that often drowned out the faint echoes of seismic signals bouncing off the bedrock. But more serious was the slowness of making a seismic survey, the drilling of deep holes for the charges and the setting out of seismometers requiring sometimes an entire day for a single depth measurement.

A simpler, faster method of sounding the ice sheet emerged in the 1960s—airborne radio echo sounding. Whereas in 1958 it required two days of hard labor by a seismic crew to produce seven spot soundings, in 1967 an aircraft with a radio echo system took three minutes to make a continuous depth profile over the same ground.

The system was developed by the Scott Polar Research Institute at Cambridge University in England under the direction of S. Evans and Gordon de Q. Robin. Early tests in Greenland and Canada indicated that radio echo sounding, a type of radar, could do for sub-ice mapping what sonar had done for sea-floor mapping and side-looking radar had done for jungle mapping.

Between 1967 and 1975 the British scientists, working with United States Navy aircraft and National Science Foundation support, conducted 345,000 kilometers of radio echo profiling. The aircraft was outfitted with a special radio transmitter and receiver, photographic-film apparatus for recording the profiles, and an antenna mounted beneath the starboard wing. The transmitter sent out a fan-shaped beam that, in part, penetrated the ice and echoed off the sub-ice terrain. In most early surveys, Robin reported, the depths from radio echo soundings agreed with depths from accepted seismic soundings to within approximately 100 meters.

The earliest flights had only dead-reckoning navigation, however, and errors in position of 30 to 35 kilometers were not uncommon. The problem was money, not technology. For the 1971–72 season funds became available so that the Navy was able to install a Litton Inertial Navigation System, which allows a continuous readout of aircraft position from a self-contained digital computer, thus enabling the pilot to follow a much more precise flight track. In addition, the navigation system records the aircraft's latitude, longitude, and ground speed every 20 seconds. When these records were compared with the radio echo profiles, it was possible to plot the soundings on a topographic map. Navigational errors were reduced to no greater than two or three kilometers.

During radio-echo-sounding operations, the height of the ice surface above sea level is obtained from the difference between the pressure altitude of the aircraft and the terrain clearance between the aircraft and the ice sheet. The former is determined by barometric instruments; the latter, either by a conventional radio altimeter or from the time delay between transmitted pulse and ice-surface returns on the radio echo records themselves. With this knowledge, plus the profile of echoes from the bedrock, Robin reported, it was possible to reduce the possible depth error over the whole network to about 30 meters.

Thus, with airborne radio echo sounding providing a rapid reconnaissance over wide areas, the bedrock relief of more and more of Antarctica began to take shape in the minds of scientists and to be translated to maps.

In a 1975 article in *Polar Record,* David J. Drewry, a scientist at the Scott Polar Research Institute, reviewed a number of the discoveries attributed to radio-sounding surveys. Among the discoveries were features suggesting quite a varied topography beneath the ice of part of East Antarctica—subglacial basins, a buried mountain range of alpine proportions, and one deep basin that "may form part of an elaborate continental rift structure." The full extent of one buried mountain range and its structural connection with the Transantarctic Mountains,

Drewry said, "could have been realized only from continuous radio echo soundings."

Drewry's article accompanied a subglacial map that had been compiled from the radio-sounding surveys. A comparison between the British map and one produced by the Soviet Union in 1966, a subglacial map in *Atlas antarktiki,* points up the advantages of radio sounding. The Soviet map, an excellent piece of work, covers the entire continent at a scale of 1:20,000,000 with a sub-ice contour interval of 500 meters. But it is based on only 934 seismic points and 6,655 gravimetric points—one data point per 1,850 square kilometers. The British map of 1975 covers only about one-fourth of the continent, East Antarctica, at a scale of 1:5,000,000 with a sub-ice contour interval of 250 meters. But, more important, it is based on more than 50,000 radio echo soundings—one data point per 65 square kilometers.

Other radio and seismic probes of Antarctica's glacial depths added an intriguing phenomenon to the continent's map. These surveys located more than seventy lakes buried under thousands of meters of ice. The largest of these, 45 kilometers long and 10 kilometers wide, is a body of apparently fresh water more than three kilometers beneath the surface, next to the Russian Vostok research base in East Antarctica. Scientists suspect that Lake Vostok's pristine waters contain living "fossil" microbes, bacteria, and viruses essentially unchanged over as much as one million years.

As more of the Antarctic ice is thus probed, more of the seventh continent's hidden topography should become part of the mapped world.

Mountains of the Sea

The greatest mountain range on Earth is not the Himalayas or the Rockies, the Andes or the Alps. The widest and deepest chasm is not the Grand Canyon of the Colorado. The broadest plain is not the steppes of Russia or the Great Plains of North America. It may come as a surprise to landlocked minds, but the greatest mountains, chasms, and plains all lie beneath the oceans, a grandeur unseen and until recent times unmapped and unmappable.

Although it represents nearly three-fourths of Earth's surface, almost nothing was known of the land beneath the sea until a century ago. To earlier people it must have seemed beyond comprehension, the last real *terra incognita,* and even beyond imagination. They could glimpse enough of the Moon and the planets to inspire dreams of discovery; but not the ocean floor. In their flights of fancy, they could strike out with reckless abandon for the planets; but, with rare exceptions, never to the ocean depths. Since near-shore soundings invariably turned up nothing more interesting than mud, pebbles, and unexceptional slopes, they assumed that everywhere the sea floor was a flat, barren expanse of dark ooze. Who could get very excited about a place like that?

Few, indeed. Not until a century ago did ships attempt to probe and measure the mid-ocean depths. Scientists on these ships had no idea what they would learn, for no one could be sure if the ocean depths ran to the hundreds of meters or the tens of kilometers—or if they were measurable at all. The methods of these scientists were crude, and so were the results, which seemed to leave everyone unimpressed and incurious. As recently as 1942, a comprehensive textbook, *The Oceans,* reflected a common attitude: "From the oceanographic point of view the chief interest of the topography of the sea floor is that it forms the lower and lateral boundaries of the water."

The discovery of the real sea floor began after World War II. With the new developments of technology and the perseverance of inspired oceanographers, it became possible to chart the submerged landscape in all its unsuspected magnificence.

Two of the people most responsible for the new map of the ocean floor were Bruce C. Heezen and Marie Tharp, both of Columbia University's Lamont-Doherty Earth Observatory. Their maps projected a new world into human minds, revolutionizing geology and our understanding of the entire planet we live on.

After years of discovery by many scientists and painstaking effort by Heezen and Tharp, a series of maps was issued by the National Geographic Society and the Geological Society of America, beginning in the late 1950s. The maps show the Earth more or less as it would look if someone were to pull the plug and drain all the oceans. No wash of cartographic blue conceals the rugged undersea world. Instead, the maps show in considerable detail and striking relief the Mid-Ocean Ridge in all its magnificence. Next to the continents and the ocean basin itself, the ridge is the Earth's largest topographic feature, whose scope was hardly suspected until Heezen and Tharp began their mapmaking. Its mountains are sometimes as much as 1,500 kilometers wide, with some of the peaks rising higher than Everest would be if it were part of the sea-floor landscape. From Iceland, where it surfaces, the ridge extends more than 65,000 kilometers down the middle of the Atlantic and around Africa, across the Indian Ocean, and up through the Pacific. Every few hundred kilometers the ridge is severed and shifted somewhat by a deep fault, producing an effect much like a long French loaf that has been sliced and then moved so that the slices no longer line up together. What looks like a deep canyon, or rift, runs along the crest of the mountains, the discovery of which proved important in explaining how the oceans and continents got where they are today.

To either side of the ridge are broad abyssal plains, where the terrain is generally so flat that it varies only a few meters every 100 kilometers. The plains in the Atlantic are beyond question the flattest places on Earth. In the Pacific, the relief is a little more varied, with more deep trenches and a scattering of flat-topped seamounts, most of which are probably drowned volcanic islands. Then there are the continental slopes, where the deep ocean basins end, and it is a steep rise to the continental shelves; the escarpments along this boundary can average 4,000 meters high. Finally, there are the continental shelves, those submerged extensions of the adjacent land. They are under relatively shallow water and are grooved with deep canyons running perpendicular to the continents.

Like other pioneering maps, the one by Heezen and Tharp is not complete and not always completely accurate. It is, nonetheless, one of the most remarkable achievements in modern cartography. It is the graphic summary of more than a century of oceanographic effort.

A t first, in probing the sea floor, there was little to go on but guesswork, educated and otherwise.

When, in 1521, Ferdinand Magellan lowered a sounding line to 750 meters without reaching the bottom, he is said to have proclaimed the Pacific Ocean immeasurably deep. Indeed, some sailors, in their fears and fantasies, were sure the ocean was bottomless. Natural philosophers in the mid-eighteenth century offered reasoned speculations wherein the average ocean depths ranged from 20 to 35 kilometers. Others tended to agree with the early Greek philosophers. On purely logical grounds, they contended that the seas must be as deep as the mountains on land are high. It was a variation on that ancient and often misleading theme of nature's symmetry, but in this case it turned out to be a fairly good guess.

As far as anyone knows, the first general estimate of ocean depths by scientific methods was made in 1856 by Alexander Dallas Bache, a great-grandson of Benjamin Franklin and for more than twenty years superintendent of the United States Coast and Geodetic Survey. It was an indirect but ingenious measure, in the best tradition of Eratosthenes.

From a Russian naval officer's account of the destruction of his ship in the Japanese earthquake of 1854, Bache established the time of the earthquake's first shock wave. By means of tidal-gauge records he determined how long it took the earthquake's first seismic sea waves, or tsunamis, to cross the Pacific and reach the California coast—an interval of some twelve hours. Bache then made his calculations, using a recently discovered physical law that related the velocity of very long waves such as tsunamis to the depth of the ocean. In this way, Bache concluded that the average depth of the Pacific along the wave path must be about 3,600 meters. It was a reasonably accurate estimate. The average depths of the Pacific, Atlantic, and Indian oceans are, respectively, 4,280, 3,930, and 3,960 meters.

This left still unrevealed the nature of the submerged landscape. How could anyone ever possibly map that which had not been seen or reached? The question posed one of the greatest challenges for cartography.

First, there was the problem of sheer scope. The deeper regions under the sea—those covered by 2,000 meters or more of water—total 312 million square kilometers, an area eight times as large as the surface of the Moon. Who could begin to survey such a domain? And who could conceivably want to? To early navigators the ocean beyond the shallow coastal waters was simply "off sounding." They cared not how deep the ocean so long as it was deep enough.

Yet the scope of the unmapped terrain was not the most formidable obstacle. There was also the problem of pressure, a problem surveyors had never before faced. In the deepest parts of the ocean, the pressure of all the water overhead is so crushing that surveyors could not think of going there and, in the established practice, scanning the terrain from some undersea hill. At one time it was thought that the pressure acted as an impenetrable barrier even to probes by sounding lines. In 1844, the anonymous author of *The Ocean, a Description of the Wonders and Important Products of the Sea* expressed such a concern:

"Heavy bodies, which will sink rapidly from the surface, do at length apparently cease to descend long before they have reached the bottom; the pressure of the water being such as to cause them to remain at certain depths, varying in proportion to their weights. Thus it is that the plumb line will not act beyond a certain length, and we have no means, of course, of extending our inquiries deeper."

Even if surveyors could descend into such depths, using the modern pressure-resistant submersible vessels, they would run into a third problem—the absence of light. In water, light can penetrate only about 200 meters from the surface, precluding the use of cartography's important new tool, aerial photography. Cameras have been dropped to the bottom of the ocean and with artificial light have taken photographs. But their range is too limited for the purposes of general mapping.

Oceanographers were, therefore, left with no alternative but to feel their way across the ocean floors, relying on one of the oldest and simplest techniques of depth measurement, called "heaving the lead." They would lower a hemp line—or, in a later refinement, steel piano wire—

with a lead weight attached to the end. When the weight was felt to hit the bottom, the depth could be roughly determined by the length of the line that had been paid out.

The first truly deep-sea sounding by line and lead was achieved by James Clark Ross in 1840. While on an expedition to the Antarctic to locate the magnetic south pole, Ross stopped his ship, the *Erebus,* a number of times in the South Atlantic to take deep soundings.

"We have made several fruitless attempts to obtain soundings as we passed through the tropics," Ross wrote in December of 1839. "These repeated failures were principally occasioned by the want of a proper kind of line, but they served to point out to us that which was most suitable. I accordingly directed one to be made on board, 3,600 fathoms, or rather more than four miles in length, fitted with swivels to prevent it unlaying in its descent, and strong enough to support a weight of seventy-six pounds."

In less than a month, on January 3, 1840, the new sounding line was ready. It was wound on a huge drum, which was fitted on one of the ship's ocean rowing boats. While the line was lowered from one boat, another boat, tied to the first and manned by strong oarsmen, held position against winds and currents. Ross timed each 100-fathom (180-meter) section of line as it went over the gunwale, and as soon as a significant change in the time of unreeling was noted, it was assumed that the bottom had been reached. By this improved but still primitive technique, Ross determined the depth of the South Atlantic, at a point about halfway between the lower halves of Africa and South America, to be 2,425 fathoms, or 4,365 meters—"a depression of the bed of the ocean beneath the surface," Ross observed, "very little short of the elevation of Mount Blanc above it." The whole operation took four hours, clearly too long for soundings to become a routine part of oceanographic expeditions.

Improvements were not long in coming. Because the most onerous and time-consuming chore was the hauling in of the heavy wet line, one early innovation was the use of a disposable thin twine. A cannonball was attached to the end of the twine. When the ball reached the bottom, the twine was cut and the depth calculated from the length of the string left on board.

But the best guarantee that the bottom had indeed been reached was to retrieve a sample of sediment. This was made possible with a device developed in 1854 by John Mercer Brooke, then a midshipman in the U.S. Navy. For a weight Brooke also used a cannonball, but one with a hole through the middle. Into the hole was inserted a hollow brass rod about twice as long as the diameter of the cannonball, and this was attached to the thin line. When the cannonball hit the bottom, the impact drove the rod into the sediment and then released it from the cannonball, so that the rod filled with the sample could be pulled back to the ship. The Brooke sounding device, or variations thereof, became standard equipment on survey ships for several decades, providing chartmakers with more accurate depth measurements and scientists with coveted samples of ocean sediments. Even later, steam engines were employed to drive the winches that hauled in the sounding lines.

During this time, in the middle of the nineteenth century, Matthew Fontaine Maury initiated the first systematic program of sounding the oceans. A Navy lieutenant of irrepressible drive, Maury almost single-handedly led the United States into oceanographic exploration, all the while operating from the rather obscure post of superintendent of the U.S. Navy Depot of Charts and Instruments, to which he had been relegated because a stagecoach accident rendered him unfit for sea duty. Maury made himself useful compiling wind and current charts from the logs of mariners and preparing maps of whale migrations. Eventually, as the commercial value of such charts became apparent, he persuaded Congress to authorize research ships for his work.

From the cruises of the *Taney* and the *Dolphin,* his two research ships, and soundings made by other Navy vessels, Maury gathered enough data to produce, in 1854, the first contour map of the Atlantic sea floor. The map, encompassing an area of 52° N to 10° S, was called the *Bathymetrical Map of the North Atlantic Basin with Contour Lines Drawn in at 1,000, 2,000, and 4,000 Fathoms.*

It was a rather presumptuous publication, but then Maury never let modesty stand in the way of promoting himself and ocean research. Fewer than 200 soundings deeper than 1,000 fathoms served as raw material for the entire chart. Maury himself suspected the accuracy of many of the measurements, and he therefore placed question marks after some depth figures. Nonetheless, Maury's chart did provide substantial evidence overturning traditional notions of an essentially smooth ocean floor. Down the middle of much of the North Atlantic basin ran a broad elevated feature that Maury labeled Middle Ground. It was also known as Dolphin Rise, after one of Maury's ships. Vague as it was, this was the first recorded suggestion of the Mid-Ocean Ridge.

Exploration of the ocean floor spread as efforts to lay a transatlantic cable gathered momentum, culminating in success in 1866, and as scientists began mounting round-the-world oceanographic expeditions. The first of these, by the British, was the famed voyage of the *Challenger* in 1872–76.

During a journey of more than 125,000 kilometers, which set the standard for all oceanography to follow, the *Challenger* made 364 stops to gather marine life and sedimentary samples and, in many cases, to take sea-floor soundings. In the Pacific, the *Challenger* made one sounding at a depth of about 10,000 meters, a record at the time.

In the last two decades of the nineteenth century, after ships of other nations followed the *Challenger*'s lead, a number of sketchy topographic maps of the Atlantic, Pacific, and Indian oceans were produced from the accumulated sounding data. Major seafaring nations felt, however, that the time had come for a comprehensive chart of the ocean basins using a common scale and terminology. The coordinating task fell to the least of the seafaring states, the Principality of Monaco.

Prince Albert I of Monaco, an avid oceanographer himself, put his "scientific office" to work plotting 18,400 soundings and, in 1904, completed the first *General Bathymetric Chart of the Oceans* in twenty-four separate sheets. The work of updating the chart with new soundings every few years is carried on to this day by the International Hydrographic Bureau, which is based in Monaco. Member nations bear responsibility for furnishing data on the sea floors with which they are most familiar.

Despite the proliferation of sea-floor charts at the turn of the century, no one could be entirely satisfied with any of them. The problem of sheer scope had not been overcome. Ocean regions as large as Mexico were still untouched by a single line-and-lead probe. There were many gaps of hundreds of kilometers between measurements, for the methods of sounding remained all too slow and unreliable. Current and drifting ships were always drawing out the sounding lines into great S-shaped curves, producing significant errors on the deep side. And even with steam-driven Brooke sounding devices, a single reading still meant stopping a ship for several hours.

It is no wonder that, more than seventy-five years after Ross, deep-sea sounding remained a sufficiently uncommon practice as to arouse the curiosity, and profanity, of passing sea captains. In an account of work off Alaska during the early twentieth century, the commanding officer of a Coast and Geodetic Survey ship, a Lieut. R. R. Lukens, wrote:

"Occasionally, while sounding at night, a passing steamer is attracted by our strange behavior and comes close to inquire the trouble and to ask if help is needed. When the well-meaning captain is told that we are only sounding, he lets loose a few nautical oaths and slams his telegraph full speed ahead, wondering why anyone would be so crazy as to sound in the middle of the ocean."

The term *sounding* in the depth-finding sense is derived from the verb "to sound," meaning "to probe." In its original usage, sounding had nothing to do with acoustics, though early oceanographers did cast a wistful eye at the possibilities. Maury wrote of "exploding petards and ringing bells to attempt to hear an echo when the wind was hushed and all was still, but no answer was heard."

The technological breakthrough ocean cartographers were waiting for came, as such breakthroughs often do, in response to disaster and war—the sinking of the *Titanic* in 1912 and the introduction of the submarine in World War I. The breakthrough was the acoustic depth finder, an early example of remote sensing, known variously as the echo sounder, fathometer, and, in its more modern form, sonar, which is the acronym for Sound Navigation and Ranging.

After the *Titanic* struck an iceberg and sank, inventors in the United States and Europe began experimenting with ways to detect large objects under water by means of echoes. Reginald A. Fessenden, a former assistant to Thomas Alva Edison, was one of the first and most successful. While working for the Submarine Signal Company in Boston, Fessenden developed an oscillator, an electric device capable of producing a penetrating underwater sonic wave. A pulse of electricity set off the vibration of a crystal that produced a sharp "ping." The signal's echo, returning from some underwater object, was detected by a listening device known as a hydrophone. In a test in 1914, Fessenden obtained an echo return by bouncing the sound off an iceberg. In France, Paul Langevin and Constantin Chilowsky further perfected echo sounding for submarine hunting.

After the war the new technology was applied to surveys of the sea floor. With echo sounding, depths are determined by measuring the time required for a sound wave, or rather a train of sound waves, to travel from a survey ship to the ocean bottom and back to the ship. The ocean bottom reflects sound just as a wall or mountain face reflects sound in air, producing an echo. In fact, echoes are much less difficult to obtain from the ocean bottom than from objects in the air. The reasons are several: First, there is always a reflecting surface, the sea bottom. Second, sound in water is not subject to as much absorption or to such large variations of attenuation as sound in air. Third, because of less absorption, sound in water will travel many times as far as sound of the same initial intensity will in air. And finally, the properties of sea water make for reasonably constant velocities of sound—about 1,440 meters per second.

As in radar, sonar distances are calculated by clocking a signal's round-trip travel time, dividing by two, and then multiplying by the velocity of sound in sea water. By this method, the measurement of great depths, which once took hours, can be obtained in seconds.

In 1922, an acoustic depth finder incorporating the Fessenden oscillator was operated by the *Stewart* during an eight-day passage from Newport, Rhode Island, to Gibraltar. The system's first extensive deep-sea trial disproved the pessimistic predictions that echo sounding would be limited to the relatively shallow waters of 40 to 50 fathoms. For the *Stewart,* without stopping once, was able to record a track of 900 measurements, one as deep as 3,200 fathoms, across the breadth of the Atlantic. A rough profile of the Atlantic basin was then drawn. But, since there were still gaps of several kilometers between measurements, it was not possible to round out the true shape of the sea floor's mountains and valleys. The profile thus looked more like the zigzag graph of a sharply fluctuating stock market.

Other ships began using the echo sounder for surveys across the Atlantic, along Panama Canal routes, off California, and across the North Pacific. Fathometers became standard equipment on commercial liners, which searched out the already known submarine ridges and valleys to help them pinpoint their positions at sea. The most thorough test, however, was run by a German research ship, the *Meteor.* On the German Atlantic Expedition of 1925–27, the *Meteor*'s two echo sounders took about 33,000 duplicate soundings. The two devices usually agreed to within 15 meters in deep water. The measurements revealed for the first time the general ruggedness of the mountains rising from the floor of the Atlantic, but unfortunately the data were destroyed in the bombing of Berlin during World War II.

Greater precision seemed beyond the means of the early equipment, primarily because of the many opportunities for human error. With the first echo sounders, the operator pressed a telegraph key to initiate the sounding signal and simultaneously set a stopwatch to ticking. He listened with a set of earphones to detect the return signal. It took considerable skill and concentration to distinguish the attenuated echo and immediately mark the time interval, usually no more than six seconds, by stopping the watch. He then recorded the echo time and the ship's position in a log for computation later. As a common practice, this was repeated only about once every 16 kilometers.

The most experienced observers could make mistakes that threw off calculations by tens of fathoms, and sailors who made many of the measurements as an extra duty while standing watch were not always the most conscientious observers. H. William Menard, then of the Scripps Institution of Oceanography at La Jolla, California, tells of a sailor's confession in a bar in San Diego. The sailor, according to Menard, said that his "normal practice was to ignore the sounder and just record the same depth as the sailor who had been on the previous four hours."

It took another war, World War II, for the U.S. Navy to bring into being the continuously recording deep-sea echo sounder. Since the Navy was not particularly interested in the deep-sea floor, only in submarines and navigation hazards, the first machines had a range of less than adequate for most deep-ocean mapping. But after the war one innovation followed another, improving the resolution and precision of echo sounders, and ocean cartographers finally had the instruments they needed.

Scientists at Woods Hole and Lamont-Doherty developed machines that sent, received, and recorded signals automatically and continuously. They could record differences in bottom relief with an accuracy of better than 2 meters at a depth of 4,000 meters. The sounding machines produced an ultrasonic "ping" every second, received the echo, measured the intervening time, converted the time interval into a depth reading, and registered this in an electronic computer's memory bank. At the same time, a stylus on a rotating arm swept across a moving strip of paper, sketching a continuous profile of the bottom relief along the ship's track. A fleet of research ships could record as many as 10 million soundings in a month.

The addition of graphic recorders to the echo sounder revealed more features of the ocean floor—thousands of volcanic peaks, for example, some with their tops mysteriously flattened at depths far below present sea level (guyots, discovered by Harry H. Hess of Princeton University, while a Navy officer in World War II); and long, straight, parallel troughs, ridges, and escarpments (fracture zones) discovered by Menard and Robert S. Dietz, a government scientist, in the eastern Pacific.

A test comparing the modern echo sounding methods with the old line-and-lead techniques was conducted in 1968. A research ship of the Coast and Geodetic Survey, the *Discoverer*, paused during an expedition and resurveyed the site of the Ross soundings in 1840. Nowhere within 8 kilometers of the site could the *Discoverer*, using the most sophisticated acoustic sounders, find a depth as great as Ross's sounding of 2,425 fathoms. Ross had been in error by about 325 fathoms, nearly 600 meters, because he failed to recognize the initial moment of bottom contact or because he made a mistake in his ship's position.

"The older spot soundings simply do not give the same kind of information as modern, recorded echo soundings," wrote Menard, an oceanographer with considerable experience in mapping the undersea world of the Pacific, in his book, *Anatomy of an Expedition*. "Older methods yielded a generalized map. New ones produce a continuous picture of the sea floor under a moving ship. The difference in what can be learned is about the same as the difference between looking at the moon with the naked eye and taking pictures of it with a rocket which flies just above it."

Unlike many new technologies of remote sensing, which remove men from intimate contact with what they are surveying, the continuous echo sounders offer a new sense of immediacy and drama, a new sense of reality, to marine cartographers. There is no waiting now to get back on shore and plot a few lead-line measurements and ponder the meaning of a rise—is it a plateau, the peak of a hill, or the slope of a greater peak?—or the meaning of a depression—is it the bottom of a valley or the slope leading to a vast chasm? Instead, oceanographers often look at the echo sounder's unfolding profile of the submerged landscape and, forgetting themselves, believe they are actually seeing the hills and valleys below.

"As a marine geologist surveys a new range of undersea mountains," Menard wrote, "he senses them around him. I was once surveying with a captain new to the game. He remarked, as we headed again toward a peak on the map we were making, that he could not suppress a captain's feeling that we would hit it, even though he knew perfectly well it was a mile below the ship."

The "hypnotic fascination" of watching the stylus arm of the echo sounder making its tracings and revealing new lands for mapping was described by David B. Ericson and Goesta Wollin, both of the Lamont-Doherty Observatory, in their book, *The Ever-Changing Sea*.

> At each sweep [they wrote], the profile of the bottom advances by a fraction of a millimeter. For miles the bottom had been level and smooth, but now each new mark left by the stylus is higher. As you watch the rising

curve you wonder whether the ship is approaching a new seamount. To qualify as a seamount it will have to rise to at least 3,000 feet [1,000 meters] above the surrounding plain. Now the trace is rising more steeply; perhaps it will be a really big one. But no, before reaching more than 870 meters, the trace turns down. However, you are not to be misled by that; the trace of echoes is strictly two-dimensional; it tells nothing about what may lie on either side. The trace before you may represent the flank of a much higher mount. So you fetch the officer whose watch it is, and between the two of you, you decide to turn the ship back and recross the region one mile to the east of the first profile. Now the trace of echoes keeps on rising until it has reached a height of 1,400 meters. As a real seamount it deserves more detailed exploration. By this time it is three o'clock in the morning; although the chief scientist was at work until midnight he must be called. After some plotting of positions and calculations a series of crossings of the region are laid out in such a way as to determine the general form of this newly discovered feature of the earth. By sunrise the survey is done.

Shortly after World War II, equipped with the new echo sounders, the ships embarked—Navy ships on routine patrol, ships carrying supplies to overseas bases and to Antarctic explorers, survey ships, and oceanographic research ships. Each ship, plotting depths as she went, contributed to the mapping of the land beneath the sea.

A primary objective of the postwar exploration was the mountains of the Atlantic, whose existence had been known in a vague way for years. Charles Wyville Thomson, chief scientist of the famous *Challenger* expedition in the 1870s, had discovered the mountains through a brilliant deduction. In comparing water temperatures taken at great depths in the eastern and western Atlantic, Thomson noted an almost unvarying difference of more than one degree between the two, the eastern Atlantic being slightly warmer. He concluded that there must be a barrier of some sort dividing the eastern and western waters. Soundings taken by the *Challenger* in 1873 bore out the prediction. Although more romantic laymen back in London greeted the news as confirmation of the legend of the lost continent of Atlantis, Thomson simply called it "a long sinuous ridge," which came to be known as the Mid-Atlantic Ridge.

During research cruises in 1947 and 1948, Maurice Ewing, with Heezen as his assistant, crossed and recrossed the Mid-Atlantic Ridge. They took soundings from the latitude of Newfoundland down almost as far as the bulge of South America. They exploded depth charges of TNT to penetrate the sediment with sound waves and reflect echoes off the bedrock. They found sediment layers much thinner than had been expected, one of the first important clues to the relatively young age of the ocean basins and a discovery that would take on even greater significance in a few years. Ewing's team also discovered deep ditches that seemed to run parallel to some of the steep slopes of the ridge. Ewing correctly observed that the trenches "may mark the location of the great faults" responsible for the many submarine earthquakes that were known to shake the region.

Could it be that earthquake zones were somehow related to ocean ridges? British scientists, in mapping the Arabian Sea and part of the Indian Ocean in the 1930s, had identified a "deep gully" between parallel crests of the Carlsberg Ridge, which extends southeast into the Indian Ocean from the mouth of the Gulf of Aden. They suggested that the gully, more than 300 meters deep, might be an extension of the Great Rift Valley of eastern Africa, a known earthquake fault. In 1949, American seismologists prepared a revealing map showing all earthquake epicenters recorded between the years 1922 and 1944. Beyond doubt, a zone of earthquake activity coincided not only with the Great Rift Valley but also with the Mid-Atlantic Ridge and the Carlsberg Ridge.

A pattern seemed to be evolving, but it was not yet clearly recognizable—not, that is, until Marie Tharp in 1953 began plotting the sounding data for a physiographic diagram of the Atlantic basin. "Physiographic diagram" is the term she and Heezen borrowed from land geology to describe their undersea maps. Such diagrams, sometimes called land-form maps, resemble a landscape as it might be sketched in perspective by an observer from a great height, but rendered more maplike with the addition of coordinates and scale.

While she was tracing the contour lines for a preliminary sketch, Tharp kept noticing a cleft in the echo-sounding profiles made across the crest of the ridge. The more closely she examined the data, the more she realized the dimensions of the cleft were far from insignificant. She found herself drawing a deep valley down the center of the entire mountain range, and the valley averaged 2,000 meters in depth and ranged in width from 15 to 50 kilometers. It was as deep as the Grand Canyon, and considerably wider.

Tharp showed the sketch to Heezen. "He was skeptical, and he stayed skeptical for a few years," Tharp recalled. "He wasn't sure we had enough to establish a broad pattern."

"You can't imagine how primitive our knowledge was then," explained Heezen.

Tharp had to agree. "We were working with only six complete transatlantic profiles at that time," she said. "But with new profiles coming in we always found this crack in the middle of the Mid-Atlantic Ridge. We called it a gully, a ditch, a crack. But it was always there."

This moved Heezen to make some independent investigations. He had a team of draftsmen plot all the known breaks in transatlantic cables. They all coincided with Tharp's ditch. He also ordered a new plotting of the earthquake epicenters in the Atlantic. They too coincided with the cable breaks and with the valley bisecting the length of the Mid-Atlantic Ridge.

Seismology had undergone major improvements as a result of advances in detecting tremors from far-off nuclear-bomb tests, and so seismologists could map with even greater precision the Earth's earthquake belts. New and more sensitive seismometers established in South Africa soon enabled seismologists to link the mid-Atlantic earthquake line with another line dividing the Indian Ocean. They also plotted a well-defined extension of the submarine earthquake belt running to the south of Australia, through the Pacific, up off the west coast of South America, and into the Gulf of California, where it cuts through western North America. There were other earthquake regions, but the tremors that occurred along this submarine belt seemed to have a kinship: they were all of relatively shallow origin, less than 35 kilometers below the surface. Moreover, in the Atlantic, seismologists determined with even greater precision the epicenter belt. It ran not just roughly along the Mid-Atlantic Ridge, but as a narrow, coherent line less than one-tenth the width of the ridge—and it ran right along the rift valley that split the length of the crest of the ridge.

Ewing and Heezen studied those maps and pondered the inevitable question: If a range of rugged mountains coincided with the mid-Atlantic and at least part of the mid-Indian earthquake lines, was it not possible that such mountains might be found almost everywhere there was a belt of shallow earthquake epicenters? It was a staggering idea. If true, it would mean the discovery of the Earth's largest geographical feature since Columbus came upon the New World.

In 1956, Ewing and Heezen felt sure enough of themselves to announce to scientific colleagues their daring prediction. There must be,

Section of the floor of the North Atlantic,
by Bruce C. Heezen and Marie Tharp, 1957

they said, a range of undersea mountains like the Mid-Atlantic Ridge extending through all the oceans. They called it the Mid-Ocean Ridge.

Evidence to support their prediction soon poured in. During and after the International Geophysical Year in 1957–58, research ships of many nations went out to look for sections of the ridge wherever there was a pattern of earthquakes. Invariably, they found the mountains they were looking for, a range 65,000 kilometers in length running through the Arctic, Atlantic, Indian, Antarctic, and Pacific oceans. And, as the data came in, Heezen and Tharp under the supervision of Ewing began the monumental task of preparing the map that would do justice to such a discovery.

In an explanatory note accompanying the physiographic diagram of the North Atlantic, Heezen, Tharp, and Ewing wrote: "There is a fundamental difference between the preparation of a terrestrial and a marine physiographic diagram. In the former the major problem is to select from more-detailed maps the features to be represented. Except in unexplored, inaccessible areas, the shape of all land features is a matter of recorded fact; the problem is to abstract and artfully draw the features in question. In contrast, the preparation of a marine physiographic diagram requires the author to postulate the patterns and trends of the relief on the basis of cross sections and then to portray this interpretation in a diagram."

In short, even with the wealth of new sounding data, cartographers of the land beneath the sea still had to guess a lot. The cross sections, or slices, were the depth profiles derived from a series of closely spaced echo sounding readings. These showed the depth at each peak, trough, or change of slope, and each was matched with longitude and latitude readings from the survey ship's position reports. But each profile represented the cross section of terrain only every few kilometers apart, at best, or possibly only every few hundred kilometers.

"You go from a plain and suddenly jump to the crest of the ridge," Tharp explained. "This called for a lot of interpolation. We [Heezen and Tharp] had lots of fights and arguments over this, but eventually we settled them. I did the drawing, but that takes less time than anything else. It takes longer to figure out what you are going to draw, weeks to interpret the data and ten minutes to draw it."

It took 70 working sheets, each encompassing an area 10° by 10°, to complete the North Atlantic sea floor. In the beginning, the sketch from each working sheet was transferred by hand, with pen and ink, to the master sheet. Now, to save time and money, this is done by photographing each working sheet, fitting each picture into a mosaic, and then rephotographing the whole. Heezen and Tharp thus produced eight basic maps—the North Atlantic, South Atlantic, Indian Ocean, and five of the Pacific Ocean. These maps became a primary reference work in a revolution in science.

O nce, while pondering an ordinary map of the Earth, long before the mountains of the sea were charted, Alfred Wegener became "impressed by the congruency of both sides of the Atlantic coasts." Africa and South America, in particular, looked as if they could fit together like giant pieces of a jigsaw puzzle. Was this more than mere coincidence?

Wegener, a German meteorologist, explorer, balloonist, and university lecturer of the early twentieth century, could not take his mind off the way Brazil bulged out just where West Africa, some 6,000 kilometers distant, was indented. Making some measurements on a globe, he found that South America could fit snugly with the curving indentation of West Africa along the Cameroon coast. Likewise, North America could be made to join the coastlines of North Africa and Western Europe, with Greenland filling the gap. This suggested to Wegener that Africa, Europe, and North and South America might have once been joined as one massive supercontinent. He called it Pangaea, meaning "all lands."

Although Wegener suppressed the thought for several years because, he said, "I did not consider it probable," he never dismissed it. In 1912, after he learned of fossils indicating that similar plants once flourished simultaneously in Africa, South America, and India, Wegener finally announced his theory. He hypothesized that the continents had begun as a single landmass surrounded by one ocean not unlike Homer's circumfluent ocean. The landmass then split apart and its pieces moved slowly into their present positions. Furthermore, basing his theory on comparative longitudinal measurements of Greenland, he suggested that the continents were still moving, ever so slowly. This became known as the continental drift theory, and a controversial one it was.

Scientists of the day scoffed. A president of the American Philosophical Society dismissed Wegener's idea as "utter, damned rot." To think that the Earth's crust was moving horizontally was completely alien to Wegener's contemporaries. They were accustomed to thinking only in terms of vertical motion—mountains thrust up from below and then eroded away.

Moreover, the skeptics correctly pointed out that Wegener had failed to develop a satisfactory physical explanation for how continents could so move. By the time the last of several revised editions of his book, *The Origin of Continents and Oceans,* appeared in 1928, Wegener had marshaled considerable evidence to support his theory, but he had to concede, "The Newton of the drift theory has not yet appeared." For want of an adequate physical explanation, therefore, Wegener's idea remained in disrepute. The prevailing theory of the permanence of continents and ocean basins won out—for the time being.

But the arguments raised against Wegener suffered a fundamental defect. They were based on the assumptions of geology of that day, assumptions based only on a knowledge of the dry land. For geology then, and until recently, was a land-based science that did not, could not, seriously probe the sea floor for what it could reveal of the true nature of the Earth.

With the discovery and mapping of the Mid-Ocean Ridge, however, scientists began to resurrect Wegener's ideas and suggest that he might have been essentially right after all. "You have shaken the foundations of geology," Bruce Heezen was told after he described the discovery to scholars at Princeton University in 1957. Perhaps that double row of towering mountains with a vast canyon between them represented the crack where Pangaea broke apart. Perhaps this was the source of new material from the interior of the Earth, renewing and reshaping the face of the Earth. Out of this new thinking came the latter-day version of the continental drift theory, called the sea-floor-spreading theory, and this in time was embraced in the more general theory of plate tectonics— the idea that Earth's crust consists of large, dynamic segments, or plates, that are created at one edge and destroyed at the other.

In 1962, Hess and Dietz independently reached an astonishing conclusion. The only way to explain the newly mapped undersea ridges and rift valleys, the deep trenches, and the relatively young age of the ocean sediments, they argued, was to posit that the ocean floors were splitting. Said Hess: "The ocean basins are impermanent features and the continents are permanent, though they may be torn apart or welded together and their margins deformed. The Earth is a dynamic body with its surface constantly changing."

Although a cautious Hess called his paper on the subject an "essay in geopoetry," the idea of sea-floor spreading gained wide acceptance among scientists. According to the theory, as expanded and refined, it now appears that the Earth's surface is divided into a dozen or so thin, rigid crustal plates, which are in constant, restless motion like massive ice floes. Continents are "islands" of lighter crust riding on the denser plates, with the oceans opening and closing around them. Where the plates diverge, as at the Mid-Atlantic Ridge, they carry the American continents away from Europe and Africa at a rate of a centimeter or two a year. Where the plates converge they form deep ocean trenches as one edge slides beneath the other and is reabsorbed into the Earth. The grinding and thrusting between the plates, it appears, account for most of the great features of terrestrial geography—the location of the continents and ocean basins, the earthquake faults, and the mountain ranges.

The plate tectonics theory explains why so much of the Earth's volcanic and seismic activity occurs where it does—along the undersea ridges where the lavas erupt and along the trenches and continental edges that are the end of the sea floor's spreading process.

The new theory also offers a plausible explanation for the symmetrical pattern of the slopes and terraces and broad plains on each side of the major undersea ridges: the outflowing lava from the rift valleys of the ocean ridges must be moving at the same rate in both directions from the ridge.

The theory seems, moreover, to clear up another of oceanography's puzzles. If the oceans were permanent features, as had been thought, they should have acted as catch basins for enormous amounts of sediment over several billion years. But in their probings the oceanographers could never find any really ancient or thick deposits of sediment; they found only a thin veneer. It is now understood that sediments eventually get driven into the interior, along with the crustal material.

Such a broad and bold theory had to be further tested, however, before being accepted as anything more than geopoetry.

Geologists examined rocks on the Atlantic islands, establishing their ages by radioisotope dating methods. The islands farthest from the Mid-Atlantic Ridge, such as Bermuda, were indeed older than the islands closer to the ridge, such as the Azores. The islands were presumably formed at the ridge and then moved to their present positions by the movement of the plate.

Even more convincing tests were soon proposed. Two Cambridge University scientists, Frederick J. Vine and Drummond H. Matthews, who had been inspired by the Hess geopoetry, suggested that evidence proving or disproving sea-floor spreading should be contained in a kind of geological tape recorder.

As Vine and Matthews noted, it had already been established that the tiny grains of iron in lavas align themselves in the direction of the Earth's magnetic field at the time the lava cools and solidifies. That alignment never changes. Lava that cooled today would forever be magnetized toward today's north magnetic pole. But it had also already been established that, for unaccountable reasons, the Earth's magnetic field periodically reverses itself; in another time, for instance, a compass needle would point south instead of north and cooled lava would be magnetized accordingly. By comparing the age of rocks, determined by dating fossils and radioisotopes, with the direction of magnetism, it was discovered in the early 1960s that the Earth's magnetic field has reversed itself, flip-flopped north and south, at least 171 times in the past 76 million years.

This may not seem often, but it is often enough to have left an indelible and possibly telltale record. If basalt is welling up and solidifying at the mid-ocean ridges, Vine and Matthews reasoned, it should contain the magnetic polarity prevailing at the time of solidification. And if sea-floor spreading is correct, there should be alternating bands of magnetized rock, pointing north, then south, and then north again, running parallel to the Mid-Ocean Ridge and extending across the ocean basin on either side of the ridge. Some such magnetic "strips" on the Pacific floor had been charted by Scripps oceanographers, who noted in 1961 that they ran generally parallel to the ridges. But the explanation and the true testing were yet to come.

Once again, research ships took to the seas, this time towing long, slender torpedo-like instruments back and forth across several suspected rift areas in the sea. These instruments were seagoing magnetometers to record variations in the magnetization of the ocean floor. Aircraft with magnetometers flew evenly-spaced parallel lines across the Mid-Atlantic Ridge. The magnetometers found precisely the pattern of alternating magnetic bands that Vine and Matthews had predicted should exist. The bands appeared with remarkable symmetry on both sides of the major ridges, like mirror images. The resulting magnetic maps of the sea floors made believers of most of the world's scientists.

If further confirmation was needed, it came in ample measure from the voyages of the *Glomar Challenger,* starting in 1968. The 120-meter drill ship was built by Global Marine, Inc., an offshore oil-drilling company, for the Deep Sea Drilling Project, a government-funded project of a consortium of oceanographic institutions, headed by Scripps Institution of Oceanography. In her first five years of operation, the *Glomar Challenger* drilled into the floors of all the major oceans, except the Arctic, and recovered 29 kilometers of bottom samples. This was done by lowering sometimes more than 6,000 meters of pipe, boring into the sea floor, and then bringing up a drill pipe full of sediment and rock, called cores. When the core samples were dated, they supported the sea-floor-spreading theory. Not only were the sediments relatively young, but so also was the basement rock.

The rocks were found to be progressively older the farther the drill ship moved away in either direction from the ridges. The oldest material yet recovered, near the continental edges, was no more than 160 million years old. In contrast, the Earth is some 4.6 billion years old, and continental rocks found in Greenland have been dated at 3.98 billion years old.

From data gathered primarily by the *Glomar Challenger,* it has been estimated that the Atlantic Ocean is no more than 200 million years old. North America must have split away from Africa 135 million years ago. India separated from Antarctica and slammed into Asia some 40 million years ago, the collision causing the uplifting of the Himalayas and a multitude of faults running through Tibet and China. The Pacific Ocean may be the remnant of the original world ocean, but its sea floor has been subject to much the same conveyor-belt transformations, since the *Glomar Challenger* has thus far found no Pacific rocks more than 160 million years old.

Maps have been instrumental in many geographical discoveries, but rarely have maps been so crucial to the discovery of the very nature of the Earth.

In the plate tectonics theory, and the maps that contributed to it and derived from it, scientists hold an important tool for possible earthquake prediction and control. The San Francisco earthquake of 1906 was caused by the crushing pressures and stresses from the North American Plate grinding against the Pacific Plate. By knowing the direction and rates of crustal movements it may be possible to predict where stresses are building up to the point of triggering an earthquake. Oil and mining prospectors are interested in the new theory because it gives them insights into where to search for untapped reserves of wealth. By reconstructing how two continents once fit together, geologists have already predicted the location of ore deposits on one continent from the location of similar deposits on the other continent.

While it may be of no practical value, Robert Dietz and John C. Holden, both of the National Oceanic and Atmospheric Administration, have applied the present knowledge of plate tectonics to produce a

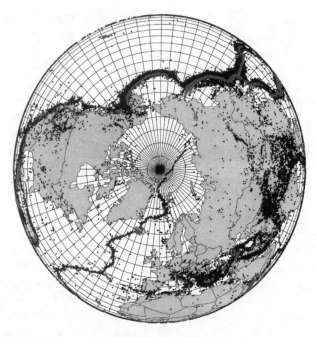

North Pole—stereographic projection emphasizing mid-ocean ridges

map of what the Earth should look like 50 million years from now. At first glance, nothing seems to have changed. No new continents. No new oceans. But closer examination reveals Los Angeles up near the Aleutian Islands, a shrunken Mediterranean Sea and an expanded Red Sea, a new gulf between East Africa and the rest of the continent, Australia overrunning much of Indonesia, Central America vanished and the Atlantic and Pacific meeting in the Caribbean Sea, and North America and Europe 1,120 kilometers farther apart. Such is the map of the distant future, if present rates of continental drift continue.

The Mid-Ocean Ridge had been discovered and mapped, it had inspired bold theories about the basic physical processes of the Earth—but it had never really been seen, only sensed and probed by remote means. Here was enough virgin geography to stir any explorer's restless heart. To descend to the submerged heights would be a venture as daring and ambitious as anything imagined by Jules Verne. But not impossible; and most assuredly, an idea whose time had come.

In late 1971 an international research project, the French-American Mid-Ocean Undersea Study (FAMOUS), was initiated to carry out just such a venture. Scientists from the two nations, along with associates from Britain, Canada, and the Netherlands, conceived of a plan to send men in the world's deepest-diving vessels down into the central rift valley of the Mid-Atlantic Ridge. It would be a journey, if not to Verne's center of the Earth, at least into a rugged chasm the French called "the navel of the Earth."

A period of intensive preparations preceded the FAMOUS dives. The sea floor at the site, southwest of the Azores, was surveyed and mapped as no other deep-sea terrain had ever been. Aircraft ran magnetic surveys that confirmed the existence of the typical pattern of alternating north-south and south-north magnetic stripes on both sides of the axis of the ridge. Ships remapped the ocean floor with a variety of sounding devices, some of which were military technologies released for the first time for use in a civilian scientific project. The first precisely navigated wide-beam (a 30° beam angle) sonar survey revealed that the floor of the rift valley was fairly flat and surprisingly narrow: whereas the entire valley is about 80 kilometers wide, the inner floor is less than 2.5 kilometers wide. Surveys with narrow-beam (2° angle) sonar produced bathymetric charts showing contour intervals as fine as 10 meters—a far cry from earlier surveys where the resolution of depth profiles was measured in kilometers, not meters.

A British research ship towed a seven-ton side-scan sonar system back and forth through the waters. Dubbed "Gloria" and shaped like a gigantic torpedo, the system directed its sonar beam obliquely and thus could map sea-floor topography at least ten kilometers on either side of its track. The outlines of submarine canyons showed up particularly well on Gloria records, as both sides of a canyon were detected simultaneously and were sharply etched. From Gloria's soundings the east wall of the rift valley was found to be less steep than the west wall.

For up-close surveys of microtopographic relief an American ship deployed Deep Tow, an instrument package developed by Scripps and towed through the depths of the rift valley like a metal fish. The "fish" emitted sonar signals whose echoes provided data for mapping more than a kilometer on either side of the track. In addition, the "fish" took photographs, recorded sediment thickness on the bottom, and examined the magnetic properties of underlying rock.

The U.S. Navy hauled out a relatively new underwater photographic system called LIBEC (Light Behind the Camera). The camera is suspended into the dark depths to within three meters of the high-intensity electronic flash lamps. By diffusing light over a large area, LIBEC reduces backscatter, which is the blurring reflection of particles suspended in the water, and thus yields reasonably clear pictures of sections of the ocean floor as broad as 40 meters. These and other photographs were assembled into photomaps and spread out on a gymnasium floor for study by the scientists before they set out on the expedition. The photomosaic resulted, said James R. Heirtzler, an American oceanographer, in "the finest bathymetric charts ever made of a small section of the bottom of the deep ocean."

American space satellites were also enlisted for the preparations, enabling the survey ships to determine their position at sea with unprecedented accuracy. By tuning in on the radio signals of the navigation satellites, ships could fix their positions to an accuracy of less than one-fifth of one kilometer, far better as a rule than fixes obtained by star sightings or shore radio signals. Knowing the ship's true position at the time of a sounding means knowing more exactly where the photographed or sounded undersea feature should be located on the map.

This method of position fixing grew out of an understanding of the Doppler shift, the variation in sound frequency that causes the oft-heard changing pitch in the whistle of a passing train. The phenomenon was first explained in 1842 by the Austrian physicist Christian Johann Doppler. Through analysis of the Doppler shift of the satellite's signal, it is possible to determine the moment when the satellite is nearest, its distance, and its orbit. Based on this discovery, Navy scientists had decided that the satellites could be used in reverse to fix positions on Earth. As the satellite approaches over a ship at sea, the frequency of the satellite's radio signals appears to rise and then fall. The ship records this Doppler shift and the orbital position broadcast by the satellite, and then computes where it is in relation to the satellite and thus where it is at sea.

In August 1973, the seven test dives began, using the French bathyscaphe *Archimède*. The *Archimède* took an hour to sink more than 3,000 meters to the bottom at the exploration site. By the light of four powerful floodlights the two pilots and Xavier Le Pichon of the National Oceanographic Research Center in Brittany, a prime mover of FAMOUS and an authority on sea-floor spreading, gazed into the maw of a small volcano whose fresh lava flows bore witness to recent eruptions. So did the absence of any sediment. The terrain was a jumble of rock. Several times the hull of the *Archimède* scraped against something hard—a nerve-racking experience, for rupture of the craft's gasoline envelope would cause it to lose buoyancy. On the first dive, the *Archimède* spent two hours at the bottom, moving about slowly and cautiously, ranging about 300 meters horizontally.

During this time the pilots navigated with the assistance of a network of anchored sonar buoys, each tethered 300 meters above the bottom. These were their "light-houses." The three buoys were laid out in a triangle along the axis of the rift valley. Each buoy was equipped with a battery-powered transponder that spoke only when spoken to. The surface ship guiding the submersible established its own position by sending out a distinctive sonar "ping" and timing the replies from the transponders. The submersibles also emitted pulses that were received directly by the surface ship and that also triggered separate responses from the transponders. The difference in arrival times between the direct signal from the submersible and transponded signals were used to compute the submersible's position. A small computer on the support ship digested these periodic calculations and plotted the submersible's position in relation to mapped features on the sea floor. This provided an invaluable record of the coordinates where discoveries were made and photographs taken.

All three research submarines—the American *Alvin* and the French *Cyana* as well as the *Archimède*—were ready in the summer of 1974, and they made a total of 44 dives, nosing about the valley depths like huge, curious fish, operating at depths (between 2,160 meters and 2,900 meters) that would have crushed ordinary submarines, taking in the strange sights where the Earth's surface is being torn apart and born anew.

When FAMOUS came to an end, the scientists had gathered several hundred kilograms of rock and sediment samples and had taken 100,000 photographs. They had, in one place at least, bridged the gap between indirect and direct exploration of the sea floor. The view from *Alvin,* *Cyana,* and *Archimède* was, of course, a limited one. Even with floodlights, objects twenty meters away were hazy.

These expeditions, and others that followed, added heretofore unimagined detail to maps of the ocean basin. The improved maps superseded in breadth and refinement the earliest Heezen-Tharp "physiographic diagrams," rendering them relics of one more pioneering stage in the history of mapping the world.

Perhaps the single most stunning discovery among the newly

mapped mountains and basins of the sea occurred in 1977. Exploring a volcanic ridge at the bottom of the Pacific, off the Galapagos Islands, American oceanographers peered out a porthole of the submersible *Alvin* at a hellish sight. Black clouds of mineral-laden water and hot gases billowed out of a sea-floor vent. Around the hydrothermal geyser stretched a biological wonderland of huge clams, two-meter-long tube worms, and swarms of tiny organisms. Life, in forms never before seen, was thriving on chemicals and heat from the planet's interior.

As similar communities of organisms were found on other mid-ocean ridges, scientists realized that these places were not only sources of new planetary crust, where oceans and continents are created, but also habitats of bizarre creatures that never know the warmth of sunlight. The findings inspired new theories about the origin of life. Beginning with Darwin, scientists had generally assumed that life's precursor chemicals made their fateful transformation into organisms in sun-drenched waters of ponds or ocean shallows. Now scientists had to stop and think: could this have happened instead in cauldrons at the bottom of the sea, without sunlight? No one knows the answer. And even if it turns out that this is not the case on Earth, the discoveries of hydrothermal vents have encouraged scientists to consider that life might originate and exist in many unfamiliar environments elsewhere in the universe.

While scientists looked into the depths, cartographers still had much to do turning sound into shape and producing better maps of the sea floor. Progress was slow as long as the means to improved mapping was controlled by secretive military establishments.

Through most of the second half of the twentieth century, the Cold War between the United States and the Soviet Union fostered the invention of new techniques and technologies for mapmaking. Nothing unusual about this; warfare and mapping have been entwined at least since early Chinese cartographers plotted the positions of fortresses on their maps. The practice of precise measurements in the Renaissance, so important in the evolution of mapmaking, stemmed in part from the invention of artillery, creating the need to construct bastions at angles to repel cannon balls and generally promoting the study of practical geometry. More often than not, remote-sensing mapping in the last century was driven by the military, which only slowly and reluctantly shared its talents with nonmilitary cartographers.

In the 1960s, for example, the U.S. Navy developed a new echo-sounding system that revolutionized sea-floor mapping. Navy engineers replaced conventional single-beam systems with a multibeam instrument that emits an array of sixteen pings once every 100 to 200 meters. Each pass of the survey ship could then map the contours of the ocean bottom in a swath 80 percent as wide as the water is deep. This development greatly accelerated the pace of data collection and produced maps showing a hundred times more detail than in previous maps. Yet nonmilitary cartographers were denied access to this technology until the early 1980s, when multibeam echo-sounding became available under the commercial name Sea Beam.

A more innovative technique for mapping the sea floor remained under tight military control until after the collapse of the Soviet Union in 1991 and the end of the Cold War. Then in 1995 the U.S. Navy declassified top-secret topographical maps of the sea floor that had been derived from an ingenious method of space satellite observations.

Space-borne remote sensors cannot peer directly through ocean depths to chart the underlying terrain. But this can be done indirectly by measuring the various heights of the ocean surface. From data showing bumps and depressions in the water surface, cartographers could reliably infer the contours of deep-ocean trenches and mountains. The data represented variations in the gravitational pull of rock at the bottom. Submerged ridges 2,000 meters tall, for instance, attract water from surrounding areas, producing a two-meter bulge in the sea surface over the ridge. Such swells, imperceptible to passengers on a ship, stand out clearly in satellite gravity measurements.

From 1985 to 1990, a secret Navy Geosat mission used radar altimeters to gather gravity data over all the world's oceans. No standard multibeam echo-sounding surveys could have provided so much mapping information so rapidly and completely, but it was for military eyes only. American submarines relied on the new maps to navigate the depths and play hide-and-seek with their Soviet counterparts. They also needed the data for aiming underwater-launched missiles, which could be nudged off course by the gravitational tug of submerged mountains.

Civilian cartographers got enough brief glimpses into the capabilities of the new gravity surveys to be envious. In 1978, NASA had tried a similar approach, at a reduced level of resolution, with its Seasat spacecraft. Before Seasat's power failed after only three months, the satellite had transmitted enough data on ocean surfaces for William Haxby, a marine geophysicist at Lamont-Doherty, to produce a new global map of the sea floor, which revealed ridges, rifts, and seamounts never before known.

Finally, when the Navy relented, all scientists and cartographers got their hands on the maps, recorded data, and other fruits of the Geosat mission. The maps had lost their strategic importance with the passing of the Cold War, and competition from the European Space Agency's European Remote Sensing-1 satellite had brought to the general public data of equivalent quality.

Working with Geosat and European data, David T. Sandwell of the University of California at San Diego and Walter H. F. Smith of the National Oceanic and Atmospheric Administration prepared in 1995 a comprehensive and detailed map of the ocean floors for nonmilitary use. False colors highlight the submarine relief: green for essentially level basins; orange-red hues for increasingly stronger-than-normal gravity, reflecting rugged ridges and mounts; and blue-violet-magenta colors for increasingly weaker-than-normal gravity, or deep depressions. These data provided the first view of the ocean floor topography in many remote areas of Earth, including one of the least known geographic regions, the land under the Arctic Ocean. With plodding surface ships making sonar soundings at the current rate, it would have taken 125 years to chart the ocean depths in such detail.

B ruce Heezen did not live to see these new maps or know of the exotic hydrothermal vents. He died in 1977, at the age of fifty-three. A few weeks before, he had won the American Geophysical Union's Bucher medal for "original contributions to the basic knowledge of the Earth's crust." When death came, from an apparent heart attack, Heezen was aboard a Navy research submarine preparing to explore at first hand the mountains of the Reykjanes Ridge southwest of Iceland, a planned voyage down to the mountains of the sea, which he had put on the map of the Earth.

Base Lines Across a Continent

Not all twentieth-century mapping operations were directed toward unexplored worlds of jungle or ice, bedrock or sea floor. Those foot soldiers of mapping, the land surveyors, still fanned out across the more familiar terrain, but they were measuring and remeasuring at precisions never before attainable, with technologies that have made museum pieces of chains and tapes.

Everywhere they went, these surveyors put their mark on the land much as they had been doing for almost two centuries. They left monuments—not the kind of monuments that memorialize great deeds (although they might have been justified in some instances) or draw large and respectful crowds, but simply small bronze disks, no bigger than saucers, set in concrete or rock. You have probably come upon them along a road or railroad, in a field, or under a clump of poison ivy. They bear the name of a government agency and usually a warning: "$250 Fine or Imprisonment for Disturbing This Mark." A string of such disks, at almost a kilometer apart, runs along the Hudson River from New York City to Albany. One is embedded in the north portal of the Golden Gate Bridge. The tip of the Washington Monument is a marker itself. And if you ascend Mount Whitney there will be a disk at the peak noting the elevation in no uncertain terms—14,494.777 feet.

These small monuments say to the passerby or, more meaningfully, to some future surveyor that once, a few years or more than a century ago, some surveyors stopped here and did some measuring and took some star sightings and then determined to the best of their ability the location of this place in relation to the rest of the world.

Consequently, Joseph H. Hirshhorn in 1976 could stand in the living room of his Connecticut mansion and be certain of his precise place in the world. Not as a wealthy financier and art collector, but his place there on the living-room hearth—41°6'12.75657"N and 73°40'25.88079"W. He knew this because one of the two oldest existing geodetic survey stations in the United States was there, marked by a bronze disk on his hearth. The station was established in the summer of 1833, when this site was open country and known as Round Hill. It was one corner of a triangle in a survey that eventually extended from Maine to New Orleans. Another corner of the triangle was fixed during the same summer at a place called Buttermilk Hill, several kilometers away in New York State. By coincidence, the site is on the property of another wealthy family, the Pocantico Hills estate of the Rockefellers.

As part of the bicentennial celebration of 1976, the National Geodetic Survey erected historical plaques at Round Hill and Buttermilk Hill. In dedicating one of the plaques, Leonard S. Baker, the survey director at the time, said: "For a geodesist, this could be considered, well, holy ground. From this meager beginning we now have 200,000 such triangulation stations throughout the country."

The meager beginning was the work of Ferdinand R. Hassler, a skilled Swiss engineer who initiated American geodetic land surveying. He was the first superintendent of the Coast Survey, which was founded in 1807 and ordered by President Jefferson to make "a complete triangulation survey, including the determination of latitude, longitude, and azimuths—direction from Polaris—of the principal places and bases, measured with the greatest possible accuracy."

Europe was well ahead in geodetic surveying, building on French achievements going back to the seventeenth century. By 1800, most European countries had plans to establish triangulations or were in the process already; by 1842 these triangulations would span the continent from the Mediterranean in the south to the Arctic and from Ireland and England in the west to the interior of Russia. Despite Jefferson's initiative, comparable American surveys had a slow start. The parsimony and political opposition of Congress virtually paralyzed Hassler at first and led to his abrupt dismissal. Nothing much was accomplished in the fifteen years he was out of the job. Rehired in 1832, he was finally able to muster the modest resources and manpower to begin the first major triangulation in the United States.

Even then, Hassler's work was exceedingly slow, though meticulous. It took him forty-five days, for example, to measure a 13.8-kilometer base line on Fire Island, a sandy barrier off the south shore of Long Island. At every 400 meters and 1,000 meters, strong pegs were driven into the ground to mark the distance, as measured laboriously by four two-meter iron bars laid end to end. Monuments of red sandstone marked each end of the base line. Hassler's methods were acclaimed by scientific societies, and subsequent surveys attested to the base line's accuracy. But at the rate he was proceeding, it is no wonder that when Hassler died in 1843, following an injury sustained while trying to protect his instruments in a storm, he had progressed no farther than southern New Jersey.

Those who followed Hassler struck out on projects establishing triangulation arcs from the Atlantic to the Pacific coasts, down the Pacific Coast, along the Gulf Coast, and through the Mohawk Valley of New York to the Great Lakes—work that took the rest of the nineteenth century to complete. The Coast Survey became known as the Coast and Geodetic Survey, reflecting its extended responsibilities, and today is the National Geodetic Survey, a branch of the National Ocean Service of the National Oceanic and Atmospheric Administration.

The purpose of the ubiquitous geodetic markers has not changed since Hassler's time. Now as then, each marker is a "known" starting point for all conscientious surveyors in the vicinity, whether they are laying out a highway, defining a property line, or preparing a local map.

In the United States the points in the geodetic network are generally spaced 3 to 8 kilometers apart in urban areas, 8 to 13 kilometers in rural areas, and 16 to 25 kilometers in mountainous and other remote regions. For much of the twentieth century the zero point for North America, the point to which all the bronze disks are tied indirectly, was a single disk set in concrete in a field on Meades Ranch in north central Kansas. The Meades Ranch disk appears to be no different from the others; the survey upon which it is based was no more painstaking than most geodetic measurements. But Meades Ranch happens to lie close to the center of the conterminous forty-eight states and at the crossing of two major geodetic survey arcs. One arc, the first transcontinental triangulation, completed at the end of the nineteenth century, runs along the 39th parallel from Cape May, New Jersey, to Point Arena, California, linking the Atlantic and the Pacific. The other arc was measured from Canada to the mouth of the Rio Grande along the 98th meridian. Thus, in a sense, when a local surveyor checked the corners of your property and tied them to the nearest Geodetic Survey marker, he was positioning your lot in relation to Meades Ranch.

Surveyors who ignore these reference points risk professional embarrassment, as has happened in a number of cases the Geodetic Survey likes to publicize by way of emphasizing the practical benefits of its work. There was the instance in 1969, in Pennsylvania, when the state highway department used its own reference points on each side of a river, instead of the Survey's; construction of a bridge started from each shore, and in midstream the two sections were four meters apart. And several S curves in the New Jersey Turnpike are there because of positioning errors resulting from similar inattention to geodetic references; the curves were the only way for the turnpike to connect to some of its cloverleaf turnoffs.

Beyond its local applications, moreover, each geodetic marker represents a point in an invisible web encompassing North America and, to one degree or other, all countries and continents. This web, or network, is the common mathematical basis for all mapping operations—topographic, geologic, and hydrographic surveys; soil and cadastral surveys; the rigid location of political boundaries and property lines; and studies of earth-crustal movements due to earthquakes, sea-floor spreading, and other shifts in the tectonic plates. The accuracy of all mapping is dependent on the accuracy of the underlying geodetic control network.

The fundamentals of establishing geodetic reference points are likewise little different from those of Hassler's time. There are still two types of geodetic land surveys—horizontal and vertical.

Vertical-control surveying determines heights above a reference surface that approximates sea level, which is done by means of measuring differences in elevation between consecutive benchmarks. This has usually been done by leveling, a type of surveying that remained essentially the same for several centuries. Directly or indirectly, the survey must be tied to the nearest shoreline. The leveling party starts from a tidal benchmark, which records the average sea level at that location based on years of tide-gauge readings, or from another benchmark already established in relation to approximate sea level. Then the crew moves inland, sighting on a level rod step by step. A surveyor holds upright a rod with a measure of heights in inches and feet or centimeters and meters clearly marked on its side. Another surveyor, standing on a point of previously determined elevation, looks at the leveling stick in the distance through a small telescopic instrument and notes exactly how much higher or lower the stick shows the distant site to be. After some distance, the surveyors retrace their steps and take readings all the way back to check their figures.

For the most precise work, known as first-order vertical control, the leveling data are adjusted mathematically to take into account the true shape of Earth, which involves knowing the actual gravity values at each benchmark. For some remote regions, especially in mountains, surveyors do well getting any kind of measurement; elevations there are often determined by barometric readings (a less accurate means) or trigonometric calculations (better, but still not as good as step-by-methodical-step leveling) based on long-distance angle measurements.

Until recently, leveling parties were a familiar sight on the landscape, generally working along highways or railroad rights-of-way, wearing their bright orange vests to alert oncoming traffic and leaning into instruments on tripods shaded by orange umbrellas. But leveling, as we will see, has all but disappeared in favor of space-age technology. The few such surveys today are aimed at refining previous work, usually for the purpose of some construction project. Or they are dealing with "problem areas," given the impermanence of the Earth's surface. Elevations in some areas, such as the California desert near Palmdale, rise because of subterranean forces associated with earthquake zones. Elsewhere, in the San Joaquin Valley of California and the Galveston Bay area near Houston, the land is sinking several centimeters a year and must be frequently resurveyed.

All told, the United States is dotted with more than a half-million benchmarks, points of known elevation, most of which were measured the now old-fashioned way.

Much more effort has been expended on the task of establishing and maintaining points of known latitude and longitude—horizontal control. Here, too, Hassler would recognize the fundamentals, if not some of the methods.

If the leveling parties were conspicuous by their orange vests and instruments mounted on tripods, the horizontal-control surveyors used to announce themselves by their steel towers, some of which rose higher than a ten-story building. These were Bilby towers, one of the few innovations in the fieldwork of geodesy introduced in the first half of the twentieth century.

Considering the tower's simplicity in concept and design, it is a wonder someone did not have the idea before Jasper S. Bilby did in 1926. The steel towers resembled frail oil derricks, and were portable. A crew of six or seven, with the nerve and agility of acrobats, usually erected a tower in six or seven hours and, after the surveyors finished their observations, dismantled it in even less time. Each tower was really a tower within a tower, neither of which touched the other. The outer structure (its parts color-coded blue) supported an observer and recorder and the signal lamps or reflectors to be observed from other towers in the vicinity. Since the structure they were standing on was not connected to the other, there was no likelihood of their disturbing the surveying instruments, which were mounted on the inner tower. To make sure that the top of the inner tower was centered on the ground mark (a previously implanted geodetic disk or the point where a new one was to be established), the chief builder set up the collimator, an optical device that aligned the two points (the mark and the top of the inner tower) vertically.

Bilby, a signalman for the old Coast and Geodetic Survey, designed the tower and, in a book published in 1929, *Bilby Steel Tower for Triangulation,* explained the need for his invention:

> In many regions it is not possible to select stations for a scheme of triangulation and have the stations intervisible from the ground, as trees, buildings, and other objects obstruct the line of vision between adjacent points. On geodetic surveys, covering wide expanses of territory, the curvature of the Earth must also be taken into consideration. Towers are therefore necessary to elevate above intervening obstructions the observer and his instruments at one station and the signal lamp or object on which he makes observations at the distant station. Wooden towers were used exclusively for triangulation towers for a great many years, but in recent years the cost of lumber and labor has become so high as almost to prohibit their use.

From the first deployment of these structures in 1927 to 1931, the pace of surveying accelerated, and about 14,000 kilometers of first-order work was accomplished, including completion of the Atlantic coast arc from Providence, Rhode Island, to Key West, Florida. At one time, more than two hundred Bilby towers were being used by field parties establishing horizontal control. They did their work mainly by triangulation but also by trilateration and traverse procedures—all well known to generations of surveyors before the time of the Bilby towers.

Triangulation, as we have seen earlier, is a system of joined or overlapping triangles in which the length of an occasional side, known as a base line, is measured and the other sides are computed from angles measured at the triangle vertices; this was Hassler's technique and, before him, that of the French arc measurers. Trilateration is a method of surveying in which the lengths of all the triangle sides are measured. Traverse is a method in which a sequence of connecting straight lines between points on the Earth are measured and used in determining positions of the points; it is more straightforward than the preceding two methods, but until recent times did not lend itself to rapid, accurate measurements over national or continental expanses.

The standard for a first-order horizontal survey is an accuracy of at least 1 part in 100,000. That is, there can be tolerated an error of no more than 1 meter over a distance of 100 kilometers. Such accuracies are required for surveys that develop the national geodetic network and support scientific and extensive engineering projects. Lesser accuracies— 1 part in 50,000, 1 part in 20,000, and 1 part in 10,000—are acceptable in lower-order surveys.

These accuracies, and even greater ones, were achieved by people who thought nothing of spending their days and, more often, nights atop Bilby towers measuring angles and distances and living a nomadic life.

Geodetic survey parties used to be sent out from Washington, D.C., whenever funds were available and weather permitted. They did specific jobs and then returned home. A system of permanent, year-round mobile parties began in the early 1930s and grew in number in response to public-works programs of the New Deal. This changed the lifestyle of surveying from its all-male expeditionary character to one of migrating households. The parties, including wives and children, numbered as many as fifty people. Their caravan consisted of trucks and office trailers of the Geodetic Survey and the mobile homes owned by the surveyors. They often found themselves north in the summer and south in the winter; they were seldom more than three to six months in any one place.

Edward L. Word worked as a theodolite observer with a typical field party in the 1970s. Among the party's many jobs was a five-month survey of Broward County in Florida, including Fort Lauderdale and Hollywood, to establish the latitude and longitude of more than eighty geographic positions. It was a bread-and-butter type of geodetic survey—encompassing no unsurveyed lands, striving for precision but not any new threshold of precision, simply providing more and better data to be used for mapping, locating permanent boundaries, planning the alignment of highways, utility lines, and other construction.

Each afternoon, around 3:00 or 3:30, Word would leave the trailer park near Pompano Beach, an encampment of forty vehicles (twenty-three trucks, thirteen mobile homes, two office trailers, and two supply trailers) that was Field Party G-23's base of operations. He would head for the site from which he would operate his theodolite. Once he hauled his equipment up the 150 steps to the top of a lighthouse, where there was a survey monument cemented into the circular platform. Other stations were set up atop a university library and a hotel. A few times, where no surface obstacles interfered and the measurement distances were short, he worked on the ground. But usually Word's work took him to the top of a Bilby tower.

After reaching the tower, Word and his assistant, the data recorder, hoisted their instruments to the top, some by pulley, some on their backs. By the time he was ready to observe, it would be dusk or dark. Triangulation parties work primarily at night because the atmosphere is clearer; the tremulous heat waves rising from highway pavement on a hot summer's day bespeak the visual distortions possible in daytime surveying. As he adjusted the theodolite, which was mounted on the inner tower, Word could make out the light of the first of his targets for the night.

Other surveyors on other towers or on building tops were getting into position and setting up target lights for Word's theodolite observations. Those at the targets were five to ten kilometers distant, in all directions, and communicated with one another by walkie-talkie. Word pointed his theodolite at the first light and measured the angle of direction; following the Survey's usual practice, he measured each angle thirty-two times to ensure accuracy. A normal night's observations lasted until 9:00 or 9:30, but sometimes Word did not get back to camp until 2:00 a.m.

Like most such field parties, G-23 was also equipped with a new device that was probably doing more in those years to facilitate and improve geodetic surveying than any other land-based (i.e., not dependent in some part on spacecraft support) instrument. This was the electronic distance-measuring instrument, a means of using the speed of light for determining linear distances.

The advent of electronic distance-measuring in geodetic surveying was an illuminating example of the dynamics of twentieth-century technology: a case of one new technology (rocketry) establishing a need that, in an amazingly short time, could be met by another new technology (solid state physics and electronics), which was then applied by an old science (geodesy) to not only satisfy the original need but also solve a broader set of problems. Surveyors had already started using electronic computers to free them from the drudgery of calculating the mathematical equations required in processing geodetic data. Now technology was throwing the surveyors a new challenge and handing them the tools to meet it.

The Air Force came to the Geodetic Survey in late 1959 with a problem—it wanted to install a system of missile-tracking devices in the vicinity of Cape Canaveral, but it had to know the exact locations of the ground stations in order for the tracking data to be of sufficient accuracy. This was the problem: if you do not have an accurate ground-position reading of the lift-off point, and the exact distance and direction from it to each of the ground stations in the area, you will never put a missile into the proper orbit or trajectory to deliver it to the intended target. And all the experts agreed that the triangulation network in the Cape Canaveral area in 1959 was insufficiently reliable—probably no better than an accuracy of 1 part in 50,000.

At first, the Air Force requested a resurvey to meet ordinary first-order standards—1:100,000. A few weeks later, it doubled its requirements. Then those requirements were redoubled. An impossible request? Accuracies of 1:400,000 over many hundred square kilometers had never been achieved before.

As Lansing G. Simmons, a Survey mathematician, wrote at the time: "To assure such results would require the measurement of many long base lines to control the triangulation lengths. Those sides of the triangles which form sort of a traverse from the Cape directly to the camera sites would have to be measured. This, in the area under consideration, would be highly impracticable if not impossible. The clearing of trees and brush and the staking of so many long base lines across country presents virtually insurmountable problems."

So the Air Force's problem became the Geodetic Survey's challenge. The Survey's best hope was the geodimeter, but could anyone be sure of this relatively new instrument? This was before similar instruments were used successfully in the Antarctic surveys.

As the first of the commercial electronic distance-measuring instruments, the geodimeter promised to be the most important improvement in base-line measuring technology in more than half a century. The kind of iron rods used by Hassler had long since been replaced by steel tapes, and the thermal-expansion problem of steel, which caused deviations in the length of the tapes, was resolved by a French discovery at the turn of the twentieth century: with the French invention of invar, an alloy of nickel and steel, tapes and wires became feasible, and measured distances faster. Hassler had averaged less than a kilometer a day in measuring his base line by rods; with long invar tapes, eight kilometers or more a day became routine. But the geodimeter was expected to reduce the time for measuring base lines from weeks to hours, without any reduction in accuracy.

The geodimeter sprang from experiments to determine the velocity of light. After they were successful, Erik Bergstrand, a Swedish scientist, decided in 1947 to adapt the procedure to measuring distances. He projected a highly collimated light beam to a distant reflector. When the reflected light returned to the device, the time the light took to make the round trip was calculated electronically, and from this calculation it was possible to figure the distance between the two points.

For several years, at least some skeptical surveyors scoffed at geodimeters, considering them nothing more than gadgets. But the Geodetic Survey decided it had little choice but to use them in its Cape Canaveral project.

Surveyors worked from March through July 1960. Their triangulation network included sixty-three stations, some of which were as much as 80 kilometers from the Cape. In any lesser survey, they might have contented themselves with measuring off only two or three base lines, but this network comprised two hundred lines, forty-three of which were done by geodimeter. Extraordinary care was taken. The mean value of two nights' geodimeter readings at a single station had to agree to within 2.5 centimeters or the measurements had to be repeated several more times. Operating geodimeters from Bilby towers proved no problem, despite early doubts; the instruments could be operated effectively under higher wind conditions than could theodolites.

Both the Air Force and the Geodetic Survey were pleased with the results. The missile-tracking stations were installed at positions with a probable error of less than one part per million relative to the Cape. In a report, Simmons said that the accuracy of the scale attained "was made possible only by the use of many geodimeter measurements." An important new tool in geodetic surveying had passed one of its critical first tests.

But Simmons, reflecting the conservatism of his profession, ended his report on a cautionary note. "It does not follow, as a result of this work, that comparable accuracies can be attained everywhere," he wrote.

Nevertheless, an attempt was soon made in an ambitious project known as the High-Precision Transcontinental Traverse. It would be the last of its kind.

The high-precision traverse produced the indispensable fieldwork for the revision of the North American Datum. A datum, in this sense, is any numerical or geometrical quantity or set of quantities that serves as a reference or base for mapping and other geodetic research. The last time the North American Datum, including Canada and Mexico as well as the United States, had been adjusted was in 1927. Based on five years of surveys, the 1927 datum consisted of about 200,000 points described by their longitudes and latitudes and about 500,000 points described by their elevations as well. But this datum had long been out of date. New measurements each year had to be made to fit into the old and relatively imprecise network, and in the process some of the very precise work achieved with new surveying technologies had been forced to accommodate distortions of as much as 1 part in 15,000.

At the start, in the latter part of 1961, the Geodetic Survey's primary purpose was to establish more satellite tracking cameras at various places along the east coast. Then the agency, its confidence in the new electronic distance-measuring equipment rising with each completed triangle, decided to resurvey the entire country. Much of the financial backing came from the Defense Department, one of many cases in which the Cold War would contribute to geodesy and cartography.

The super-framework for the traverse consisted of seven major loops—one small loop in Arkansas, Louisiana, and Mississippi; several small loops in California; and spurs from Aberdeen, Maryland, to Bangor, Maine, and from Jacksonville, Florida, down through Cape Canaveral to Homestead, Florida. By the end of 1976, at completion of the traverse in northern Michigan, field parties had determined the geographic positions of more than 2,750 sites, spaced at intervals of about 15 to 20 kilometers in forty-four states. This result, officials of the Geodetic Survey said, constituted a tenfold improvement in the precision of the geodetic base for United States mapping.

Extreme care was taken to maintain the highest possible standards of accuracy. Because slight temperature variations affect the geodimeter's light, balloons were lofted to get exact temperatures near the midpoint of each line at the height of the geodimeter ray path. Humidity readings were obtained with a psychrometer. Altimeter readings were taken at the beginning and end of each completed measurement of a line. All these data were factored into the calculations. Moreover, each side of the elongated triangles standard for the survey was always measured on different days with different geodimeters. The two measurements had to agree to within 1.7 centimeters. Very seldom, surveyors said, was it necessary to remeasure a line to meet the required agreement.

The ultimate test came when the surveyors returned to the starting point of one of the traverse loops. By how much did they miss closing the loop? One reason for the small Arkansas-Louisiana-Mississippi loop, 1,200 kilometers in length, was to provide a check on the accuracy of the methods before proceeding to the much larger loops out west. The results were encouraging: a miss of only 1.2 meters in 1,200 kilometers, or the desired one-part-in-one-million standard.

While the traverse was in progress, technological advances contributed to further improvements in the speed of high-precision surveying.

In 1966, a self-taught electronics engineer with the Geodetic Survey, George B. Lesley, experimented with modifications of the geodimeter that incorporated laser light. It was believed that laser light might extend the range of each measurement and, unlike the light beams then in use, be effective in daytime surveying. A test by a traverse party in Nebraska showed conclusively that laser geodimeters could measure lines up to 40 kilometers in length, double the range of the standard geodimeter, and up to 80 kilometers under certain conditions. Laser light became the favored means of electro-optical measuring by surveyors.

Another innovation facilitated the transmission and calculation of data collected by field parties. Until 1975, a field party observed an angle or measured a line, recorded the measurements in a record book, then mailed a list of measurements to the Survey headquarters in Rockville, Maryland. There the data were recorded on magnetic computer tapes for processing and analysis. With new portable equipment, data could be fed into a computer right at the field site, processed, and communicated by telephone directly into the headquarters computers—probably the very day the measurements were made.

Indeed, technology was now outrunning the pace of surveying. Triangulation by space satellite was also introduced during this time, a technique that would lead to a new world geodetic network (see Chapter 21). If geodesists had been sure of satellite triangulation in 1961, they later said, they probably would not have initiated the Transcontinental Traverse based on electronic distance-measuring techniques. On the other hand, without the traverse measurements, they would have been unable to evaluate the accuracy of the satellite measurements. The one, it seems, had to precede the other.

In announcing the completion of the traverse, the Geodetic Survey director Leonard Baker described it as "the most accurate survey of its size ever done in any country on Earth, and its results will serve surveyors, engineers, and scientists for many years."

On the basis of the new measurements and thousands of complex mathematical calculations beyond pre-computer capabilities, geodesists constructed the North American Datum of 1983, replacing the one of 1927 and embracing Canada, Central America, Greenland, and Mexico. The revised datum did not mean changing anyone's property lines or shifting any of those brass geodetic markers. But the determination of your precise place on Earth, the point of longitude and latitude, may have become slightly different—only slightly, for the measurement of the distance between New York City and San Francisco changed by only a few meters. And Meades Ranch in Kansas lost its central place in the scheme of North American geodesy. The new network was tied to one common point, the mass center of the Earth.

But Baker's optimism proved unfounded; the 1983 datum was by no means the last word in accuracy. Soon it had to be upgraded with position and distance measurements using space satellites in what is known as the Global Positioning System, to be described in Chapter 21. Tests had shown that surveying with satellites was certainly easier and quicker, and it routinely produced accuracies of one part in a million.

Looking back in retirement, Joseph F. Dracup, a geodesist, said that with the announcement of the test results "the geodetic world was turned topsy-turvy." Theodolites, Bilby towers, and electronic distance-measuring instruments became obsolete overnight. Versions of geodimeters may still be used in less extensive surveys. But the last Bilby tower was erected by the Geodetic Survey in 1984 at a station appropriately named Bilby, near Hartford, Connecticut.

The High-Precision Transcontinental Traverse marked the end of an era in land geodetic surveying. From then on, surveying and mapmaking moved resolutely into the space age.

PART *Four*

The Flight Out

From Bright Angel Point we flew down Bright Angel Canyon, passing over ancient cliff dwellings, to Ribbon Falls. We swung over the falls and spotted a wooden footbridge spanning the rushing stream. This was our landmark. Marion "Rags" Connaway, the pilot, picked out a landing area amid the bush and boulders and set the helicopter down there, on a beach of sparkling white gravel.

Across the bridge Wendell Mason and I found the boulder into which Brad Washburn had already drilled a hole. We re-established radio contact with Washburn and Buddy Cutshaw at Yavapai. As expected, they were unable to find the Ribbon Falls station through the Rangemaster's telescope. The place, off in a narrow side canyon, was too secluded and the background a bewildering mass of rocky cliffs. So I stood on the boulder and burned a signal flare that gave off bright orange smoke, while Mason installed the reflectors. Cutshaw saw the flare and got the station in focus.

We strained but could not see the laser beam this time. The morning light had become too bright, and from where we were we could not be sure exactly where Yavapai was, only its general direction. But over the radio came assurances that Cutshaw was making one good measurement after another.

Two hikers with heavy packs happened by and asked what we were doing. Helicopters were not usually permitted to land in the Canyon, and our radio and prisms understandably aroused curiosity. The hikers heard our explanations, and we heard theirs, about the trails taken, the last campsite, and the next. We wished each other well.

Then, as quietly as they had appeared, the two hikers departed down the dusty trail, never looking back, never suspecting the fleeting image they evoked. They were Matthes, the old topographer in the Canyon on foot. They were Matthes, come to remind us of the slow, plodding ways of mapmaking before we had not only the modern instruments but also the vehicles for reaching the most remote or hitherto inaccessible regions of the Earth—and beyond.

When the word came by radio, Mason and I packed up again and left Ribbon Falls. We had one more station to visit, one more set of laser measurements to make, and then the helicopter would have us out of the Canyon by lunchtime, before Matthes could have made five more kilometers the hard way.

But the newer means of transportation, we soon found out, have not altogether removed risk and anxiety from mapping.

After we took off from our last station, the helicopter encountered stiff downdrafts. The wind often spills over the south rim like an invisible waterfall, pouring down to within 200 or 100 meters of the Canyon floor. Flying into the Canyon when the downdraft is strong can be like descending in a broken elevator. "You can lose 2,000 feet of altitude just like that," Connaway said, snapping a finger. "It leaves you literally hanging by your seatbelt."

Coming from the bottom, however, our problem was one of *gaining* altitude, not losing it. The helicopter was no match for the Niagara of south wind spilling around us.

Connaway, knowing the vagaries of Canyon winds, maneuvered the helicopter over toward the north wall. Often, after the wind levels off near the bottom, it runs into a butte and rises somewhere near the north wall. Often, but not this time. Next, he flew down into the Inner Gorge, along the shadowed walls, hoping the greater heat there would create some updraft. Again, no luck.

Finally, Connaway flew over as close as he could safely get to the south wall, about 30 meters away, and found a slight updraft between the wall and the downdraft. The helicopter struggled up and out of the Canyon. We could relax.

Washburn was already back at the lodge recording the laser data and plotting preliminary distances on the aerial photographs. It was surprising, he said, how closely the new data agreed with the older maps of the Canyon. "When one has worked in this magnificent wilderness," Washburn remarked, "one's respect for these old-timers never ceases to continue to rise. The details are new and exciting, but the basic facts are still the same as they were at the turn of the century."

I thought of the two hikers and of Matthes. We had accomplished more in a few hours, with the laser and the helicopter, than the earlier cartographers could have done in a week—or ever. We had seen and recorded the Canyon whole, from the perspective of an aerial camera. We had reached places where no old-timer on mule or horse could go. Was there any limit to the new cartography?

Geodesy from Space

Flight into space has been, to borrow from that historic phrase, one giant leap for geodesy and no small step for cartography.

Before the space age, geodetic surveys nearly always stopped at the water's edge. The oceans were seemingly insurmountable obstacles to world-wide surveys. The relative positions of the continents could not be determined with precision, and islands in the oceans between the continents were, in a geodetic sense, adrift. Some islands in the Pacific were mislocated on maps by hundreds or thousands of meters. Geodesy, the global science, suffered from its lack of a global view.

Geodesy had nonetheless made substantial progress toward providing an understanding of the shape of Earth. We have seen that the French geodesists of the eighteenth century confirmed Newton's idea that the Earth was flattened slightly at the poles and bulging at the equator. The discovery shifted geodetic thinking from the ancients' spherical Earth to one that was spheroidal, or ellipsoidal. The problem then was to define the ellipsoid, to measure the figure of the Earth. Between 1750 and 1950 the value of the flattening was refined until, by the time of the space age, the value generally accepted was 1 part in 297.1. This represented the presumed difference between the equatorial diameter and the polar diameter, divided by the equatorial diameter. It was arrived at by a combination of four kinds of data: measurements of arcs of the meridian, like those the French expeditions made; records of gravity over as much of the world as possible, since the surface gravity field is an indirect expression of the Earth's form; analysis of motions of the Moon, which reflect irregularities in the Earth's gravitational tug; and calculations of the precessional movements of the Earth's axis in relation to the stars.

But knowing the shape of the ellipsoid, scientists had realized for years, solved only part of their problem.

Early in the twentieth century geodesists moved into what is called the geoidal epoch of geodesy—the third epoch following spherical and ellipsoidal. Geodesy uses several reference "figures of the Earth," depending on the purpose and desired accuracy. For maps of small geographical areas, such as city maps, the Earth may reasonably be considered a plane. Many astronomical and navigational computations use the sphere as a surface representing the Earth; a sphere is mathematically easy to deal with. And, as we have seen, there is the ellipsoid, a mathematically defined regular surface with specific global dimensions. One other figure of the Earth involved in geodetic measurements is the geoid. This has nothing to do with mountains and valleys, the familiar surface, but rather is a hypothetical surface to which the ocean would conform if free to adjust to the Earth's gravitational attraction and the centrifugal force of the Earth's rotation. Because of the uneven distribution of the Earth's mass, the geoidal surface is irregular, and because the ellipsoid is a regular surface, the two do not ordinarily coincide.

Determining the Earth's geoidal surface has proved to be quite an undertaking. Gravity surveying with pendulums and gravimeters, even with the aid of airplanes, was tedious and a seemingly endless task. The science of geodesy clearly needed a new tool if it was to refine further knowledge of the shape of the Earth and the positions of places on it.

Soon after the opening of the space age, with the launching of the Soviet Union's Sputnik 1 on October 4, 1957, the artificial Earth satellite was recognized as a "star" that could be used for navigation, mapping, and geodesy. The next evening, researchers at the Lincoln Laboratory at the Massachusetts Institute of Technology were able to determine Sputnik's orbit precisely by observing how the frequency of its radio signal increased as it approached and decreased as it departed—an effect known as the Doppler shift. The demonstration that a satellite's orbit could be easily determined from the ground also suggested that positions on the ground could be determined from the signals broadcast by satellites.

The application of this knowledge would transform geodesy's ability to learn more accurately the shape of Earth and the dynamics of its shifting crust. Before long, it led to the development of one of the most practical and indispensable tools of surveying and mapping, the Global Positioning System.

One way to use satellites in geodesy was to treat them as points in the sky for geometrical triangulation. Another use was to analyze the effect of the Earth's gravitational attraction on satellite orbits and thereby determine with greater accuracy and coverage how gravity varies over the Earth. These two uses reflect the two primary divisions of the science—geometrical geodesy and physical or dynamical geodesy.

One of the first discoveries made with a spacecraft had to do with physical geodesy. It was a discovery about the shape of the Earth, and it was quite a surprise.

Among the early explorers out in space was a grapefruit-size spacecraft named Vanguard 1, a mite out circling the Earth, but being sensitive to the pull and warp of Earth's gravity, a mite with a message for geodesy. The message was in Vanguard's very orbit, in the subtle ways that the tiny instrumented ball drifted in its orbit.

If the Earth were a completely homogeneous spherical mass, the orbit of a spacecraft should look the same day after day, apart from small perturbations caused by the Sun and Moon. Geodesists expected some variations, for they knew the Earth to be ellipsoidal and to be an irregular geoid. But they did not expect what they saw.

Vanguard took some strange dips. By radio tracking, it was determined that Vanguard's perigee (low point of orbit) migrated slowly around the Earth. This was expected, but the surprise was that the perigee plunged lower when it was north of the equator than when it was to the south. This indicated a lack of symmetry in the Earth's shape. If the Earth were symmetrical, though flattened at the poles and bulging at the equator, the perigee would remain at a fixed distance from the Earth's center. But the distance of Vanguard's perigee from the Earth's center became about 10 kilometers smaller when the perigee was at its farthest north than when at its farthest south. Scientists could be reasonably sure of this because their radio tracking system was known to be effectively accurate to about 200 meters for a satellite in low orbit. After further observations, moreover, they convinced themselves that the orbital perturbations were much larger than could possibly be caused by the irregularities in the geoid.

The only conceivable explanation, American scientists believed, was that Earth is pear-shaped, with the "stem" at the north. More precisely, the North Pole is about 19 meters above the symmetrical figure, and the South Pole is 26 meters below. If, in other words, someone bored a hole through the ice at the North Pole and fell into the sea, he would be 45 meters farther from the equator than a person who met the same misfortune at the South Pole.

The discovery raised questions. How did the Earth get deformed like this? Why is the stem at the north rather than the south? Neither question has been satisfactorily answered.

Almost as puzzling to the scientists was the fact, of which they were promptly reminded, that Columbus had predicted just such a pear-shaped Earth. In the last years of his life Columbus speculated that Earth "was not round in the way that is usually written, but has the shape of a pear that is very round, except in the place where the stem is, which is higher." He must have been guessing, but how right the explorer was.

The orbits of the early satellites told another story, which led to yet another refinement of the value of the Earth's flattening. Photographic tracking of Sputnik 2 and radio tracking of the American satellites Explorer 1 and Vanguard 1 indicated that the westward drift of their orbits was slightly less than it ought to be if the previously accepted value of the flattening was correct. After studying the results, British and American geodesists suggested that the flattening of the Earth was closer to one part in 298 than to one part in 297.

The value now established is one part in 298.25. This means that the equatorial diameter exceeds the polar diameter by 42.77 kilometers, which is about 170 meters less than in pre-satellite calculations. The difference has little if any practical significance. But if you were a geodesist, and liked to have measurements to an accuracy of within 30 meters, the revised value was a source of satisfaction and the beginning of the use of orbital analysis for determining Earth's true shape.

Other surprises were in store for geodesists observing the orbits of the new satellites. Before the space age it had not been possible to obtain gravity information in any detail for more than 20 percent of the Earth's surface, primarily because it was impossible to survey gravity at sea and in many remote lands. But since the existence of satellites geodesists were no longer so restricted. They learned that they could survey the Earth's gravitational field from orbital perturbations and thus map the many humps and valleys of the geoid. Such observations, beginning in 1961, revealed the geoid to be more irregular than had been generally thought and showed even the equator, once assumed to be a perfect circle, to be slightly lopsided. These discoveries of satellite geodesy were derived from both photographic and radio tracking techniques.

At the start of the space age the Smithsonian Astrophysical Observatory established a worldwide network of 12 satellite-tracking cameras. These were remarkable instruments, the Baker-Nunn cameras, capable of photographing objects in the sky more than a hundred times fainter than the faintest star visible to the naked eye. Pictures of the streaks of light of the satellite were timed, processed, and carefully analyzed to determine variations in the Earth's gravitational field as reflected in observed perturbations of the satellite's orbit.

Another technique developed in the 1960s was the radio Doppler tracking of satellites, as already described. By precisely timing the radio signals from a satellite and measuring the Doppler shift it is possible to calculate changes in the range and velocity of the satellite with respect to the observer. While the Navy began using Doppler tracking to fix the positions of ships at sea, geodesists became interested in what the satellite signals told them about the shape of the geoid. They noticed that sometimes, for no apparent reason, the satellite speeded up, and at other times it slowed down. Upon analysis scientists saw that the changes in velocity occurred over and over again at the same places in the satellite's orbit. From this they learned that the equator had its ups and downs. Further surveys over the decade, covering much of the globe with even more refined Doppler measurements, as well as the Baker-Nunn optical observations, led to a number of maps showing in unprecedented detail the contours of the geoid.

One of the earliest such maps was prepared by William M. Kaula of the University of California, Los Angeles. Kaula followed a practice common to all recent geoid maps. In this map the basic regular shape of the Earth is taken as an ellipsoid with a flattening of 1/298.25 and an equatorial radius of 6378.165 kilometers. The map's contour lines give the amount in meters by which the geoid differs from the basic ellipsoid. And the differences turned out to be surprising. South of India, in the Indian Ocean near the Maldive Islands, there is a deep depression in the geoid. Over near New Guinea is the highest hump on the equator.

In a 1976 article in the journal *Science*, Desmond King-Hele cited the Smithsonian Standard Earth II as the best geoid map for summarizing the achievements of satellite geodesy in the 1960s. The map represents the results of more than 100,000 photographic observations as well as Doppler surveys and some Earth-based gravimetric work. The depression south of India, a dip of 110 meters, is by far the greatest departure of the actual geoid from the reference ellipsoid. The hump north of New Guinea is 81 meters high. The other two major humps on the geoid, each about 60 meters high, are centered near Britain and south of Madagascar, while there are major depressions between 45 and 60 meters deep southeast of New Zealand and off California and Florida.

Even more accurate models of Earth's gravitational field—hence the shape of the geoid—were produced in the 1970s through a combination of satellite and laser technologies. A number of satellites were equipped with special quartz reflectors. When the satellite passed within range of certain ground stations, a powerful pulse of coherent light, a laser, was transmitted into space, reflected off the satellite, and returned to the ground station. By measuring the time it took the light to make the round trip, it was possible to determine the distance between the station and the satellite to an accuracy of 10 centimeters. The ultimate accuracy, according to King-Hele, is likely to be 2 or even 1 centimeter.

One result was a new geoid map prepared by the National Aeronautics and Space Administration's Goddard Space Flight Center. The map, GEM (Goddard Earth Model) 8, was derived from 562,000 optical, radio, and laser observations of 27 different satellites and another 1,600 surface gravity measurements. These were plotted with a contour interval of 2 meters, compared with 10 meters for earlier geoid maps.

This new and more precise knowledge of the shape of the Earth, both the pear-shaped ellipsoid and the bumpy geoid, gave geophysicists new insights into the Earth's dynamics. The increasingly accurate description of the Earth's gravitational field should eventually lead to a better understanding of the internal structure of the Earth, including the underlying forces responsible for earthquakes and volcanoes. The geoid maps have already been compared with the tectonic maps and found to have some possible significant resemblances. But the most immediate incentive for satellite geodesy was not so much intellectual as military. The trajectory of an intercontinental ballistic missile is influenced by those variations in the Earth's gravitational field that cause the satellites to speed up or slow down. The physical shape of the Earth is thus of prime importance to those who worry about the political and military shape of the Earth.

While physical geodesists used satellites to map the shape of the Earth, others looked to space and saw a way to determine very precisely the relative positions of various points on the surface of the Earth and particularly to interconnect points a continent or even ocean apart. These are the geometric geodesists.

Their idea is to extend traditional triangulation to a point in space. If a satellite is observed at exactly the same time by tracking cameras in several different places on Earth, a triangulation network can be built up that would be just as good as, if not better than, the ones previously used to establish latitude and longitude control points for mapmakers. The satellite would serve as one of the three points of the triangle. The other points on the Earth could be several thousand kilometers apart, making possible triangulation surveys that span the oceans and fix the positions of remote islands in relation to each other and to the large continents.

The possibility of satellite triangulation occurred to a few American geodesists soon after the launching of the first Sputniks. Lawrence W. Swanson and some of his colleagues at the Coast and Geodetic Survey (now the National Ocean Survey) had been working with Hellmut H. Schmid at the Army's Aberdeen Proving Ground in Maryland to learn his methods of analytic aerotriangulation in ballistic tracking. Out of curiosity Schmid photographed the first two Sputniks with ballistic tracking cameras at Aberdeen. When he showed the results to Swanson, the two men pondered the idea of determining geodetic positions by photographing a passive (sun-illuminated) satellite simultaneously with three cameras at different positions.

This could be the long-sought solution to one of geodesy's primary problems: triangulation across large bodies of water, beyond the range of the theodolite. Some progress had been made since World War II, but not enough to satisfy geodesists.

Shoran held the greatest promise prior to the space age. This was an electronic navigation method by which distance measurements are made by the determination of the travel time of a radio signal. For trilateration an airplane equipped with the shoran transmitter-receiver flies across a line about midway between the distant ground stations equipped with shoran transponders. The airplane thus serves as the third point in a surveying triangle. The method was tried successfully in extending geodetic triangulation from Florida over the Bahamas and Cuba, and was considered fairly accurate over distances of about 800 kilometers, although variations in the atmosphere—temperature, pressure, and humidity—raised uncertainties about the precise velocity of light during the operations.

In 1946 Yrjö Vaisala of Finland proposed the use of high-altitude flares as targets to extend geodetic control beyond points intervisible on the ground. The idea was to drop parachute flares from an airplane so that they could be sighted on by theodolites at several triangulation stations. Uncertain weather proved to be a nuisance, and all the waiting for a clear night greatly increased surveying costs.

The satellites seemed to afford a much more reliable light in the sky. Another advantage, besides reliability and the potential for intercontinental surveying, was equally appealing to geodesists. The great strength of the geometric satellite method, Schmid asserted, was that it "allows the determination of the three-dimensional positions of a certain number of selected stations on the physical surface of the Earth without reference to any geophysical hypothesis, specifically without reference to either the direction or magnitude of the force of gravity."

Schmid and his associates went to work developing the equipment and techniques for satellite triangulation. In 1963, after almost two years of field tests and experimentation, its feasibility and its accuracy in establishing geodetic control had been amply demonstrated. Most of the experiments were performed with Echo 1, the communications balloon satellite, providing a target suitable for station separations of about 1,500 kilometers. French geodetic surveyors, using the same satellite and similar techniques, were able in 1964 to determine the relationship between Europe and North Africa within an accuracy of 10 meters.

In another demonstration, American surveyors used lasers to measure a base line between Aberdeen, Maryland, and Cape Canaveral, Florida. Then cameras at Aberdeen, Cape Canaveral, and Bermuda simultaneously photographed a satellite. With the known Maryland-Florida base line, it was possible to calculate the distances from both points to Bermuda and to fix more precisely Bermuda's position.

Thus encouraged, the National Ocean Survey made plans for surveying a new world geodetic network. The centerpiece of the effort was a 30-meter balloon called Pageos (Passive Geodetic Earth Orbiting Satellite) that was launched by NASA in June 1966. Pageos followed a near-polar, circular orbit about 3,700 kilometers above the Earth. Sunlight reflected off its silvery skin made the satellite visible at night to sharp-eyed tracking cameras.

As in any triangulation survey, traditional or satellite, the preparations included running highly accurate base lines. The Australians measured two lines: an east-west line connecting stations at Culgoora and Perth, and a north-south line between Culgoora and Thursday Island. A line was run roughly along the 12° parallel in Africa, from Dakar through Nigeria into Chad. Several European countries cooperated to measure a line from Tromsö, Norway, down through Denmark, West Germany, and Austria to Catania, Sicily. The North American line ran from Beltsville, Maryland, to Moses Lake, Washington—the two primary stations in the United States other than those in Alaska and Hawaii.

A maximum of 17 camera stations were operated at any one time over the four years of fieldwork, from 1966 to the end of 1970. Wild BC-4 ballistic cameras, being more portable than the Baker-Nunns, were deployed for photographing Pageos as it passed in view. Since 3 or more cameras had to snap simultaneous photographs, timing was critical to the operation. The camera shutters were controlled by electronically regulated clocks at the field sites. And each of those clocks, though separated by thousands of kilometers, had to be synchronized to an accuracy of plus or minus 50 millionths of a second.

The necessity for such accuracy called into being a new specialist: the traveling timekeeper. There were three keepers of the clocks assigned to the project, each based in a different part of the world to maintain a so-called atomic clock that was synchronized with a master clock at the Naval Observatory in Washington. Even though light-years beyond any of John Harrison's designs, these clocks still had to be adjusted every so often. About once every six weeks, each of the timekeepers would take an eighteen-kilogram portable crystal clock and fly to the field parties in his region. At each stop, he synchronized the party's clock with respect to his portable clock to within a millionth of a second. In addition, each field party received regular radio time signals from the Navy and the National Bureau of Standards, enabling them to compute time lost or gained. But poor reception or occasional power failures reduced the radio's reliability, and so the traveling timekeepers would be called in.

All this effort produced thousands of stellar and satellite photographs. Because Pageos was photographed against a background of stars whose positions are known and catalogued, it was possible to determine the satellite's exact position. And because it was photographed simultaneously from several points on Earth, it was possible to calculate through trigonometry—and the indispensable electronic computer—the exact positions of the cameras on Earth. This established a series of measured triangles by which it was possible to develop the worldwide geodetic network.

For those interested in knowing exactly where they are—missile launchers, spacecraft trackers, navigators, cartographers, geodesists, and other sticklers for precision—Hellmut Schmid said that the new measurements achieved an accuracy on the order of plus or minus 3 or 4 meters. That is, the longitudes and latitudes of the 45 camera stations, as well as many positions embraced by the triangulation network, were now known with a precision hitherto unthinkable. The measurements, for example, showed a 100-meter change in the relative mapped locations of Europe and North America, a 400-meter difference between North America and Hawaii, and displacements of 3 kilometers or more for some islands in the Pacific.

Eventually, as the data were computed and added to, geodesists made further refinements in the Earth's reference ellipsoid and established a common worldwide mathematical base, or datum, for all mapping.

O ther extraterrestrial means of measuring Earth were introduced from the 1960s through the '80s. They proved to be especially suitable for measuring movements of Earth's crust, and to be exceedingly precise.

The dynamic behavior of the crust required another important dimension in geodetic measurements—time. After the discovery of tectonic plates, those continent-size moving pieces of Earth's crust, geodesists began to measure the rate of movement along fault lines between the plates. They wanted to record changes affecting their geodetic control networks, and thus their maps, and also any subsurface shifts and stresses that could presage an earthquake. At first, they used traditional ground-surveying techniques. Then they looked to space.

In 1966, astronomers working with the big radio telescopes decided that, in viewing distant objects in the heavens, two widely spaced telescopes would be better than one. This was the beginning of very-long-base-line radio interferometry (VLBI). When two or more radio antennas, separated by miles or continents, listen simultaneously to signals from an object in the sky, such as a quasar, they are, in effect, widening the observing aperture by the distance between the antennas, and this produces much better resolutions. Irwin I. Shapiro, a Massachusetts Institute of Technology scientist, heard of VLBI for astronomy and got the idea to apply it to measuring Earth crustal motions. A quasar is for all practical purposes a fixed object. When two or more antennas receive the same radio signal from a quasar, and register the reception times by precise atomic clocks at each antenna, scientists can see that the signal reached the antennas at slightly different times. The difference in the arrival times can be computed to determine the distances between the antennas. If such measurements were made periodically, Shapiro reasoned, any changes in the distances between antennas—hence changes in the Earth's crust—could be detected and measured.

Using MIT's Haystack Observatory in Massachusetts and the National Radio Astronomy Observatory in West Virginia, Shapiro made his first measurements to demonstrate the technique. Later, experiments were extended to include an antenna in California. By 1979, the technique had been improved so that it was possible to obtain geodetic measurements within an accuracy of one centimeter.

Another technique was introduced in 1976, with the launching of the Laser Geodynamic Satellite, Lageos, into a 6,000-kilometer-high orbit. Thereafter, geodesists relied more and more on extraterrestrial techniques. NASA positioned mobile laser systems in California along the San Andreas Fault—the contact zone between the Pacific and North American plates—and bounced laser pulses off Lageos. Analysis of the triangulation data over time indicated an observed movement of the plates of several centimeters a year, more rapid than had been predicted.

A network of more than a dozen laser satellite-tracking systems was established around the world. Signals between a station and the laser-reflecting Lageos permitted scientists to pinpoint the location of the station to within 3 to 5 centimeters and, through further computations, determine if it had shifted—and the direction of the shift—over time due to the movement of Earth's crustal plates.

Other experiments in the 1970s, a most fruitful time in geodesy, involved another application of Doppler satellite tracking. The result was yet another means of establishing geodetic control points for mapping.

The Navy's navigation satellite system, as already discussed, had become the principal satellite method used to determine Earth's gravity field and thus to map the geoid. If the Doppler tracking of such satellites could fix the position of a ship at sea—their original and primary purpose—why could it not be used to fix the longitude and latitude of a point on land or of the several control points required for the mathematical framework of a map?

In the first tests to compare the accuracy of Doppler positioning with that of optical systems, the Defense Mapping Agency (since renamed National Imagery and Mapping Agency) operated Doppler radio receivers at some of the same sites where the Wild BC-4 cameras were photographing Pageos. The Doppler method had certain readily apparent advantages. The receivers, designed by the Applied Physics Laboratory of Johns Hopkins University and built by Magnavox Research Laboratories under the trade name Geoceiver, were more portable and simpler to operate than the Wild ballistic cameras. A Geoceiver weighs about 450 kilograms, compared to the 5 to 6 tons of equipment associated with the BC-4 system. Under normal conditions, satellite observations with the Geoceiver require no more than ten days at each site—and half the time in higher latitudes, where the navigation satellites pass within range of a site more frequently. The normal occupation of a single BC-4 station during the world-wide triangulation project had been more than six months.

But were the results of Doppler positioning as accurate as those of other methods? The tests were so encouraging (agreement to within one meter) that the National Geodetic Survey began sending Geoceiver teams into the field. The first operations were in Alaska, particularly in remote, heretofore unsurveyed places, and in the Gulf of Mexico, where it was necessary to extend geodetic control farther out on the continental shelf to improve maps for oil exploration and drilling. Accurate control points were thus established in a relatively short time and at lower cost than had been possible. Mappers of the Antarctic were impressed and decided that satellite Doppler geodesy might help them establish control whereby they could convert new space photography into maps of the icecap.

Starting in 1974, the Geological Survey dispatched Geoceiver teams on Antarctic traverses, going in by aircraft to occupy a site for a few days of observations and then moving on. Their objective was to fix the positions of three or four features for each space photograph. In a report on the work, Rupert B. Southard, Jr., of the Geological Survey praised the Doppler technique as "a solid achievement" and further said: "Refinement of one-pass solutions from primary Navy satellites (assuming good data) will yield an accuracy within 10 m. With three or more passes of a primary satellite, the accuracy can be increased slightly. Although greater precision is obtained with more observations, field parties must move swiftly because of the short austral season. The accuracy of positions derived from short-duration observations more than meets the current standards for medium- and small-scale mapping, as well as the immediate requirements of ice-movement investigations."

Indeed, the Geological Survey reported in 1975 that calculations made from Doppler positioning experiments showed that anything placed on the Antarctic ice cap at the exact spot of the South Pole will move in the direction of Rio de Janeiro at the rate of eight to ten meters a year.

Further experimentation has convinced more and more geodesists of the utility and accuracy of Doppler satellite tracking in determining the longitude, latitude, and elevation of just about any place on Earth. "At each point of change," Southard said, "a newly developed instrument or technique was applied to the survey operations, resulting in greater accuracy, better control distribution, and a better map." Southard was speaking of the evolution of geodetic surveying in Antarctica, when the methods changed from solar observations to helicopter-supported electronic traverses to Doppler positioning, but what he said could be applied as a one-sentence summary of twentieth-century, particularly space-age, geodesy and mapping.

The culmination of these technologies was the development of the handiest device for mapping and perhaps the greatest advance in navigation since the invention of the compass for direction, the sextant for determining latitude, and John Harrison's chronometer for fixing longitude. Nothing before could match this device's convenience and reliability for telling you where you are and giving directions and distances to where you want to go—anytime, night or day, and anywhere, on land or at sea. Like the compass, the device is inexpensive, portable, and easy to operate. Today's surveyors never go into the field without one of these electronic instruments, and even hikers, weekend sailors, explorers, truckers, and some drivers of ordinary cars are using simplified versions and wondering how they ever got along without them. These little instruments are Global Positioning System (GPS) receivers.

The system grew out of a brainstorming session at the Pentagon over the Labor Day weekend of 1973. As successful as the Navy's navigation satellites had been, there were too few of them, leaving the system vulnerable to breakdowns or enemy attacks. Officials and engineers of the

Department of Defense were looking for a foolproof system, one that would ensure continuous and instantaneous global service, and could not be incapacitated by the malfunctioning of a few satellites. After long hours, the Pentagon brainstormers settled on the concept of GPS, which was both elaborate and essentially simple.

The space components of GPS are 24 Navstar satellites, each the size of a large automobile. Each satellite orbits Earth every 12 hours in a formation that ensures that every point on the planet can always be in radio contact with at least four satellites. The first of the Navstar satellites was launched in 1978, and the system reached full capability in 1993, at a cost of $12 billion.

The simple part is the system's operating principle. Each satellite continuously broadcasts a digital radio signal toward Earth. The signal includes the satellite's position and the time, exact to a billionth of a second. A GPS receiver on Earth registers the arrival time of the signals, usually from four satellites, and uses the information to calculate the distance between the satellites and the receiver. By checking its own time against the time of the four satellites whose positions are known, the receiver can pinpoint its longitude, latitude, and elevation.

At first, GPS was reserved for the exclusive use of the military. The receivers were, in any case, large and too expensive for widespread application. But with the miniaturization of electronic circuitry, hand-held receivers became available for about $3,000 each, then for less and less; some simple models now cost no more than $100. As the manufacturers looked hungrily at the enormous civilian market, they pressured the Pentagon to make GPS available for commercial use. The Pentagon eventually relented in 1988. The most accurate GPS signals, though, were reserved for military and other authorized users, including some government mapmaking agencies. In a practice known as selective availability, the satellites broadcast two signals: a civilian signal that was accurate to within 30 meters and a second signal, encoded for special users, that was accurate to less than 20 meters. In 1996, the White House announced that an even higher level of GPS accuracy would be made available to everyone worldwide and free of charge.

By then, GPS was established as a technology with many applications, commercial and recreational, scientific and cartographic as well as military. The National Academy of Sciences cited GPS as an example of how basic research leads to practical technologies that were "virtually unimaginable at the time the research was done."

With GPS, a survey that once took days can now be completed in a few hours, with 10 to 100 times more accuracy.

Inset: Like calculators, GPS equipment is becoming less expensive with time. Some hand-held devices are available today for under $100.

In the 1991 Gulf War, the academy noted, American troops used GPS for navigation on land, sea, and in the air, for targeting of bombs, and for guidance systems on missiles. Ground troops with handheld receivers could move swiftly and accurately through the featureless desert of the Arabian Peninsula.

GPS probably saved the life of an American pilot shot down in 1995 by Serbian forces in Bosnia-Herzegovina. Stranded behind enemy lines, the pilot knew almost exactly where he was because his life vest contained a tiny GPS receiver. He radioed his position to Air Force pilots overhead and then to Marines who were sent in to rescue him. Captain Scott Brady was found and brought back to safety, and hailed as a hero, but it was GPS that told the rescuers where to look for him.

The science academy also enumerated some of the "almost limitless" nonmilitary applications:

- Emergency vehicles use GPS to pinpoint destinations and map the most expeditious routes.
- Vessels lost or stranded at sea can be located with GPS.
- Trucking and transportation services use GPS to keep track of their fleets and to speed deliveries.
- Shipping companies equip tankers and freighters with GPS receivers for navigation and to record and control the movements of their vessels. Pleasure boaters increasingly rely on GPS for navigation.

- Civilian pilots use GPS for navigation, crop-dusting, aerial photography, and surveying. Airlines save millions of dollars by using it in making and following flight plans. (The first round-the-world balloonists, in 1999, navigated by GPS.)
- Some automobiles are now equipped with GPS so that drivers not only can find out where they are but also can get directions to their destinations. In Japan, where it is especially popular, 500,000 automobiles have been fitted with GPS-based navigation systems.

Among the first nonmilitary users of GPS were geodesists, who adopted it as a replacement for the huge and immobile deep-space radio antennas used in measurements defining exactly where some place is on the geoidal map. GPS was especially convenient in measuring changes in Earth's surface associated with earthquakes and the shifting of tectonic plates.

Even if GPS came too late for its Transcontinental Traverse, the Geodetic Survey put the system to several tests in 1983. Those tests left no doubt that this would soon become the technology of choice in establishing horizontal survey networks and in time would be used for conducting vertical surveys too. Before the system was fully operational, GPS receivers became standard equipment with surveying field parties. Joseph Dracup of the Geodetic Survey exuberantly proclaimed GPS "the most important development in the history of surveying."

Without the portability and increasing accuracy of GPS measurements, surveyors for the U.S. Geological Survey and the Army Corps of Engineers could not have mapped Florida Bay and the Everglades in the time and at the precision demanded by the Everglades Restoration Project in the 1990s. The surveyors started from benchmarks previously measured to a high degree of accuracy, one at the town of Flamingo on the mainland and another in the backyard of a bungalow in Tavernier in the Florida Keys. From the brass disks at these "control points," the surveyors fanned out to make GPS measurements at other sites nearby. Each measurement recorded the distance (GPS in this instance replacing the laser distance measuring instruments, which had not long before replaced the traditional surveying chain) and direction of the site from the control point. In this way, the positions of outlying sites—their longitude, latitude, and elevations—were determined with the same accuracy as the positions of the control points.

Although each reading with the GPS receiver took only a minute or two, the work demanded patience and fortitude. On Lower Matecumbe

Key, the surveyors had to wait out a shower and brace themselves against winds that seemed to stop pelicans in mid-flight. In afternoon heat, they slogged through mangrove thickets and mosquito swarms to fix one more control point in one more line of sites measured with signals from satellites in high orbit.

"Surveying is like a spider spinning a web," one of the surveyors, Brian Shannon, said. "You're moving back and forth, building something one line at a time until you have a complex network that captures a place."

Not that GPS has the solution for every question of location, as a paleontological expedition learned in the Gobi Desert of Mongolia. Thanks to their small GPS receiver, the paleontologists always knew the coordinates of each fossil bed they discovered and the distance and direction to the next destination. Still, they got lost. In a land without roads or good maps, they could never be sure whether the best way forward, to avoid uncharted obstacles like ravines, was to the south, north, or straight ahead. All they could do was head for the nearest dwelling off in the distance, perhaps a few hours away. They would have to engage the Mongol herder in palaver about his horses and sheep before finally getting to the point. Although the herder could only shake his head in wonderment over the peculiar device in the paleontologist's hand, he alone could tell the expedition the surest way to get where it was going.

Knowledge of location, no matter how precisely specified, either by GPS or traditional surveying, is of limited value in cartography unless it is linked to knowledge of context—the things to be put in their proper place. As GPS technology reached maturity, it so happened that other technologies had taken enormous strides. Remote sensing from space and computers with their prodigious memories were collecting and processing all manner of data, as never before—information about people, places, and environments, the stuff of maps.

Knowing the shape of Earth and its dynamic crust, knowing precisely where you are—all this, and more—contributed to a revolution in mapmaking and the study of geography at the end of the twentieth century.

Mapping from Space

The geographer Joel L. Morrison, reflecting on the sweeping changes in mapmaking today, looks back to 1972 as a turning point in the application of remote-sensing technology to mapping from space. That was when the National Aeronautics and Space Administration initiated the Landsat program for surveying Earth with orbital multispectral scanners, another military byproduct. "Landsat," Morrison said, "represents the beginning of the systematic mining of this new source of raw materials for cartographers."

The new space technology passed some of its first and most demanding tests in Antarctica.

In February 1972, while navigating the British Antarctic Survey's DHC-6 Twin Otter, Charles Swithinbank noticed that Latady Island, a desolate thumb of ice and snow west of Alexander Island, seemed to be mismapped. Latady appeared to be longer—about 50 kilometers longer, Swithinbank estimated—than depicted on the latest maps. But he had to be sure. The weather and the bleak terrain had played many a trick on earlier polar explorers. They were often confounded by the size and distance of things, a mountain far inland seeming to be right on the shore, a rock towering 300 meters seeming to be a mere 50 meters high. Not until 1940 had they proved that Alexander Island was indeed an island and not a part of the Antarctic mainland. So Swithinbank looked again, more carefully. He had not confused what he saw with Charcot

Island to the north or Beethoven Peninsula to the south. It was Latady Island, all right, and the maps were wrong.

Months after that flight, Swithinbank returned to the Scott Polar Institute at Cambridge, England, and was able to confirm the discrepancy. Examining very high-resolution radiometer pictures from the United States weather satellite NOAA-2, whose remote sensors could delineate the sharp boundary between sea and land surfaces, he traced the outline of Latady Island as it really is. Later his attention was drawn to another set of pictures, also taken from a spacecraft. The revealing new pictures were made between January and March 1973 by an unmanned, butterfly-shaped spacecraft known as Landsat 1. Thus he was able to prepare a new map of Latady Island and environs.

Swithinbank had little trouble assembling the eight frames into a mosaic, superimposing terrain features in overlap areas, and matching features at all the edges of the pictures. Longitude and latitude graticules were positioned with reference to known survey points on both sides of George VI Sound, which separates Alexander Island from Palmer Land. This gave Swithinbank confidence that there were no detectable scale errors in that part of the map. For the rest of the map, including Latady Island, he and his associates checked the principal point coordinates derived from the spacecraft's position and orbit at the time each picture was taken. It was not always a perfect fit, but close enough for Swithinbank to figure the map's accuracy to be plus or minus 10 kilometers.

In the *Polar Record* of May 1974 Swithinbank published the results and his new map based on the Landsat imagery. According to the map, Latady Island is almost twice as long and somewhat wider than it had been originally mapped. Charcot Island is about 30 kilometers south of its previously mapped position, and Beethoven Peninsula a little south of where it had been on the map. Some of the mountain ranges on Alexander Island were also realigned. "This may be the last time in the long history of exploration," Swithinbank wrote, "that any coastline on Earth will be found so far from where it was supposed to be."

Polar experts in the United States were equally impressed by Landsat images of the rest of Antarctica, including many places that had never been surveyed by aerial photography because they were too remote from staging bases. Topographers of the Geological Survey demonstrated the feasibility of using Landsat images to revise other Antarctic coastlines, reposition islands and the Ross Ice Shelf, and discover uncharted mountain ranges. The imagery, coupled with the new

Doppler control-establishing technique based on tracking Navy navigational satellites, promised to fill many gaps in Antarctic cartography.

Landsat 1 was launched in July 1972 by NASA for the express purpose of determining if orbiting observatories could help improve our knowledge of the Earth. From its orbit 900 kilometers above Earth, Landsat had a magnificent perspective, which scientists sometimes called a "God's-eye view." More technically, the view was synoptic, by which is meant observations of a wide-area distribution of conditions are possible at the same time. By whatever name, the view from Landsat produced impressive results pointing to new ways of preparing and revising maps, not just of Antarctica but of anywhere else on Earth.

Mapmaking from space was thus shown to be feasible. But the possibilities were first glimpsed as early as 1963, when L. Gordon Cooper orbited Earth in a Mercury spacecraft.

T he early advocates of space flight had overlooked cartography and Earth surveys in their predictions and planning for the new age. An authoritative report, *The Next Ten Years in Space,* published in 1959 for the Select Committee on Astronautics of the House of Representatives, made no mention of orbital photography of the Earth's surface except as a military reconnaissance technique. It was assumed by many experts that, apart from the obvious problem of cloud cover, the Earth's dense atmosphere would probably have a blurring effect and preclude photography of the clarity required for scientific and cartographic observations. Those experts were soon proved wrong.

In 1962, Paul M. Merifield, a graduate student in geology at the University of Colorado, began a doctoral thesis on the geologic uses of hyperaltitude photography. He happened to examine a photograph of Morocco's Atlas Mountains, which had been taken a year earlier by an automatic camera aboard the first unmanned Mercury spacecraft to orbit the Earth. He immediately realized the exciting possibilities. In that single picture Merifield could readily distinguish major geologic features, from folded mountain belts to vast expanses of sand dunes.

Merifield suggested to Paul D. Lowman, Jr., a NASA geologist whom he had known as a fellow graduate student at Colorado, that he try to get the Mercury astronauts to take more such photographs during their forthcoming flights, which they did. As far as Lowman and other geologists were concerned, the most fruitful of the Mercury missions was the last, the 22-orbit flight of Gordon Cooper in May 1963. Cooper was equipped with a hand-held 70-mm Hasselblad and with incredibly acute vision, 20/12 instead of the normal 20/20. Because of weather conditions below and the configuration of his orbit, nearly all of the best color pictures Cooper took were of Tibet. He passed over that country several times, traveling at an altitude of more than 160 kilometers and a speed of 28,000 kilometers an hour.

Cooper's report after the mission astonished everyone. "I could detect individual houses and streets," the astronaut said, "in the low-humidity and cloudless areas such as the Himalaya . . . area, the Tibetan plain, and the southwestern desert area of the U.S. I saw several individual houses [probably large lamaseries] with smoke coming from the chimneys in the high country around the Himalayas. . . . I saw what I took to be a vehicle along a road in the Himalaya area and in the west Texas–Arizona area. I could first see the dust blowing off the road, then could see the road clearly, and when the light was right, an object that was probably a vehicle."

Could this really be possible? Or was Cooper seeing things? Since the physiological effects of space flight were at that time largely unknown, a number of scientists suspected that Cooper, under the influence of prolonged weightlessness, had suffered hallucinations. But subsequent investigation revealed that a large white-topped truck had indeed driven along that rather deserted southwestern highway at the exact time and place Cooper had reported. To test their visual capabilities, astronauts on future flights were asked to identify large white markers on the ground. Man and camera, it turned out, could see the Earth quite clearly from space.

When Cooper's pictures of Tibet were developed, Lowman and other scientists at NASA's Goddard Space Flight Center remembered some old maps of Tibet, the work of Sven Hedin, that had been captured in Germany at the end of World War II. Hedin, a Swedish explorer, had mapped Tibet and much of Central Asia the hard way in expeditions between 1893 and 1908 and between 1927 and 1933. During the Sino-Swedish Expeditions of 1927–33, Hedin carried out extensive topographic and geologic reconnaissance surveys. He had sophisticated equipment—theodolites and steel tape, radios for obtaining longitude fixes, cameras for some photogrammetric work, barometers for determining elevations. Sometimes in the deserts, however, with few or no well-defined landmarks, triangulation and plane-table work became impossible. The surveyors had to resort to compass traverses, the distances being measured by pacing. And, with rare exceptions, they still had to rely on animals, often camels, for transportation.

Lowman's group inspected the Hedin maps and began comparing them with Cooper's photographs from space. The geologists could identify in the pictures unmapped mountains, locate more precisely a number of large lakes and the dry beds of ancient lakes, and thus were able to produce major revisions of the maps. They applied Hedin's survey sightings on mountain peaks to the space photographs as a frame of reference; a picture from space, as from the air, cannot be translated into a true map without some measurements on the ground. Using one particular picture and a plotting table, Lowman was able to map the Tibetan Plateau, about whose geography and geology practically nothing was known. Geologic folds and domes, glacier lines, and snow-covered highlands were charted accurately for the first time.

The Tibetan photographs were also brought to the attention of some of the surviving members of Hedin's 1927–33 expedition, notably N. P. Ambolt and Erik Norin. The photographs, Norin said, "opened extraordinary possibilities for closer study of the physical geography of High Asia." With the assistance of Goddard experts, Norin and other Swedish geographers used Cooper's photographs and others from subsequent missions of the Gemini Project to fill in the map with new detail. Combining measurements from camelback expeditions and photographs from space, the old and future cartographies, they constructed a new map of remote Tibet.

This, Lowman said, marked the beginning of "serious planning" for Earth resource surveys from space.

In a 1965 paper in *Photogrammetry Engineering,* Lowman outlined the "unique capabilities" of space photography. He aimed much of his discussion at those critics who argued that it was an unnecessary and expensive duplication of aerial photography. To Lowman the advantages of space photography were the following: greater perspective, wider coverage, greater speed, and rapid repetition of coverage.

The perspective afforded by the high altitude of spacecraft, Lowman wrote, permitted one to see entire mountain ranges, drainage basins, and fault systems at a glance—something aerial photography cannot usually do. Worldwide coverage would be especially important, he pointed out, in surveying areas not readily accessible to regular aircraft, such as the polar regions and the oceans. The advantages of a spacecraft's speed and repetition of coverage were self-evident, as they permitted a survey of the world in a few days, not years.

Geologists like Lowman were not alone in their interest in space photography. Robert N. Colwell, a forestry professor and associate director of the Space Sciences Laboratory at the University of California, Berkeley, had learned aerial-photo interpretation during World War II and after the war had sought to apply aerial-survey techniques to mapping the extent and severity of crop disease. Using various film-filter combinations, Colwell began photographing cereal crops from the air in the early 1950s, trying to discover recognizable tone or color differences between healthy and diseased plants. As he studied his cereal photographs, he became increasingly excited. Not only could he detect blight in wheat and oat fields weeks before it was apparent on the ground, but his photographs also revealed soil types, mineral deposits, different plant species, water depths, and blight in timber stands. If this could be done with aerial surveys, could it not be accomplished more efficiently from space?

NASA asked Colwell and other university scientists to recommend what specific remote-sensing wavelength bands, used individually and in various combinations, would work best for mapping the Earth's resources from space. After five years of research, they selected four wavelength bands, two in the visible bands of the spectrum (red and green) and two infrared bands, just beyond what the human eye can see. The test came on the flight of Apollo 9 in March 1969, and the results caused a sensation.

The photographs, taken from 200 kilometers in space by a multi-spectral camera combination, were so sharp and clear that the freeways of major cities were visible. The spectral signatures of 14 different field crops and 17 vegetable crops could be identified. Land with no visible crops was accurately divided into fallow, plowed, leached, abandoned, harvested, and seeded. University of California geographers took a particularly clear color infrared photograph of one million acres of the Imperial Valley and produced a land-use map that has since become a classic, appearing in publications throughout the world. Before the advent of space imagery, land-use maps of an area that size frequently took three years to complete, were costly and inaccurate, and rapidly became out of date.

To Colwell and Lowman and many other experts the Apollo 9 test proved what could be done with space photography.

"One of the first pictures I examined after the flight was of San Diego County," Lowman recalled. "Now, you'd think that the place would be mapped up and down. It had been, with some of the best possible geologic maps. But after seeing that picture, I jumped on a plane for San Diego, rented a car there, and drove over to the area around the

San Andreas Fault. The maps show a predominance of northwest-trending faults related to San Andreas, like the San Jacinto and Elsinore faults. The surprise was to find, on the Apollo 9 picture, a number of unsuspected northeast-trending lineaments. When I got there, it was hard to find some of those lineaments on the ground. As I've said, you can't make out an extensive pattern as well on the ground or even from a plane as you can from space. Most of those lineaments did turn out to be faults, some 40 to 45 miles long. You have to rethink some of your theories and redo all the geologic maps when you do orbital photography."

Analysis of other photographs, from the Gemini and Apollo flights, revealed an unmapped, relatively young volcanic field in the Mexican state of Chihuahua. Although of no apparent geologic importance, Lowman said, the discovery "was an early demonstration of the use of orbital photography in correcting regional geologic maps."

In comparing orbital photographs with published topographical maps, Ronald G. Dalrymple of Goddard found many other examples of how the space view could benefit cartographers. Pictures of Honduras and Nicaragua revealed meander loops in the Coco River, near the town of Bilas, where the latest maps showed a much straighter course. Two lakes covering an area of about 100 square kilometers were discovered south of Addis Ababa, Ethiopia, where the maps showed only a swamp. In Peru, a river was found to be longer than it appeared on the maps, and the meanders of other nearby rivers differed markedly from the maps. A mound-shaped volcanic feature was discovered on the southeastern side of Lake Titicaca, in Bolivia; it covered 260 square kilometers, was quite obvious in the photographs, but was not shown at all on the maps.

Dalrymple concluded that orbital photographs, even the early ones, could be useful in updating obsolete maps, correcting glaring errors in portrayal, and detecting important omissions. If this was possible with random photographs taken by the astronauts of Mercury and Gemini and Apollo, how much better it would be to have a more systematic program of viewing Earth from space.

Nevertheless, despite the mounting evidence supporting the usefulness of Earth surveys from space, Landsat was a long time coming to fruition. The program ran into just about every conceivable difficulty—bureaucratic infighting, budgetary constraints, conflicts over military secrecy, delays, and postponements.

In 1968 NASA eventually began financing the development of Landsat, coordinating the effort with the Agriculture, Commerce, and Interior departments, the Army Corps of Engineers, and, after it was established, the Environmental Protection Agency. Management of the project was assigned to Goddard, a NASA center concerned mostly with unmanned satellites. Geologists at Goddard now had their chance to fly a spacecraft designed exclusively to survey the Earth from space.

The spacecraft chosen for Landsat 1 and built by the General Electric Company was a modified version of the successful Nimbus weather satellite. It had two stubby wings inlaid with photovoltaic cells for converting sunlight into electricity. As equipped for Landsat, it carried television cameras, a multispectral scanner, videotape recorders, and radio receivers and transmitters.

The television system, consisting of three identical cameras, was designed to take pictures registering not only what is visible to the human eye but also wavelengths of light the eye cannot see. It can take simultaneous pictures of the same 185-by-185-kilometer section of the Earth in three bands of the spectrum—green, red, and near-infrared. A shutter and lens system in each camera produces images on a photosensitive plate, and these images are then scanned with an electron beam to create a video output. In this way, it can produce a set of three images of the same surface area every twenty-five seconds. The images can be transmitted immediately to ground stations, if the spacecraft is within range, or be stored on videotape for later transmission.

The multispectral scanner was an array of electronic sensors capable of measuring light simultaneously in four bands of the spectrum—blue, green, red, and near-infrared. Such instruments had already returned amazing results in aircraft tests. It had been possible, for example, to fly a scanner over a region, feed the electronic readings into a computer on the ground, and have the computer print out a map showing where there was wheat and where there was rice, even where an incipient disease might be threatening a crop. Every species of plant reflects a characteristic combination of wavelengths of light, a signature, that can be used to identify it.

To make these observations, an oscillating mirror in the scanner causes the light energy from a 185-kilometer swath perpendicular to the satellite's path to be swept across the focus of a small telescope. At the focal point is a cluster of tiny optical fibers, six for each band being monitored, which carried the received energy through spectral filters to detectors that convert it to an electrical signal. The scanner was designed to operate continuously, producing data for a strip photograph of all the ground over which Landsat passed.

Even though they were a far cry from Hedin's surveying technology, the cameras and scanning device of Landsat were not the best money could buy. They were compromises dictated by the size and power of the spacecraft, the project budget, and military security restrictions. The Landsat systems were designed to distinguish at optimal conditions objects on the ground no shorter than 100 meters, the length of a football field. Cameras and film flown on classified military surveillance satellites can obtain pictures showing features less than 3 meters in length. They have the capability, it is said, of pinpointing objects smaller than a Volkswagen from altitudes of 150 to 500 kilometers. This would have been ideal for cartographers interested in large-scale mapping, though not necessary for other investigators primarily interested in synoptic observations. Despite entreaties from scientists, the Department of the Defense would not release higher-resolution optical systems for use in a civilian program.

Finally, after the long struggle and some additional delays on the launching pad at the Western Test Range in California, Landsat 1 took its position in orbit on July 23, 1972. The direction, shape, and altitude of the orbit had been carefully planned to give mapmakers every possible advantage.

By traveling in a roughly north-south direction, crossing the polar regions, the spacecraft achieved worldwide coverage. The spacecraft made a complete circle of the Earth every 103 minutes 14 times a day. Because of the Earth's rotation beneath the spacecraft's fixed orbit, Landsat covered a different swath of ground on each pass during an eighteen-day period, then covered the same ground over the next eighteen days, and so on. This repetitive coverage made it possible to view each part of the United States at least one time or another without any obscuring clouds.

The altitude of 900 kilometers was chosen for purposes of timing and perspective. A spacecraft's speed is a function of its altitude: the higher the orbit, the lower the speed. At its particular altitude, Landsat was synchronized with the Earth's rotation in such a way that the spacecraft's course followed the Sun in its east-west migration. Thus, every time Landsat crossed the equator on its north-to-south track it was 9:30 A.M. local time. This meant that any picture at a particular latitude was always taken under the same general lighting conditions.

By maintaining a near-circular orbit, with an almost constant altitude of 900 kilometers, Landsat achieved broad and uniform perspective. A single image portrayed the same expanse of ground that it would take an airplane 1,000 pictures to encompass. Not only were all the Landsat images shot from the same altitude, but from that particular altitude each point on each image appeared to have been photographed from directly above. This reduced distortions and facilitated the processing of the imagery for the purposes of making maps.

Although electrical switching trouble forced flight controllers to turn off the television cameras early in the mission, the multispectral scanner mapped more than three-fourths of the Earth's landmass, including both polar regions. A sample of this was displayed within a few months on the wall in the office of William Nordberg, the chief project scientist. This was a mosaic of 11 photographs showing a strip of land 185 kilometers wide running from Quebec down to North Carolina, taken by Landsat 1 in less than six minutes.

The most valuable results from the Landsat data came only after hours of painstaking analysis of the images through magnification, use of color filters, and other manipulations. Although the data were first processed into black-and-white images, scientists found that they could draw out more detail by assigning a different color to each of the wavelength bands observed by the multispectral scanner. They then superimposed three images, laying one color upon another, to create the false-color photograph. Individually, in black and white, each image of the identical scene appeared very much the same, a landscape of black, grays, and white with low contrast. But, converted to color, the scene fairly leaped up through the magnifying glass. The picture took on a new and more revealing dimension, mountains and valleys standing out in relief, highways and rivers and the quiltwork of croplands etched in finer detail.

A photograph of western Nevada illustrated the possibilities. There was Reno in gray and blue gray; cities and other works of man were made to show up in those and even lighter shades. There were the suburbs in pink and the farmlands in bright red, the color signatures of vegetation because of the very high reflectivity of chlorophyll-bearing leaves captured in near-infrared.

These were the kinds of patterns that land-use planners, cartographers, and agricultural experts looked for in the Landsat photographs. Upon careful analysis, they were able to spot trends in urban sprawl, revise maps, make timber inventories, and chart the various uses to which land was put, even distinguishing between pastures and croplands, vineyards and orchards. A land-use map for an area the size of Missouri, which would cost $1 million and 44 man-years with conventional aerial photography, could be produced from Landsat imagery for an estimated $75,000. A land-use map of Rhode Island, Connecticut, and Massachusetts was prepared in a matter of days.

In one of the earliest Landsat experiments, Purdue University scientists took imaging data from part of Texas and Oklahoma and determined that the area included the following: 4.1 million acres of range and pasture; 2.7 million acres of cropland; 1.5 million acres of forest; and 190,000 acres of water. A "ground-truth" survey largely confirmed the findings. Moreover, Robert Colwell took Landsat data of northern California and, by analyzing color signatures, identified the general types of crops in fields measuring 20 acres or larger.

Geologic mapping turned out to be the most rewarding use of the early Landsat data. At a meeting of project scientists at Goddard in March 1973, Paul Lowman reported the discovery by Landsat of many previously unmapped fractures branching off the San Andreas Fault in California, fractures that had not been detected by Apollo cameras. Other scientists used Landsat imagery to map more than 500 previously unknown lineaments in the Adirondacks of New York; to identify new areas of extensive fracturing in Alaska that may be associated with mineral deposits; to determine the relative ages of superimposed lava flows in Iceland; and to trace the linear-terrain features where, according to continental-drift theories, India is supposed to have slammed into Asia millions of years ago. University of Wyoming geologists prepared the first detailed map of the many cracks and other structural features of the Wind River Mountains, a job that would have taken at least five years by conventional methods.

One Landsat photograph permitted a rare glimpse into the Earth's past. A half century ago J. Harlen Bretz, a University of Chicago geologist, suggested that the barren, heavily scarred region of eastern Washington known as the Channeled Scablands had been made that way by a flood of phenomenal dimensions, an idea that stirred a long controversy among geologists. They had lacked conclusive evidence of the so-called Great Spokane Flood until they saw the Landsat photograph of western Montana, Idaho, and eastern Washington. This convinced them that not only had there been a flood but that it was by far the most massive one for which any evidence is available.

As reconstructed by the United States Geological Survey, the flood occurred during the last ice age, some 18,000 to 20,000 years ago. A lake almost 600 meters deep, containing half the volume of a Lake Michigan, accumulated behind an "ice cork" that plugged drainage of western

Montana into Idaho. On the mountainside above Missoula, beach lines left by this lake show that the site of that city was buried under 285 meters of water.

The ice cork formed when the ice sheet that was moving down from Canada filled a narrow valley where the Clark Fork River empties into Pend Oreille Lake in Idaho. When the ice dam collapsed, several thousand cubic kilometers of water rushed down the valley of the Spokane River across the site where the city of Spokane now stands, then turned south across central Washington, leaving scars clearly visible in the space photograph. The soil and often the underlying basalt were washed away. In Montana's Markle Pass, the currents (estimated at ten times the combined flow of the world's rivers today) formed ripple marks 7 to 18 meters high and 3 kilometers long.

"These gravel ridges are plainly visible on aerial photographs," the Geological Survey reported, "but went unnoticed for many years simply because their immense size makes their pattern and symmetry almost indistinguishable from the ground."

The Grand Coulee, 85 kilometers long and 270 meters deep, is the largest channel dug by the flood. It began forming when ice diverted the Columbia River south of its normal route close to where the Grand Coulee Dam was later built. Other ground scars of the flood can be traced in the space photographs for 900 kilometers. The Geological Survey concluded that "the many fragments of evidence have been pieced together to support Bretz's concept of the Great Spokane Flood."

Other scientists looked to the Landsat photographs for a better understanding of the Earth as it is today.

Vincent V. Salomonson, another investigator at Goddard, reported that the satellite photographs were making it possible to chart the gradual shifts in glaciers. This could lead to an explanation of why the shifts occur, and whether glacier ice, which contains 75 percent of the Earth's fresh water, is on the increase or decrease.

Geological Survey scientists used near-infrared images to detect shallow water-bearing rocks in Nebraska, Illinois, and New York. As a result, they saw the potential for producing more-accurate maps of underground water supplies. In Africa, Mali took Landsat photographs and made maps of remote areas as a tool in water exploration. In Chile, hydrologic maps based on Landsat imagery were used to estimate water availability in arid regions of the country.

Oceanographers examined the pictures to determine their usefulness in monitoring the biological productivity of the deep ocean. Landsat could, of course, not actually see fish, but the spacecraft's sensors could see the distinctive reflected energy from concentrations of plankton; the plankton produced a reddish hue in the pictures. Where there is an abundance of plankton as a source of food, there will be schools of fish. Space photographs from Apollo helped Taiwan find new fishing grounds, and infrared sensors on a Nimbus weather satellite traced the temperature contours of ocean currents and upwellings, another guide to good fishing.

Hydrographers experimented with Landsat data as a means of measuring water depths. Initial evaluation of the technique, using the ratio of two multispectral-scanner channels, indicated that it could be practical in plumbing coastal areas of low to moderate turbidity as a way of updating the locations of reefs and shoals on hydrographic charts.

Thus, from Landsat and other space imagery Swithinbank could remap an island in Antarctica, Norin could update Hedin's maps of Tibet and Central Asia, Lowman could produce a new geologic map of California, and others could turn out a variety of land-use, vegetation, water, and coastal charts. These were impressive results, and once-skeptical professional cartographers began to take notice.

As they gained confidence in the Landsat material, the cartographers decided to experiment with even more maplike products, the first and most striking of which was a "satellite image" map of Arizona. In contrast to their earlier work, this looked more like a map than a picture.

For the Arizona map, cartographers of the Geological Survey used 24 Landsat images fitted into a mosaic and checked for accuracy with photo-identified ground-control points. Then the cartographers superimposed the imagery on an existing base map of the state. In this way, they transferred to the imagery many of the lines and symbols so familiar to maps—political boundaries, highway lines and numbers, blue river lines, place names, circles of varying sizes to denote cities of varying populations. The result, published in sepia tone, was a combination image-and-line map at a scale of 1:500,000.

Alden P. Colvocoresses and a team at the Geological Survey, working with NASA and experts from several universities, conducted a number of experiments making general-purpose maps from Landsat imagery. At first the cartographers proceeded with some trepidation. The loss of the

OVERLEAF: *Composite Landsat photograph*
of the United States

SPACE PORTRAIT U.S.A.
The First Color Photomosaic of the Contiguous
UNITED STATES
Produced by
GENERAL ELECTRIC COMPANY
Beltsville Photographic Engineering Laboratory
From 569 Landsat Satellite Images
In Cooperation with
The National Geographic Society and
National Aeronautics and Space Administration

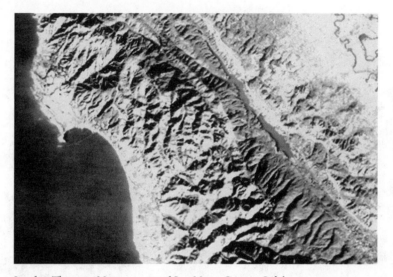

Landsat Thematic Mapper image of San Mateo County, California

television camera imagery early in the mission had been seen as a serious setback, as the Vidicon had been expected to provide better ground resolution and considerably better geometric fidelity than the spacecraft's multispectral scanner. But the scanner's imagery exceeded expectations: the image quality was excellent and the distortions were surprisingly small.

The first Landsat maps to meet cartographic standards and to be lithographed for public sale were a gridded image of upper Chesapeake Bay and color mosaics of New Jersey and Florida. They contained more detail than any cartographer could produce. The image required no rectification; a grid of longitude and latitude lines was fitted by identifying on the image discrete points, such as airport runway crossings, highway interchanges, powerline rights-of-way at highway crossings, square corners of woodland, and so forth. Through careful computations these points on the image were related to ground coordinate systems.

The New Jersey mosaic was produced from three images, which were gathered by the spacecraft in only half a minute. The Florida mosaic was prepared from sixteen separate images made during seven different orbits—but with little visible trace of the patchwork. The images were very close to being true maps once a grid was added.

In a 1975 article in *The American Cartographer,* journal of the American Congress on Surveying and Mapping, Colvocoresses discussed many of the proven advantages of satellite imagery in mapmaking. These included:

1. *Coverage.* Within two years nearly 80 percent of the Earth's land areas were imaged at least once by Landsat 1, and in most areas the coverage was repetitive—which provided some unexpected bonuses. Seasonal mapping is now feasible. Snow-covered winter scenes were found to be better for defining topography and cultural features, such as roads and urban areas, than comparable summer imagery.

2. *Speed.* An experimental photomap of the Lake Tahoe area of Nevada and California was produced within two weeks after image acquisition. The time could be reduced further to a few days.

3. *Revisions.* Although the satellite imagery is too nearly vertical for the compilation of contour maps, its fidelity to where things are has the advantage of simplifying small-scale planimetric mapping and the revision of some cultural detail. The Geological Survey used Landsat imagery as the source for revisions of the Virginia, Antarctic, and Alaska base maps, including the updating of urban sprawl and delineating of new highways. Once an area has been topographically mapped by more conventional means, the major mapping problem is not land forms (which change little) but the representation of changes in cultural features.

4. *Geometric fidelity.* Data from the satellite images, as corrected by computer processing, approach the accuracy of data from a calibrated mapping camera.

5. *Suitability for automation.* Frame cameras and Vidicon imagers record discrete scenes on a plane. Each photograph or image has its own geometric characteristics, and unless extensive analytical adjustments are made, consecutive photographs will not form a continuous map. In contrast, a multispectral scanner, as deployed by Landsat, can produce a basically continuous image on a mathematically definable map projection of negligible distortion. This characteristic provides the potential for developing an automated image-mapping system through computer-processing techniques and with only minimal requirements for ground control.

"Within the limits of the imagery capability," Colvocoresses concluded, "the age-old mapping problem of fitting together source materials of different forms and epochs is eliminated as far as the individual scene map is concerned. The automation prospect is further enhanced by the fact that Landsat repeats its coverage with considerable precision."

Colvocoresses and his team of cartographers had moved from skepticism to the firm belief that Landsat pointed the way toward the revolutionary concept of automated mapping of the Earth. All that it took to impress the layman was a huge photomap that went on display in 1974 at NASA headquarters in Washington. There, at a scale of 1:1,000,000, were the contiguous forty-eight states of the United States as they had never been seen before, certainly not by Frémont or Powell, nor by the passenger of a transcontinental jet. There lay revealed in one broad sweep the land as it really is—not a mapmaker's representation, however faithfully rendered. This photographic relief map was assembled for NASA by Department of Agriculture cartographers using 595 cloud-free images returned by Landsat 1. It was the first clear, comprehensive photomap of the United States.

Colvocoresses announced in 1974 a new projection he had conceived for maps produced from satellite imagery. He called it the Space Oblique Mercator Projection, though it became known in mapping circles as Colvo's Projection. The standard Mercator, as we have seen, is centered along the equator, with distortions of scale increasing progressively toward the high latitudes. The oblique Mercator is keyed to some other great circle, which crosses the equator at an oblique angle. Colvocoresses's new space-age projection would be based on a line running along the ground over which the spacecraft passed when the mapping images were produced.

But this was easier said than done. Because of the relative motions of a spacecraft revolving around a rotating Earth, the spacecraft's ground track, and therefore the projection's reference line, is not a true circle but a curved line. Moreover, the amount of curving varies from latitude to latitude. Historically, as Colvocoresses said, map projections have been based on static conditions. Even a mapping photograph taken from an aircraft records a scene instantaneously and the image can be rather easily fitted to a conventional map projection. But the Landsat scanning instruments produce a continuous image, which facilitates the mapping of large areas but is subject to distortion by the double spacecraft-Earth motions. The problem was to develop some mathematical equations by which the spacecraft images could be converted to accurate maps according to the new projection.

Neither Colvocoresses nor other experts under contract to the Geological Survey were able to figure out the necessary equations. At an international symposium on geodesy at Ohio State University in 1976,

Colvocoresses described the problem and said: "Here is a real challenge to the cartographic community."

In the audience that day was a chemical engineer from New Jersey, John P. Snyder, a man with no formal training in mapping technology but with a lifelong enthusiasm for the mathematical intricacies of map projections. As soon as he had learned trigonometry in high school, he had taught himself the methods and formulas of map projections. Snyder had taken a vacation to attend the geodesy symposium, and when he heard Colvocoresses state the problem he recalled thinking that he was "the last person who would be able to solve the problem."

For the next few months Snyder put in hundreds of hours in his study at home, at nights and on weekends, doing calculations involving plane and spherical trigonometry and differential and integral calculus. Using a relatively simple pocket calculator, he tested, rejected, revised, and retested equations. And still he could not figure out how to determine the curve that the ground track should follow on the map and then how to convert the images to appropriate map coordinates. The inspiration finally came to him, and it was a "straightforward calculus-trigonometry-algebra arrangement," he said.

The first set of equations could solve the problem only if the world were a perfect sphere. No one else had gotten that far, but it was not good enough. It took another month, late in the summer of 1977, to produce the equations based on an ellipsoid Earth.

Snyder mailed the equations—eighty-two of them—to Colvocoresses, with whom he had been corresponding. They were checked out with the Geological Survey's big computers. The problem had indeed been solved.

Advanced versions of Landsat continue their mapping rounds, with Landsat 7 going into service in 1999. Images from Landsat and the French SPOT satellites have been applied to many cartographic tasks, although their resolutions (30 meters for Landsat, 10 meters for SPOT) somewhat limit their utility. Other satellites offered improvements, gathering images of objects as small as 5.8 meters in width. Secret military satellites can do even better.

Some private American companies plan to operate small commercial satellites that they claim will deliver pictures at resolutions of three meters or less, at costs of a few hundred dollars each—below current charges for space mapping imagery. The companies are betting on the emergence of a robust market worth several billion dollars a year for space images to be used in urban planning, agricultural and environmental monitoring, the making and revising of maps, and perhaps cut-rate espionage.

"Some of us are skeptical that there is a big enough market," said Stephen Guptill of the Geological Survey. "But the technology is there."

Dynamic Maps:
A New Geography

According to the Turing theorem, any procedure that can be reduced to a finite number of explicit steps can be automated. Most cartographic work can be reduced to a finite number of explicit operations and thus is capable of being automated. Consequently, through the increasing use of electronic data-processing systems in the last years of the twentieth century, the ancient art and science of mapmaking came to be automated and expedited, computerized and transformed. The implications for the entire field of cartography and geography have been revolutionary.

At first, few seemed to foresee the sweep of the revolution or how rapidly it would lead to radically new forms and applications of maps. Revolutions start like that: some tea tossed into Boston harbor, an exile's return to Finland Station, or a few machines adapted for the more or less straightforward automation of conventional geographic mapmaking. The only change cartographers had in mind in the beginning was to employ computer-driven devices as faster, more efficient substitutes for human beings in handling the voluminous and tedious work of compilation and plotting that has always characterized map production.

On a visit to the U. S. Geological Survey in 1966, Jerome E. Dobson, a geographer at the Oak Ridge National Laboratory in Tennessee, saw an early demonstration of automated mapmaking. "As a blocky image slowly formed on a flatbed plotter," he said, "my host commented that publication-quality maps were conceivable, but he doubted computers would ever be powerful enough to support the massive computation high-resolution maps would require." That was, Dobson added, "the last time anyone ever underestimated" the new technologies of computerized mapmaking and geographical analysis.

A more meaningful first step was taken in 1973. After years of study and cautious experimentation, the National Ocean Survey published its first nautical chart produced almost entirely by automation. The chart covered a small part of Mississippi Sound in the Gulf of Mexico, at the harbor of Pascagoula, and it looked like any other nautical chart—blue for deep waters, pale blue for shallows, green for marshes, buff for land, magenta for pipelines and anchorages, and tiny numbers, in black, denoting depths. The only tip-off that anything was different appeared in the fine print of the legend: "This publication utilized computer assistance and machine engraving techniques."

The automation in this case began out on the hydrographic survey ship. Depth readings from echo sounders were fed into a computer on board the ship. The computer also received data on the ship's position every second, made mathematical adjustments based on the water's salinity as it affects sound conductivity, and then recorded the soundings on magnetic tape.

When the tapes were returned to the Ocean Survey's headquarters in Rockville, Maryland, they were used to send electronic signals to run a flatbed plotting table. In two hours, an automated ballpoint pen could register 2,500 depth readings on a plastic sheet that served as one of the overlays in making the printing negative for the map. The pen clicked away in rapid response to digital codes. When a signal for code 115 reached the machine, for example, the pen started drawing a dashed line, denoting a channel limit, until it stopped receiving 115. The machine "knew" where to draw the line because of a stream of coordinates that followed the code. When the machine received the code 126, it sensed that it was about to receive a depth value. The pen jumped across the board to the proper place, as directed by a coordinate signal, and began inscribing the next depth reading recorded on the tape.

"There's no sacrifice in precision," said Frank V. Maloney, an Ocean Survey cartographer specializing in computer systems. "If anything, it's better. If one man has to plot two thousand soundings by hand, you can imagine he's going to get tired and make a few little mistakes. Not the computer."

The new system sped up the production of charts from eighteen months to less than six. Only geographic names, compass roses, land features, and notes were still affixed the old-fashioned way. Once nearly all information could be stored in a computer memory bank, it was relatively easy to update charts, incorporating new data and correcting errors. A revised chart could be turned out in three months or less.

The task of converting the present nautical chart file, or any map, from graphic format to machine-readable digital form was somewhat like unmaking a map. The process is called digitizing. To convert map data into numerical form an operator moves an electronic cursor over the surface of the map. In this way features that would appear as lines on a map, such as roads, railroads, or coastlines, are translated to a digital code representing the coordinates of their positions, and recorded on magnetic tape.

More fully automated digitizing systems followed. By the 1980s, the Geological Survey was using electro-optical scanners to digitize all the content of its maps, any one of which contains more than 100 million separate bits of information.

The revolution had by then reached a new stage. Mapmakers moved beyond the use of computer-based technologies that simply replicated the products of the older cartography. With the convergence of aircraft and spacecraft remote sensing, the Global Positioning System, desktop computers, and electronic processing and display of data in graphic form, mapmakers and geographers in general found themselves with tools to make maps in ways and forms never dreamed of by Ptolemy, Mercator, or the Cassini family.

Maps can reveal more than the landscape, and have for several centuries. Remember the dictum: Anything that can be spatially conceived can be mapped. The new technologies broadened the conceivable meaning of "anything" and gave mapmakers the means to call up all manner of information about people, places, and environments—anything mappable—and visualize it swiftly and in many different guises.

The geographic distribution of short-lived phenomena, in particular, could be depicted in time, so that their study would be much more meaningful than before. Taking two years to produce a standard topographic sheet may be tolerable in light of the normal rate of change in land forms, but air-pollution maps of a metropolitan area, for example, must be available within hours if they are to be of more than historical interest. It had even become possible to collect information and have it displayed as a map on a computer screen as it was being collected.

Such capabilities especially transformed the practice of cartography known as thematic mapping. A thematic map illustrates a set of facts and figures in terms of their geographic distribution. Such a map may include some base data (coastlines, boundaries, geographic names), but only as points of reference for the phenomenon being mapped—whether it is climate or ocean currents; soil or vegetation distribution; facts of human geography such as population, health, and wealth; or just about any physical or abstract data that can be shown in their respective positions on a surface of Earth.

The practice of thematic cartography goes back at least to Edmond Halley in the late seventeenth century. Halley is usually remembered as the astronomer who correctly predicted the periodic reappearance of the brilliant comet that bears his name. But he was a man of many parts: colleague and patron of Newton, student of magnetism, fine Latin poet, and, not least, a cartographer. In 1686, Halley drew what is considered the first meteorological chart, which illustrated the directions of prevailing winds in the lower latitudes. By the use of maps, Halley observed, certain phenomena "may be better understood, than by any verbal description whatsoever."

Halley's most enduring contribution to cartography resulted from his two-year voyage as captain of the *Paramour*. The expedition embarked from England in 1698 and sailed the north and south Atlantic, everywhere observing the nature of terrestrial magnetism by noting variations of the compass needle from true north. Knowledge of these variations was of utmost importance to navigation. Shortly after Halley's return to England, he published a map based on the observations, entitled "A New and Correct Chart showing the Variation of the Compass in the Western and Southern Ocean."

Like Halley's meteorological chart, it was an early example of a thematic map, and an innovative piece of work. For Halley had to figure out a way to give graphic expression to the compass variations. He described his solution as follows: "The Curve Lines which are drawn over the Seas in this Chart do show at one View all the Places where the Variation of the Compass is the same; the Numbers of them, show how many degrees the Needle declines either Eastwards or Westwards from the true North; and the Double Lines passing near Bermudas and the Cape Verde Isles is that where the Needle stands true without Variation."

Thus came into use a valuable cartographic technique: the isoline, which is another name for Halley's curve lines, or what his contempo-

Isogonic map of the Atlantic (1701) by Edmond Halley

raries took to calling Halleyan lines. These are lines on a map connecting points of equal value.* In Halley's compass variation map, the isolines represented various zones of the Earth's surface within which magnetic variation was supposed to be uniform. Today, lines of equal magnetic declination are usually called isogonic lines.

The most familiar isolines are the contours on topographic maps, lines of equal elevation, but the technique is applied in many other ways. In 1817, acknowledging Halley's inspiration, Humboldt introduced isotherms for temperature distributions in connection with his South American mapping. Later in the century, Joseph Henry prepared the first modern weather map and originated the use of isobars to show atmospheric pressure distributions.†

Cartographers in the nineteenth century developed other methods of portraying statistical data in map form. One of the most inventive was Henry Drury Harness, a British army officer. In his maps for the Irish Railway Commission, published in 1837, Harness introduced graduated circles for cities (the size of the circle being proportional to the population of the city, as on many road maps today) and traffic flow lines (the width of the lines being proportional to the relative quantities of traffic in different directions).

In 1854, a physician named John Snow pioneered the technique of medical thematic mapping. On a map of the Broad Street area of London, Snow placed a dot at the location of each reported death from a cholera outbreak and a cross for each water pump. The map made everything clear: the incidence of cholera coincided with the people who drank from the Broad Street pump. This was a case of a thematic map not only presenting the results of an investigation but also serving as an analytical tool in the investigation, a dual function that has become even more prevalent now that computers facilitate both the display and the analysis of data.

Their prodigious capacities for storing and analyzing data made computers especially ideal for generating thematic maps, and the experi-

---

*Research by Stig Nordbeck, Lund University in Sweden, shows that Halley's maps may not have been the first to have lines connecting points of equal value. A Dutch map in 1696 used such lines to connect points having the same depths in navigation charts of river mouths. But Halley's work inspired more imitators and so is credited with launching the practice.
†A few of the dozens of other isolines are the isobath, for depth below mean sea level; isohyet, precipitation; isanther, time of flowering of plants; isopag, duration of ice cover; isodem, population; isoseismal, number (or intensity) of earthquake tremors; isotac, time of thawing; isospecie, density of a species; isovapor, vapor content of the air.

mentations coincided with and encouraged an intellectual sea change in the field of geography. A "new geography," as its apostles proclaimed, was being born.

In the 1950s, after a half century of professional stagnation, geographers began to conceive of novel ways to analyze and illustrate space-related phenomena. The themes of their thematic maps became more varied and complex, pattern-seeking explorations into areas hitherto uncharted in a quantitative sense by geographers. Not only were their early computer maps different in preparation and appearance from conventional maps, but some were also attempts to fathom deeper conceptual levels.

"We are no longer confined to conventional earth space when we think of maps," wrote Peter Gould, a professor of geography at Pennsylvania State University.

An early experimenter with computers and thematic mapping was a graduate student at Lund University in Sweden, Torsten Hägerstrand. In 1953, he began modeling processes of spatial diffusion, starting with computerized simulations showing the spread of technological innovations in a farming area of southern Sweden. In time, many of the maps used in Sweden's national planning were compiled by computers along the lines of Hägerstrand's work. For example, a map displaying the national "welfare landscape," a contoured surface showing the travel times to reach such services as doctors, dentists, libraries, and adult-education centers, could only have been compiled by computers doing the millions of calculations.

Taking spatial diffusion mapping further, geographers prepared maps of relative accessibility of places in Western Europe. If one left home at 6 a.m. and wanted to return by midnight, how far could a person go by certain means of transportation and how long could the person spend at various places? These questions were answered in map form, the answers suggesting ways to change airline or train schedules. Such analytic techniques, when programmed for computers and linked to a video screen, could be displayed in time-lapse films of dynamic simulations.

Waldo Tobler, a geographer at the University of California at Santa Barbara, advanced the concept of winds-of-influence maps. These maps illustrate relationships of a most ungeographic nature. One example was a map of the flow of ideas and information through psychology journals. Tobler coded the number of times one journal referred to or footnoted another. Not unexpectedly, journals with similar intellectual traditions and perspectives, and many contributors and advisers in common, appeared close together on the map; those with different viewpoints were spaced far apart. Tobler also produced a winds-of-influence map of botany in the South Pacific. When the islands were plotted by botanical similarity, the chart differed considerably from the more conventional geographic map, raising significant questions about the mechanisms of plant dispersal.

The look as well as content of thematic maps underwent change. A leader in developing new techniques for portraying statistical data in geographic formats, starting in the early 1970s, was the Harvard Graduate School of Design's Laboratory for Computer Graphics and Spatial Analysis. As in the case of the first computerized nautical charts, Harvard's first thematic maps produced with computers were automated versions of familiar manual thematic cartography. These include choropleth maps, on which sections determined by civil boundaries or other arbitrary divisions are colored, shaded, dotted, or hatched to make darker or lighter tones in proportion to the density of a given subject's distribution. Others are perspective maps, with peaks and valleys representing not topographic relief but variations in some phenomenon keyed to geographic distribution, and also isopleth maps, with contour lines resembling Halley's.

These earliest computer-generated maps were produced with standard line printers. Just as a typewriter could create crude pictures by taking advantage of different letters, numbers, and signs, so could a computer line printer. Maps made in this way were simple and inexpensive, but without esthetic appeal. Later, continuous-curve plotters, attached to computers, provided more complex and attractive maps. Even so, cartographers were still trying to replicate conventional mapmaking, a practice computer scientists would soon make obsolete.

Thematic mapping—and mapping in general—had come a long way from Snow's dots and crosses or even the clunky line printers of first-generation computers.

Out of the ferment of computerized mapmaking emerged a new breed of cartographers and geographers. They speak the language of electronics and computer software as well as of geography. They are more mathematical and analytical than their predecessors. They explore spaces and phenomena that often lie beyond pre-computer capabilities. They are expanding and enlarging the very definition of a map. And

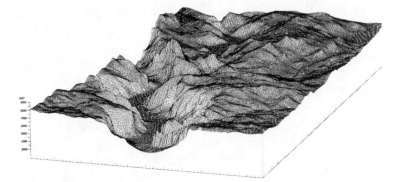

A computer-generated map of Yosemite Park

more and more of the new mapmakers come from outside the profession of cartography; some are everyday computer users in the home, office, or laboratory. The line between the producer and the user of maps, once sharp, has blurred with each advance in cybercartography.

Computerized thematic cartography depended on the merging and electronic manipulating of what are known as databases. One of these, a cartographic database, contains the digitized coordinates that describe details of geographic areas, such as a city by blocks, a country by states and counties or provinces, or the world by countries.

It used to be that the printed map functioned as its own database, the primary storage medium for spatial information. "When we had only printed maps," said Joel Morrison, formerly chief geographer of the U.S. Census Bureau, "one product had to do two things: store information and at the same time be a visualization of how the area looks. Today, the map as visualization is freed from the map as a multipurpose database."

Now that the storage function has largely been transferred to computer disks (which had replaced magnetic tapes), the maps themselves—on computer screens or on printouts—can be customized in size, scale, and perspective. Unburdened by archival responsibility, individual maps can be more pictorial. Their content can be streamlined to include only certain chosen information, such as waterways draining a region, average personal income by the postal zones of a city, ethnic populations of a particular place, or a route map for the family vacation trip. In a minute or two, for example, one can produce a map from a house in Manhattan to Jackson Square, New Orleans: distance 1,338.4 miles door to square, with the route and its required 24 highway changes clearly charted.

With the new technology, a map's content is drawn from separate databases. These are much like any other set of facts recorded on computer disks, except they are keyed to geographic territories in some way. Each base is easily updated as new information is gathered or the political geography changes, as happened with all the name and boundary revisions after the collapse of the Soviet Union. Because data and images from satellite remote-sensing systems are collected and transmitted in digital form, they can be readily stored in computers (as new or revised databases), then automatically processed and converted to mapping formats.

As often as not, these essential ingredients of computerized mapping, cartographic and content databases, are combined on disks produced by government agencies and an increasing number of private software companies. They are easy to use with nearly all desktop computers. They are particularly suitable for what is known as interactive mapping, which can be a little like playing computer games. Any number can play.

Once a map is displayed on the screen an operator can delete, add, or correct the map's contents, change the projection or scale, zoom in on certain sections, and otherwise rework the map. "Gee, what I want to see is behind that clump of trees," Morrison said, imagining an interactive mapping game. "Just rotate the visualization to behind the trees. Or I want to see the other side of the mountains. Subtract some data, and there you are. Take out trees and see what the bare land surface looks like. All these things you could not do before."

A cartographer, or anyone familiar with computers, can thus experiment until he or she finds the most effective format for the intended purpose of the map. Then the map can be recorded for later reference and also produced on paper. And ever-inventive computer specialists keep nudging cartographers to explore new methods of making and using maps.

Altogether, these new technologies are called geographic information systems (GIS). In the strictest sense, a GIS is defined as a computer system capable of assembling, storing, manipulating, and displaying data identified according to their location—the software mastermind of cybercartography. The software links geo-

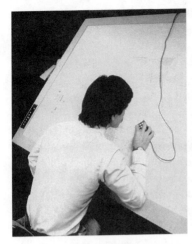

Converting map information
to digital form using a hand-held
computer mouse

An electronic scanning device
will convert some types of map
information to digital form

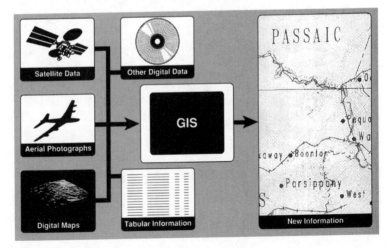

Linking information in different forms through a GIS

graphic information, or where things are, with descriptive information, what things are like. Practitioners regard the total GIS as including the data that go into the operation.

One of the most comprehensive of the new GIS databases is the U.S. Census Bureau's TIGER (for "topographically integrated geographic encoding and referencing"). It is the bureau's master address book. Every highway and street in the country is in the file, along with the addresses on each side. All new census results are keyed to this file. Much of the data is available to anyone, as are demographic and geographic databases of other government agencies. Software companies increasingly obtain such data, then process and sell them for use in making maps flashed on desktop computer screens. Such maps are used for everything from planning new shopping centers and marketing campaigns to routing deliveries by trucks equipped with GPS receivers.

In GIS mapmaking, think of each layer of data incorporated as a sheet of clear plastic bearing a distinct set of information, each representing a particular feature of the area being examined. The bottom sheet might be a map of political boundaries and streets, the geographic framework. On top of that could be placed other layers keyed to the base map: education levels, buying power, just about any demographic information one could want.

"We overlay one set on another, sometimes things you couldn't conceive of mapping before," explained John A. Kelmelis, a geographer at the Geological Survey. "If done by hand, this might have taken a whole career. And it was virtually impossible then to keep up with changes in the data over time."

At the survey's offices in Reston, Virginia, Kelmelis explored and tested the uses of digital cartography's dynamic maps in seeing facts, conditions, and probabilities almost literally spread across the landscape. He demonstrated the critical role such maps can play in decision-making and, as an example, already had in the aftermath of the 1993 flood of the upper Mississippi River.

"See that levee." Kelmelis pointed to a map on the computer screen. "What if I took it away or built it higher; what would be inundated?"

One by one, maps filled the screen with answers. A swath of blue spilled over the land, covering fields of corn, even whole towns. The GIS program was drawing on detailed models of the area's elevations and river volume and flow rates, as well as on the nature and uses of land along the banks. Such maps helped identify sites on higher ground for relocating the town of Valmeyer, Illinois, which was almost destroyed in the flood.

Next, Kelmelis called up green-and-blue Landsat images of the Missouri and Mississippi valleys as the water rose by stages and stood from bluff to bluff. Maps made from such satellite digital images alert people to an approaching crisis and help guide engineers and policymakers concerned with river-basin management. "This is the future of mapping," Kelmelis said. "For many uses it no longer suffices to make a static map."

The multiplying applications of GIS cartography brought maps and mapping to nearly every aspect of government, business, and everyday life. "Things people thought about doing in the Sixties and Seventies became perfectly possible in the Eighties," Kelmelis said. "They became commonplace and inexpensive in the Nineties."

Starting in 1994, officials and precinct commanders of the New York City Police Department would meet twice a week to study computer maps projected on a huge screen. These were up-to-the-minute maps of a dark side of urban geography. With a system called Compstat, for computer statistics, everything about any crime of murder, rape, robbery, or drug dealing was recorded and then displayed by location. This often yielded revealing patterns and possible clues and connections. The maps, which could also be called up on laptop computers at precinct stations, guided the deployment of forces and crime-solving strategies. The police looked at one map, for example, and realized that several shootings in a Brooklyn neighborhood had been committed by young men on bicycles. So the police swept into the neighborhood and stopped bikers violating any law, like going the wrong way on a one-way street. In a short time, they seized eighteen handguns, and the shootings stopped.

An early example of the application of GIS to the kind of geographic analysis pioneered by Hägerstrand, Tobler, and other thematic-mapping experimenters was an assessment of the implications and resource requirements of President Jimmy Carter's proposals for a new national energy plan. Jerome Dobson and his Oak Ridge colleagues modeled projections of electrical-generation expansions and changes to evaluate their impact on air quality, water quality, labor supply, and other economic and cultural factors. Using economic and population databases, they projected, as far as the year 2000 and county by county, energy demand and fuel use by type in each major region of the country. Even though little of the energy plans was ever implemented, the exercise encouraged the exploration by an increasing number of geographers of the analytical potential of GIS.

In a more ambitious application of the maturing techniques of GIS, the Oak Ridge group produced a worldwide 1998 population database for identifying people at risk from natural and human-made disasters. Such knowledge is essential for quick and effective response to the release of hazardous chemicals into the atmosphere by industrial accidents or radiation from nuclear power plant explosions, the spread of toxic gases from a terrorist's bomb or chemical and biological warfare, and the eruption of ash and gases from volcanoes. The analysis of the effects of these hazards was conducted at data resolutions on the order of one square kilometer or even finer. For this the Oak Ridge geographers developed a global population database called LandScan.

The collection of the best available census counts (usually at province level) for each country was only the start. Remote sensing by aircraft and spacecraft provided insights into two population variables. Images of land cover and nighttime lights revealed where people live and where they work and the routes they travel between residence and workplace; most censuses count population based primarily on where people sleep rather than where they work or travel. These variables and other probabilities, involving simulations of contaminant transport by air and water, were then collated and assessed through GIS software. It is what proponents of GIS affirm that the technology does best: integrate spatial databases to produce maps and inventories incorporating multiple phenomena. LandScan, the Oak Ridge geographers concluded, "appears to be the most suitable currently available global database for estimating populations at risk."

Applying GIS, cartographers have also become adept at converting a satellite image into a map showing specific land use, property ownership, or underlying geology. Agronomists map rainfall patterns over space and time. Environmentalists simulate the course pollutants would take if dumped into a river. When war breaks out, planners of bombing raids use highly detailed maps in selecting targets with pinpoint precision, as practiced by NATO forces in the 1999 conflict over Kosovo. But one incident there, the mistaken bombing of the Chinese embassy in Belgrade, was a sobering reminder that even maps with high-tech pedigrees are not infallible; the embassy's move to the targeted site a few years earlier had somehow gone unrecorded in the database from which the mission map was produced.

Essentially, these applications differ little from what people have long used maps for: illustrating what was, what is, or what might be over some geographic space. But the information at hand was never greater or more accessible (though not without flaws), and the tools for graphic display and analysis were never more creative.

Mapmakers and geographers have weighed the implications of cybercartography with a mixture of excitement and anxiety, not unusual for those who find themselves in the midst of a revolution. Surely some of the claims for the new technologies will prove to be extravagant. But which ones? At this stage it is impossible to know with any certainty what role traditional mapmakers will have in shaping and using whatever electronic-processing practices turn out to endure in cartography. Who will set and maintain mapmaking standards?

Modern mapmakers have tended to be professionally conservative, ready to take on worlds but only in the measured ways of their calling, unaccepting of that peak in Darien in the absence of ascending contour lines. By training, they abjure poetry and whimsy in their work. By experience, they are unaccustomed to sharp departures in practices and technologies. Changes used to be so gradual, Morrison said, "that a person could expect to finish a career using the same technology that he or she used at the beginning of a career."

Now the very character of cartography is changing all around them. The printed map itself, as we have seen, is no longer the principal repository of the geographic information it portrays; that function has passed to computer databases, so capacious and easily updated. At their command, mapmakers at computers can call up specific information, experiment with ways to display it graphically, and produce a number of maps on the screen or in hard copy.

A different type of cartographer is entering the field, atuned to the increasingly dynamic products and practices. "With previous technology," Morrison said, "you wanted introverted individuals who worked alone and came up with the best design, almost a work of art. Today, technology allows groups of people to sit around, together or online, and interactively produce maps."

Many of the new mapmakers are not necessarily trained cartographers. The "joystick generation" will adapt readily to automated geography, Dobson predicted in 1983, when many geographers were still wary of the new technologies. And so it has. Digital cartography has empowered every person to be his or her own mapmaker, with direct access to the various databases of GIS. These user-cartographers are expected to grow in number, and to stretch the scope and influence of maps.

Not that the users of the new technology, including geographers, are spending all their days bound to their office computers. For a study of the effects of acid rain, Dobson once walked through a wilderness area in the Adirondack Mountains of New York. He came upon an area of distinctive vegetation change. From the satellite image in his hands, he could see that this was no isolated phenomenon, but a transition from one vegetation cover to another over a wide area. Back in the laboratory, Dobson fed the information from direct and satellite observations into a computer and compared the results with maps of the region's watershed. GIS, he said, "was essential at every step as we conceived and tested hypotheses."

Reflecting on the experience, Dobson could not help but feel a twinge of regret. "Geographers sometimes feel they were born too late, after everything already was discovered," he wrote in 1994. "Many of us still would like to be in terra incognita adventuring and discovering places never seen before. Instead, we try to learn more about familiar places and the processes occurring in them. To us this type of discovery often is as intriguing as a new island, river or canyon, though it is difficult to generate the same level of excitement among the public."

In this era of computerized maps and new geography, the traditional producer-cartographers are not facing extinction, but transition. The producer-cartographers still collect most of the mapping information, enter it into databases, and develop new compendiums of information for use in GIS. But gone is their control over many of the ultimate products and their uses.

"Suddenly, it's a whole different world to us," Morrison said in 1998. "Our future as mapmakers—even ten years from now—is uncertain."

Old-line government and commercial mapmaking institutions have come under increasing pressure to adapt to cybercartography, as small software companies began marketing more and more of the computer disks reaching user-cartographers at their desktop computers. The largest provider of GIS software is the Environmental Systems Research Institute, now known as ESRI Inc., which was started in the 1970s in Redlands, California, by Jack Dangermond. While at the Harvard Laboratory for Computer Graphics and Spatial Analysis, he recognized the potential of computerized geographic applications in town planning, power-plant siting, and environmental assessments. Out of this work grew the company's development of the first successful commercial GIS software, called ArcInfo systems. As more companies have entered the field, supplying GIS software has become a $3 billion industry.

Among the old-line institutions, the Defense Department's National Imagery and Mapping Agency, once a bastion of secrecy, has felt compelled to offer free mapping software for viewing many of its data files. The Geological Survey made available much of its topographic map data on the World Wide Web. It also experimented with producing printed maps at something like a money machine. Insert some money, punch in the topographic map requested, and out comes the map, almost instantly. The National Geographic's *Atlas of the World*, published in 1999, was based largely on a new digital database designed to facilitate updating for future editions and to be a source of a Web site featuring maps and geographical data. This "interactive atlas," created with the assistance of ESRI, was promoted as "the gateway to digital maps of value not only at home, school, and office but, as the site develops, also to professionals in conservation, forestry, real estate, agriculture, health care, and many other fields."

The creation and the use of maps at the end of the twentieth century were indeed in the throes of revolution. So profound were the changes going on that historians and cartographers looked as far back as the Renaissance, in the fifteenth and sixteenth centuries, for precedent. Scholars note that in 1400 few people in Europe used maps, except for Mediterranean seamen with their portolan charts; by 1600, artistic innovation, an emphasis on quantification and measurement, new knowledge from global exploration, and the practices of warfare and statecraft had combined to make maps a common conceptual tool essential to a variety of professions. From the Renaissance sprang a widening map consciousness.

But in pace and magnitude, with technologies changing by the year and with extraordinary amounts of mappable information (some of it from worlds away) now at hand, the current revolution seemed to be truly incomparable.

Extraterrestrial Mapping:
The Moon

On the day before they made history the astronauts of Apollo 11 were far from Earth, 380,000 kilometers, orbiting the Moon and adjusting their eyes and minds to a new world, blinking at how well reality compared with the maps in their own hands. Neil A. Armstrong, the commander, remarked over the air-to-ground radio: "The pictures and maps brought back by Apollos 8 and 10 give us a very good preview of what to look at here."

Mission Control asked them to search out Aristarchus, the most mysterious of the observed craters. Brighter than any other crater, it had shown signs in the past of emitting a strange glow. Edwin E. Aldrin, Jr., another of the astronauts, radioed to Mission Control: "We'll get out the map and see what we can find out around Aristarchus."

In a few minutes the Apollo 11 spaceship would move around behind the Moon and out of radio contact, so Michael Collins, the third man in the crew, asked Mission Control: "Could you give us a time of crossing the prime meridian 150 west?"

This radio exchange between Apollo 11 and Mission Control at Houston on July 19, 1969, suggests the central role of cartography in the first human exploration on a world beyond Earth. Only a world that had already been charted—extensively and, in some places, well—could be discussed in such language. Places had names. Their positions were located on a grid of latitude and longitude. There were landmarks the astronauts, with map in hand, could look for and be guided by.

The Apollo 11 astronauts had with them a clear plastic pouch filled with more than one hundred folded maps: maps of the Moon as it would look to them from lunar orbit, lined with coordinates and the spacecraft's predicted orbital path; the lunar landscape under their approach path for landing, with landmark craters carefully plotted; the Earth they flew around before embarking on their lunar trajectory; and the oceans of Earth where they could splash down after their adventure was over. The lunar orbital maps were Mercator projections at a scale of 1:11,000,000. The surface exploration charts included photomaps and geology maps at scales from 1:100,000 down to 1:5,000. Identical maps were spread out on the walls and tables in the back rooms of Mission Control, for everywhere the astronauts went their progress was being traced on maps by hundreds of scientists and engineers.

It is hard to say which would have more startled Columbus or Cook: the news of men reaching the Moon or the sight of most of the world following their every step by sound, sight, and map. Explorers used to go to a remote place and then map it. But the reverse was true in the case of the Moon. Through telescope, photography, and remote-sensing technology cartography was able to precede lunar exploration *in situ*.

Those who planned and flew the Apollos drew on three and a half centuries of lunar cartography, the story of which evokes a sense of *déjà vu*, for it was in many ways a rerun of what happened in Earth cartography. Myths and misconceptions had to be dispelled. A world represented on primitive maps of little value gradually came into focus through new technologies and voyages of exploration. Geodesists (or, more correctly, selenodesists) struggled to learn the true shape of the Moon, to determine a prime meridian, and to establish reliable control points. But the Moon was a much different world and so far away, although close enough to be the object of the first extraterrestrial mapping.

No one seems to be sure to whom to give credit for the first map of the Moon. The honor used to be accorded to Galileo without question, but in recent years evidence has been advanced in behalf of either William Gilbert or Thomas Harriot.

Gilbert, physician to Elizabeth I and discoverer of the nature of terrestrial magnetism, is known to have produced a sketch of features that could be seen with the naked eye when the Moon was full. This was not published until 1651, although it could have been drawn no later than 1603, the year of his death. Gilbert expressed regret that no one had drawn such a sketch in antiquity, which he said would have enabled him to learn whether after the lapse of many centuries the face of the Moon had undergone any change.

Harriot, an English mathematician and one-time tutor of Sir Walter Raleigh, drew a lunar sketch based on telescopic observations in July 1609. Thus, his lunar sketch predated Galileo's by several months, but it was somewhat less detailed than Galileo's. However, an undated diagram of the full Moon by Harriot, presumably drawn in 1611 or 1612, has features both numbered and lettered; it can thus perhaps be called the first telescopic lunar map.

Galileo, the great Italian scientist, built his first telescope and focused on the Moon in the winter of 1609–10. He found the Moon "everywhere full of vast protuberances, deep chasms, and sinuosities." His examination of the roughness of the terminator—the line between illumination and shadow—convinced him that the Moon could not be Aristotle's perfect sphere. He referred to the dark areas as "the ancient spots" and revived, without actually committing himself, speculations that they might be lunar seas. By noting when a certain lunar peak began to be illuminated by the Sun, and roughly gauging its distance from the sunrise line, Galileo calculated the height of the peak to be about seven kilometers, which is about 50 percent too tall. Galileo published these discoveries in his book *Sidereus Nuncius,* which contained four different phase sketches of the Moon, each eight centimeters in diameter. Unfortunately, Galileo was not a particularly good artist and the crude woodcuts used in reprinting his original book diminished among some lunar authorities his reputation as an astronomical observer. In fairness, however, it should be remembered that he was using an instrument with streaky glass and a magnification of only twenty, and his ranging curiosity diverted his attention from the Moon rather too soon and led him to discoveries of Jupiter's four largest satellites and a study of sunspots.

Nonetheless, the excitement generated by Galileo's lunar observations led to the widespread use of the telescope and the drafting of a succession of lunar maps during the rest of the seventeenth century. For that reason, whether his sketches were first or third, whether they could be described as maps or not, Galileo certainly should be credited with inaugurating lunar cartography. Gilbert's map had not yet been published, and Harriot had an aversion to publication and generally kept his work to himself. A step forward in science is of little value unless it is made known to others, especially other scientists, so that they can examine it critically, duplicate it, and eventually improve upon it.

One of the new observers of the Moon was Michael F. van Langren, or Langrenus, a Flemish astronomer to the Spanish court. Langrenus produced one of the early reasonably accurate maps of the Moon. His purpose, in the tradition of classical astronomy, was more to solve an earthly problem than to learn about a new world. He was interested in determining geographical longitude—this was a century before John Harrison—through timing the appearance or disappearance of lunar features. The Moon because of its rotation was recognized by Langrenus as a potential timepiece, correct time being essential in knowing longitude. Langrenus, therefore, started observing in 1628, but his map did not appear until 1645. On it he identified 325 features and named them after members of royalty, saints, scientists, astronomers (including himself), and geographers. This was the first attempt to describe the lunar face in terms of a comprehensive nomenclature.

Many of the lunar names still in use were applied by Giovanni Battista Riccioli, a Jesuit priest who published a map in 1651. Like Langrenus, Riccioli assumed that the dark smudges on the lunar surface, Galileo's "ancient spots," were seas, and so each was given the Latin name *mare*, or sea. Those so identified were given such fanciful names as Sea of Fertility, Sea of Serenity, Sea of Tranquillity, and Ocean of Storms. He also adopted and expanded on Langrenus's practice of naming craters for scholars, scientists, and other prominent people.

As the quality of the telescopes improved so did the lunar maps. The foremost lunar cartographer of the seventeenth century was Johannes Hevelius. A patrician brewer and city councilor of Danzig, Hevelius studied astronomy and mathematics, learned drawing and copperplate engraving, and acquired some of the best telescopes of his time. He built an observatory in one of the rear buildings at his home and there, from 1641 to 1647, made regular observations and turned them into the first comprehensive lunar atlas. The three charts comprising the Hevelius atlas were, by the standards of the day, all true maps. They were drawn to scale. His height measurements of some of the lunar mountains were more accurate than Galileo's estimates.

Robert Hooke, a contemporary and rival of Newton, attempted the first detailed drawings of lunar craters in about 1665, and his depiction of the crater Hipparchus was remarkably good. While he was at this work, Hooke pondered the forces that could have caused such a scarred landscape and proposed for the first time the theories of meteorite impact and volcanic origin of the craters. He also surmised that the lunar surface was made of "an earthly or sandy substance."

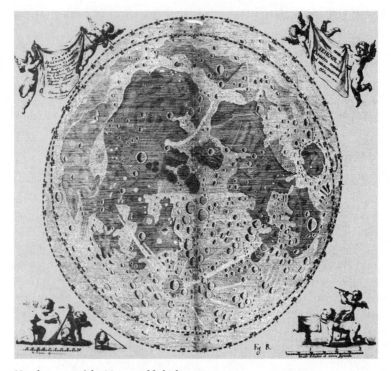

Hevelius map of the Moon, published in 1647

The most handsome map of the seventeenth century was by Jean Dominique Cassini, the founder of the Paris Observatory and the man who initiated the topographic survey of France. His full-face drawing of the Moon, published in 1680, was the first to resemble the photographs with which we have now become so familiar.

The next lunar chart of significance was produced by Johann Tobias Mayer, the German astronomer-cartographer who grappled with the longitude problem. According to Zdenek Kopal, a modern lunar authority, Mayer's work around 1750 "represents a veritable landmark in the history of selenography, which has since become a more exact branch of lunar studies."

Mayer tackled some of the fundamental questions of selenodesy. If geodesy is a science still capable of surprise a couple of millennia after Eratosthenes, then imagine the challenge of selenodesy in the time of Mayer: to establish some mathematical frame of reference for a remote, unreachable object that permits only one side of itself to be seen from

Earth. Mayer did the best he could. By observing the rotation of the Moon over a long period of time, Mayer found its axis of rotation to be inclined to the ecliptic by 1°29′ (modern angle: 1°31′). This was an important step toward finding the Moon's equator (the great circle perpendicular to the polar axis of rotation) and thus toward a system of coordinates. Mayer established a basic control point at a visible crater near the center of the apparent disk of the Moon. With a glass micrometer, an instrument for measuring very small dimensions, attached to his telescope, he fixed the positions of 23 secondary control points and estimated an additional 65 points. The Moon was getting a geodetic—or selenodetic—network, and, beginning with Mayer's work, maps of the Moon were assuming the appearance of modern maps.

But efforts to establish a more reliable coordinate system were complicated by the fact that, from the standpoint of viewers on Earth, the Moon did not seem to sit still; it seemed to rock east and west and roll north and south, phenomena known as lunar librations. The apparent east-west shift of plus or minus 8° is due to the combined effect of the Moon's consistent rotation and variable orbital motion. There is also a plus or minus 6.5° apparent north-south shift due to the inclination of the Moon's axis of rotation with respect to its orbit about Earth; when the axis is tilted toward Earth, for example, it is possible to see 6.5° beyond the south pole. These librations allow viewers from Earth the advantage of getting periodic glimpses over the top and around the edges to the Moon's far side—to the point of being able to see 59 percent of the Moon's surface. But the disadvantage, which made things tough on early selenodesists, was that the positions of features on the lunar surface seemed to shift over time, the result of the librations. This made it difficult, but all the more important, to fix the lunar equator and a prime meridian as the foundation for lunar mapping by longitude and latitude.

Two Berliners established what has become, in a sense, the Greenwich Observatory of the lunar coordinate system. In the course of preparing one of the best lunar maps of the nineteenth century, Wilhelm Beer, a wealthy banker and amateur astronomer, and Johann H. von Mädler, a professional astronomer, identified a small circular crater near the center of the lunar disk. It was near the equator, as it had now been defined from Mayer's work on the polar axis rotation, and near the longitudinal center, as identified after all the librations had been taken into account. Thus the crater, Mösting A, was used to locate the prime meridian on the Beer-Mädler maps. To this day, though subsequent observations have refined its calculated position by several kilometers, Mösting A remains a visible reference point for lunar longitude akin to the 0° longitude line drawn in cement at Greenwich.

Later, after the invention of photography, it was possible to improve and extend selenodesy through Earth-based lunar triangulation by photogrammetry. The first useful photographic triangulation of the Moon was performed about 1900 by Julius H. Franz. He fixed 150 points rather precisely and then interpolated an additional 1,150 points from measurements of five photographic plates made by the Lick Observatory. The first map made using the expanded triangulation network was the work of Walter Goodacre, a British amateur astronomer.

Lunar mappers were slow to exploit this capability. In fact, lunar studies slumped into a general decline in the first half of the twentieth century as professional astronomers set their sights on the universe beyond the solar system, leaving the Moon primarily to amateur observers. Some observatories published impressive photographic atlases, which included some photomaps, but they made no effort to convert the pictures to topographic maps.

When the prospect of traveling to the Moon became more real than fictional, however, cartographers went to work as never before. What had been the pursuit of astronomers and amateurs became—in the late 1950s—the work of large government mapping agencies. In the United States, the Army Map Service addressed itself to the problem of producing relief maps by resurrecting lunar stereoscopic photogrammetry. The Army cartographers selected eight pairs of photographs taken at the Paris Observatory between 1896 and 1907, dates indicative of the long hiatus in professional lunar astronomy.

The photographs were printed on glass plates and analyzed through a special photogrammetric plotting instrument. Variations in elevation were determined and delineated by contour lines. In the course of their work the Army cartographers had to decide on a vertical datum. What should be considered, for mapping purposes, the lunar equivalent of mean sea level?

At first, it seemed logical to measure the elevations of features as additions to the mean value of the lunar radius, which was determined to be 1,738 kilometers. But this would have resulted in too many features with negative elevation—too many lunar Death Valleys. Accordingly, the mappers elected to assign zero elevation for the Moon to the floor of the crater Aristarchus. This seemed to be the lowest place on the visible hemisphere. Lunar elevations were thus calculated from an assumed radius of 1,732.4 kilometers, which worked out to a conveniently round number for the elevation of the fundamental control point at Mösting A—7,000 meters. Unfortunately, many of these elevation figures have proved to be useless; plate grain, emulsion movements, and poor resolution were insurmountable obstacles.

The Air Force Aeronautical Chart and Information Center, working with the National Aeronautics and Space Administration, gradually moved to the forefront of lunar mapping. The Air Force improved on the Army maps by assembling new photographs taken by more modern cameras attached to telescopes at Pic du Midi in France and at Flagstaff, Arizona. Stereoscopic effects were obtained from photographs taken at different librations. Time-lapse photography increased the variety of perspectives and, therefore, the height-measuring capability. The result was a series of 1:1,000,000 topographic maps with shaded relief and contours. For subsequent work the Air Force depended heavily on the *Orthographic Atlas of the Moon,* published in 1961 by the Lunar and Planetary Laboratory of the University of Arizona. According to Ewen A. Whitaker, one of the laboratory's most respected lunar mapping authorities, the atlas drew on "the best photographs of the Moon, to a scale of 100 inches to the diameter" and many of the measured control points of Franz and S. A. Saunder, an English astronomer. The grids "were mostly accurate to one-half a kilometer, an improvement of 10 to 20 times over all previous maps."

Robert W. Carder, who was chief of the Air Force lunar mapping unit in St. Louis, recalled the early difficulties experienced by cartographers as they sought to draw the surface of the Moon. "We used conventional techniques at first," Carder said. "But the Moon doesn't have terrestrial features. There's no air or water erosion on the Moon. We were making it look too much like Earth. In our first attempts we made the ridges very sharp and the valleys pronounced, but on the surface of the Moon hills are really rounded, with no sharp peaks, and the valleys are sloping. We had to find a way of making it look natural."

It was decided to have a contest. Eight cartographers were each given a Moon photograph and asked to draw a map from it. The drawings were taken to Gerard P. Kuiper, a lunar astronomer and director of the Yerkes Observatory, who picked out the most realistic one based on his years of telescope observations. The winning drawing was by a young woman with an art degree, Patricia M. Bridges.

A lunar cartographer had to overcome several handicaps. The photographs of the Moon were of non-uniform quality because of illumination variations. There were no topographic measurements or contour lines to define slope and, hence, provide a basis for hill-shading efforts. Bridges's approach was to develop a detailed mental image of the land form to be drawn, based on meticulous examination of lunar photographs taken under a wide range of illumination and on personal observations through telescopes. The feature was then drawn to scale with a tiny spray gun called an airbrush, which can be controlled precisely by a skilled operator. She accentuated highlights by removing dark tones with an electric eraser.

The Bridges technique, refined later by Jay L. Inge, a Lowell Observatory scientific illustrator, was used throughout the 1960s to make the Lunar Aeronautical Chart (LAC) map series that aided the Apollo project and the geologic mapping of the Moon. Until these maps were produced, Carder said, the best map of the Moon he had seen was one drawn by J. F. Julius Schmidt in 1878.

To land safely on the Moon the astronauts needed maps so faithful to the lunar terrain that they would not be in danger of tumbling into uncharted craters or off misplotted mountains. The Apollo 11 astronauts, and those who followed, had such maps of their landing sites. But it took five years, a battery of computers, hundreds of scientists and technicians, and a number of robot explorers to prepare these maps. This tremendous effort marked an entirely new phase in lunar mapping.

A step in that direction was taken on October 4, 1959, when the Soviet Union's spacecraft Luna 3 flew by the Moon and televised a sequence of photographs of the hitherto unseen far side. The quality of the individual pictures was much poorer than good telescope photography, but sufficient to reveal the far side to be much different in appearance from the near side, more rugged and with scarcely any "sea" basins. A second Soviet spacecraft, Zond 3, got another photographic glimpse of the far side in 1965. The pictures were better this time, and from them Soviet scientists produced a map at a 1:5,000,000 scale covering much of the far side.

American plans for exploring the Moon and preparing for the Apollo landings involved three series of unmanned spacecraft: one (Ranger) to crash-land, one (Surveyor) to land softly, and another (Lunar Orbiter) to circle the Moon. All were designed to collect Moon-mapping information through photography at a closer range. It was as if James Cook, years before he sailed forth himself, had sent cameras on automatic ships to scout out prospective landfalls in the Pacific.

After six disheartening failures, the final three Ranger spacecraft succeeded, in 1964 and 1965, in transmitting television pictures of the lunar surface just before crashing there. Ranger 8 took close-up photographs of the Sea of Tranquillity near Armstrong and Aldrin's Tranquillity Base. All three returned pictures showing that the Moon is pockmarked with a myriad of tiny craters, some less than a meter wide. From these photographs a few topographic maps were compiled at scales of 1:250,000 with 100-meter contours and 1:50,000 with 50-meter contours. The high ground resolution also made possible the production of detailed geologic photomaps by a new astrogeology branch of the U.S. Geological Survey, established at Flagstaff, Arizona, by Eugene M. Shoemaker.

The Soviet spacecraft Luna 9 made the first soft landing on the Moon in February 1966, proving that the surface was strong enough to support spacecraft and men. Five American Surveyor spacecraft made similar gentle landings between June 1966 and January 1968. The Surveyor's television cameras took pictures completely around the horizon and from a meter or so in front of the spacecraft out to the horizon. Surveyor 6 executed a little hop after its initial landing, providing a stereoscopic base for photography. Shoemaker's team constructed a special plotting instrument to simulate this geometry, and thus was able to draw topographic maps of the landing site at a scale of 1:50 with 10-centimeter contour lines. Mapmakers could see for the first time the type of surface they were to translate into maps.

Of greatest importance to the pre-Apollo mapping of the Moon were the Lunar Orbiter missions. Five spacecraft were launched between August 1966 and August 1977, and all five returned excellent photographs of the lunar surface, front and back.

The Lunar Orbiter spacecraft, rather than carrying television cameras, were equipped with complex camera systems containing their own developing laboratories right on board. Each spacecraft had two separate cameras operating on a single web of 70-mm film. The medium-resolution camera could take pictures of objects as small as 8 meters in diameter on the lunar surface when operated from its planned optimum altitude of 46 kilometers. The high-resolution camera could photograph objects on the surface as small as 1 meter in diameter. A device that sensed the velocity-to-height ratio of the spacecraft drove the film platen so as to compensate for the speed of the vehicle and thus prevent blurring of the pictures. A clock in the system recorded the time of each exposure, which could be compared with spacecraft tracking data to determine the location on the Moon of the photographed scene.

To transmit the pictures to Earth, after they were developed on board, the strips of film were run through a scanning process consisting of a "flying spot." The spot was a beam of light that recognized varying intensities of light and recorded them. By scanning each point on each strip of film, the flying spot produced an electronic "picture" of each real picture. These electronic signals were then radioed to Earth and reconstructed into pictures at the Jet Propulsion Laboratory at Pasadena, California.

The five Lunar Orbiters returned 1,950 photographs, achieving about 99 percent coverage of a world whose thirty-eight-million-square-kilometer surface is nearly the size of North and South America combined. The first three flew in near-equatorial orbits and were directed to photograph possible Apollo landing sites. Out of some forty candidate sites identified from Earth-based observations, NASA reduced the list to five on the basis of Lunar Orbiter photography. With the primary task successfully completed, Lunar Orbiter 4 was assigned to photograph the entire front face of the Moon from a near-polar orbit, which it did at resolutions showing features 50 to 100 meters in diameter—or about 10 times better than observations with the best Earth-based telescopes. Lunar Orbiter 5 also flew a polar orbit to obtain more photographs of the higher latitudes and to photograph selected areas on the front side at resolutions of 2 to 5 meters. All five Lunar Orbiters also transmitted many pictures of the Moon's far side.

The Lunar Orbiter photographs were eventually pieced together into mosaics, with the priority on photomaps for the so-called Apollo Landing Zone, an area extending to 45° east and west longitude and 5° north and south latitude. This involved long, tedious analysis of the photographs to develop networks of control points and to determine topographic relief more accurately.

Triangulation is a trying operation in the best of circumstances, and for the Moon there were new and more complex demands. In the triangulation of a single 16-photograph strip, for example, the lunar photogrammetrists found it necessary to measure and compare more than 12,000 separate points in order to remove imagery flaws, to tie the individual frames together, and to provide the required control network. Then, to their disappointment though not surprise, the mappers discovered that their control points derived from the Lunar Orbiter photography were incompatible with those determined by Earth-based observations. Astronomers had developed a system of coordinates based on the center of the Moon's figure, as seen from Earth. This might have been adequate for small-scale mapping, but not for large-scale maps of exploration. The Lunar Orbiters had a different and truer perspective. Circling under the influence of lunar gravity, anchored, if you will, to the Moon, they produced data for a coordinate system tied to the lunar center of mass.

The new coordinate system became the framework in which the absolute positions of the Apollo landing sites were determined. For mapping the Apollo 11 landing site Air Force and Army mappers identified fifty points common to all overlapping photographic passes and used them for a refined retriangulation. This enabled them to locate the center of the landing ellipse to an accuracy of plus or minus one kilometer. Since potential landing sites were selected for their relative smoothness, the center of the ellipse had no visibly distinctive topographic features and so it was necessary for the mappers to plot a number of nearby navigational landmarks, usually small craters that could be seen through the Apollo sextant. The positions of these landmarks in relation to the landing site had to be correct to within 100 meters.

Not only did the Apollo astronauts need to know where their landing sites were, but they also needed to know their topography. From the Lunar Orbiter surveys cartographers had to extract enough information to assure mission planners that the landing site was reasonably flat and smooth. The four-legged lunar module could safely land on slopes no steeper than 30°. Moreover, the craft's navigation system used radar to measure ground clearances and to update its position during the descent to the lunar surface. Sharp variations in the terrain of the approach corridor—a deep crater or an unexpected ridge—would thus confuse the computerized trajectory calculations. There could be no such surprise.

Remarkably detailed relief maps of the Apollo landing zone were derived from Lunar Orbiter photography through the techniques of stereophotogrammetry. On all but a portion of the first Lunar Orbiter flight the cameras were programmed to take overlapping pictures. This provided stereoscopic views of the lunar surface. Using precise measuring instruments such as the comparator, the mapmakers were able to compute distances and angles between features and thus figure out elevations.

But this did not tell them the slope of the land and absolute elevations in terms of the Moon's "sea level." To find this out the mapmakers resorted to such supplementary data as the exact attitude, velocity, and trajectory of the spacecraft when it took any particular picture. However, the data were not sufficiently exact for their intended use.

An analysis of the tracking data, which told the Orbiter's position with respect to the Moon's center of mass, enabled mapmakers to compute the radius vector from the mass center to the craft. They called this R. It was simply the distance from the center of the Moon to the camera. But they also had a small r to work with, the distance from the camera to the point of the surface they had previously measured by stereophotogrammetry. By subtracting r from R they knew the elevation of the point on the surface.

Finally, to determine the slope of the surface, the mapmakers went back to the original film strips and employed techniques of photometry, a science based on measuring intensities of light. The Sun always moves from east to west on the Moon. If a point of land is tilted toward the Sun, the picture of it will be bright; if it is sloped away from the Sun it will show up dark. Because the times when the pictures were taken were known and because the raw signals from which the pictures were reconstructed were based on a sensitivity to varying light intensities, the mapmakers could use photometric equations to figure out the lay of the lunar land.

Among the maps available to the Apollo planners, based primarily on Lunar Orbiter, were those compiled at 1:5,000,000 and 1:10,000,000 scales on Mercator projections extending to 60° north and south. But the base map for most of the planning was another Mercator map at scale 1:2,750,000 for the zone between 40° north and south. Its accuracy varied from a few hundred meters at the center of the visible face to an estimated 25 kilometers on the far side. In addition, Armstrong and Aldrin were guided by a number of larger-scale maps during their landing and surface exploration activities. There was a 1:100,000-scale photomap of the landing area by which they established their initial bearings and on which they recorded observations of distant features. There also were a 1:25,000 photomap and a 1:5,000 photomap, which was the base map for the astronauts' first traverse on extraterrestrial soil.

The Lunar Orbiters demonstrated that in an incredibly short time (one year of flight time, another year of processing) a whole world could be mapped. And it was *another* world.

H ouston, Tranquillity Base here. The Eagle has landed," Neil Armstrong announced. It was July 20, 1969, and the Apollo 11 landing craft was resting steadily on a gentle 3.5° slope, which meant the photometrists could relax. The craft had flown over a hazardous boulder field but found a smooth landing site, which meant the topographers could congratulate themselves, although they resisted giving in to the rapture of the moment. They remained, in their own way there in Mission Control, as sober and attentive to detail as Armstrong and Aldrin on the Moon.

Within ten minutes after touchdown, experts of the Mapping Sciences Laboratory at Houston were studying enlargements of lunar maps, trying to determine the exact landing point. It was not critical to the mission. But Michael Collins in the orbiting Apollo wanted to know so that he could attempt to locate his companions from where he was in orbit. Shoemaker and his fellow geologists, members of the Apollo science team, wanted to know so that they could relate the astronauts' radioed descriptions to their landing-area maps. And for the cartographers it was mostly stubborn pride. Their maps had been instrumental in the selection of the site and the astronauts' navigation to a safe landing, and so why could they not now mark on their map the precise location of Tranquillity Base? They were like mathematicians who, eschewing rounded-off figures, persist in carrying a sum out a few more decimal points. Relying on descriptions of the landing area by the astronauts, Shoemaker's geologists located Tranquillity Base at 0°41'15" N and 23°25'45" E.

Extraterrestrial mapping had passed its sternest test. Maps made from a distance, often a great distance, had proved to be accurate guides in the exploration of another world. The next step, on the Moon as it has been so often on Earth, was to improve upon the maps and to employ mapping techniques not only to get somewhere but to *know* the place where you have been.

Although the "Moon walks" made the big headlines, the command module that remained in lunar orbit was responsible for the best and most comprehensive mapping results of the Apollo Project. The last three missions—Apollo 15, 16, and 17—carried the first truly photogrammetric systems for extraterrestrial mapping. These systems were housed in the rear compartment, the so-called service module, of the Apollo spaceship. They included a mapping camera, a stellar camera, and a laser altimeter.

The 76-mm primary mapping camera photographed the lunar surface as the spacecraft orbited the Moon. Each frame covered an area of 167 square kilometers at resolutions of 20 to 25 meters. Because the orbits of most of the Apollo missions were close to the lunar equator, it was not possible to study the total surface—only about one-quarter of the Moon. The precision of the results depended heavily on the stellar camera and the altimeter. The stellar camera looked to the side of the spacecraft and photographed the star field each time the mapping camera was exposed; this established the angle at which the mapping camera was photograph-ing the Moon, while tracking data recorded the position. Each exposure by each camera contained fiducial marks, fixed points of light which were used to define camera coordinate systems for the analysis required to convert the pictures to maps. Placed next to the cameras was the laser altimeter, which obtained the range from the spacecraft to the surface by measuring the time required for a laser pulse to be transmitted to the surface, reflected back, and received by the instrument.

The altimeter measured the distance from the spacecraft to the lunar surface with 1-meter accuracy. However, since the distances from the center of the lunar mass to the spacecraft were only accurate to from 50 to 300 meters, the height of mountains and valleys could not be determined with precision—but it was better than ever before. Some highland areas on the far side were shown to rise almost 4 kilometers above the average surface level. There is a drop of about 8 kilometers from the highest lunar mountains to the floor of surrounding basins.

From the camera system lunar cartographers obtained the data for a three-dimensional control network with positions on the lunar surface defined with respect to the center of the lunar mass. The Moon was being mapped in relation to itself, not as an object observed from Earth. Apollo 15 alone supplied the positional data for more than 175 1:250,000 lunar orthophoto maps published since 1973. Using radio antennas on Earth and laser beams bounced off reflectors on the Moon, selenodesists have fixed the locations on the Moon of several Apollo scientific stations and the Soviet Lunokhods, and as a result have helped establish posi-tional relationships and expanded the control structure on which Apollo photography is converted to maps.

With the Apollo missions it became possible—and necessary—to map more of the Moon than its general topography and gross geologic features. A control network to span the Apollo area was developed soon to support geologic, seismic, chemical, and gravimetric mapping work.

One of the most ambitious of these undertakings was the chemical mapping of one-quarter of the lunar surface. Astronauts visited only six places, whose exact chemistry could be deciphered from the returned samples. But on later missions the Apollo spaceships were also equipped with three remote-sensing instruments for chemical mapping from orbit. They all depended on radiations coming up to the spacecraft from the Moon, radiations that carry chemical signatures. The three radiations were x-rays, gamma rays, and alpha particles.

The x-ray detector received radiations from the Moon that were excited by solar x-rays. When solar x-rays hit the lunar surface, they penetrate less than a millimeter deep but they excite the atoms in the soil and cause them to give off weak x-rays of their own. From these secondary x-rays, detected and sorted out by the Apollo instrument, scientists could measure concentrations of major elements such as silicon, aluminum, and magnesium.

The gamma-ray instrument detected emissions of radioactive elements in the lunar rocks and soil, by which scientists could map concentrations of potassium, uranium, and thorium in the top few centimeters of the surface and, through further analysis, also identify the presence of iron, calcium, and titanium. The alpha-particle experiment sought to find any emissions of the radioactive gas radon from vents in the lunar surface. This was an attempt to detect recent volcanic activity on the Moon.

When data from the x-ray and gamma-ray experiments were plotted on base maps, Apollo scientists made an important discovery: there are definite chemical differences between the dark maria and the lighter highland areas. The highlands are rich in calcium and aluminum. The maria are made up of rocks rich in such elements as magnesium, iron, and titanium. Generally the topographic lows are radioactive highs, and vice versa. The gamma-ray instrument detected a small but definite signal of radon near Aristarchus, the intriguing crater the Apollo 11 astronauts had been instructed to observe closely. Scientists are still not sure whether this is indicative of lunar vulcanism or simply of slow leakages of radioactivity.

In addition, the Apollos mapped a phenomenon that had been discovered by an analysis of the Lunar Orbiter tracking data. Radio signals showed that the spacecrafts failed to move regularly: over certain regions of the Moon they speeded up, suggesting extra concentrations of mass, and thus stronger gravity, in the outer layers of the Moon. The

OVERLEAF: *Map of the Moon from the Apollo missions*

INDEX MAP OF THE NEAR SIDE OF THE MOON
Number above quadrangle name refers to lunar base chart (LAC series);
number below refers to published geologic map

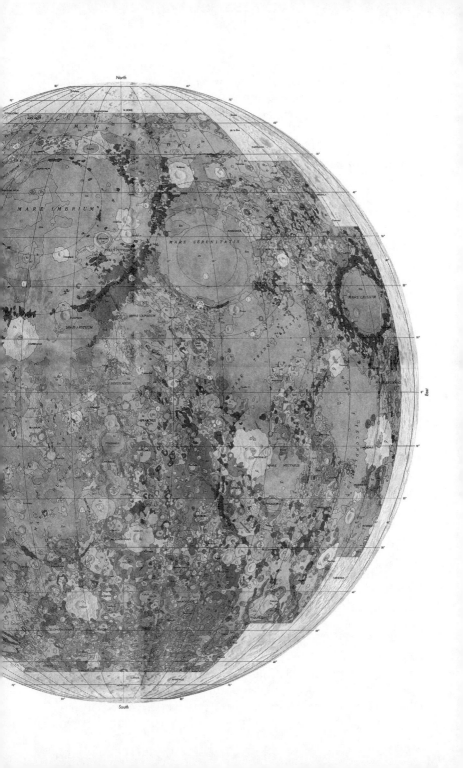

detective work was done by Paul M. Muller and William L. Sjogren of the Jet Propulsion Laboratory. They isolated the sources of the perturbations as mass concentrations (mascons) of denser and heavier rock in the circular *maria*—Imbrium, Serenitatis, Crisium, Humorum, and Nectaris.

The work of Muller and Sjogren provided the beginnings of a gravity map for the near side of the Moon. Then Apollo 15 and Apollo 16 released small unmanned satellites that remained active in lunar orbit for days, furnishing additional tracking data to confirm earlier findings and to map gravity over larger regions of the Moon, including the far side.

The seismometers left on the Moon by the astronauts gave scientists their first clues to the lunar interior—and the first diagrams of the layered subsurface structure, which were rough beginnings of lunar seismic mapping. From the records of thousands of moonquakes, meteorite impacts, and the deliberate crashes of discarded Apollo rocket stages onto the surface, it was determined that the Moon is divided into layers much as the Earth is. The outermost part of the Moon, the crust, is about 60 kilometers deep. Beneath the crust is a thick layer of rock (the mantle) that extends down to more than 800 kilometers.

O nce American astronauts had achieved the Moon, once they had triumphed over the Soviet Union in this particular engagement of the Cold War, lunar exploration was shunted off the main line of world attention. Mapmakers felt frustrated. As proud as they were of their latest lunar maps, they knew that much work remained to be done. Because modern mapping of the Moon was largely based on surveys for and by the Apollo Project, the best charts were confined to equatorial regions, leaving ample territory in need of more detailed mapping. And scientists wanted better maps of their areas of special interest. Geologists needed more comprehensive maps of the chemical and mineral composition of the lunar crust. All scientists were eager for maps of the polar regions. But more extensive lunar mapping had to await the next wave of exploration, and that was a long time coming—almost a quarter of a century. So far, it has included only two modest missions.

The first was by the spacecraft *Clementine* in 1994. A small experimental vehicle flown by the Defense Department, *Clementine* after a time was diverted to orbit the Moon. Though not designed for such tasks, the craft's radar and other instruments were used to conduct a brief but productive reconnaissance. This project yielded improved maps of the high latitudes, notably the poles. The biggest surprise was the detection of what appeared to be water ice in great quantities in the permanent shadow of the lunar south pole.

Another small spacecraft, *Lunar Prospector*, began an eighteen-month mapping mission in early 1998. Orbiting the Moon at altitudes ranging from 100 kilometers down to seven, the spacecraft collected and radioed much more detailed information about lunar composition and structure, which mapmakers turned into high-resolution geologic maps.

By analyzing shifts in *Lunar Prospector*'s radio signals, scientists identified slight variations in the craft's velocity as it orbited the Moon, and translated this information into maps of the lunar gravitational field. Scientists could infer from this data the distribution of the Moon's mass and, through further calculations, discovered that the Moon probably has a small metallic core. The nature of the deep lunar interior had eluded previous investigations. The relatively small size of the core was interpreted as further evidence of the Moon's origin in a collision between a huge protoplanet and Earth: Rocky debris scattered in the collision had clumped together to form Earth's only natural satellite.

Lunar Prospector also took a close look at both polar regions and detected hydrogen atoms at the surface, seeming to confirm Clementine's findings of water ice. The ice was presumably dumped on the Moon by comets and survived only at the poles, where temperatures never rise much above −220° C. Some scientists remain highly skeptical. But if the ice is really there, the new maps of the lunar poles may someday guide visitors in their search for prime sites for bases and fueling stations.

Extraterrestrial Mapping: Mars

B y November of 1971 cartography had come a long way from Eratosthenes and the well at Syene. Earth had been surveyed on foot and camel and horse, from ship and airplane and spacecraft, by camera and radar and laser beam. Earth's continents and ocean basins had all been mapped, if only lately and not always too well. Even Earth's Moon had been charted, and the first radar signals had penetrated the clouds of Venus. And now cartography, in another act of audacity and virtuosity, had reached out millions of kilometers to embrace Mars.

The first surveying party reached Mars on November 13, 1971. It arrived in the form of the American spacecraft Mariner 9, an unmanned vehicle weighing less than a ton and shaped something like a flying windmill, a complex robot equipped with the essential field tools for extraterrestrial mapping—television cameras, remote-sensing instruments for determining elevations and distances, and a finely tuned communications system for transmitting its findings back to Earth. After a 167-day voyage from Earth, covering 397 million kilometers along a curving trajectory, Mariner 9 fired its braking rocket, slowed down from 17,600 kilometers an hour to under 13,000, and was captured in orbit around Mars. It thereby became the first man-made satellite of another planet in the solar system.

From its vantage point, ranging between 1,350 and 19,000 kilometers above the surface, Mariner 9 conducted its historic survey for nearly a full year. The spacecraft took and transmitted more than 7,000 television pictures of Mars before its maneuvering rockets ran out of nitrogen gas in October 1972.

This was all the cartographers needed, and considerably more than they had expected. Within a month after Mariner ceased operations, the cartographers produced from a mosaic of the photographs the first detailed map of the entire globe of another planet.

The man most responsible for this achievement was Harold Masursky, chief scientist of the U.S. Geological Survey's Center of Astrogeology at Flagstaff. Studying a little-understood volcanic region of central Nevada aroused Masursky's initial curiosity about the geology of other worlds. A knowledge of the basins of the Moon, he thought, should help him understand central Nevada, and vice versa. It was 1962, the space age was young, but the Survey was already extending its franchise to the Moon—and so did Masursky. He joined the staff of the newly established Center of Astrogeology and soon found himself very much a man of the new cartography, knowledgeable in aerial photogrammetry, computers, and spacecraft television systems. He became an expert at interpreting photographs of the Moon taken by the new unmanned spacecraft, and soon he was applying the same expertise to Mars.

"We've done the same thing for Mars as the old guys of the Survey did for the American West," Masursky said. "No one really knew what was out there, and so they went out and mapped the West. Those maps were the tools for exploring the West. These maps of ours are the tools for exploring Mars."

Mars had long been the object of all manner of misconceptions because, until Mariner 9, man's view of that planet remained unclear. At its closest approach Mars is never less than 56 million kilometers away, and such viewing opportunities come only every two years and last only a few weeks. At those times, with the best of telescopes, astronomers can see little more of the Martian surface than one can see of the Moon with the unaided eye.

In 1840, Beer and Mädler, the outstanding lunar mappers, produced the first map of Mars that was more than a crude drawing. They endowed it with a network of longitude and latitude lines, but the substance of the map bears little resemblance to any later maps of Mars. They simply lacked the facts.

This and subsequent maps drawn from Earth-based observations represented more a vision of Mars than the reality. They were a shadowy blur of light and dark revealing only the broad variations in the planet's albedo, the light reflectivity of its surface. They showed the polar caps and a few other surface features, but many of them, like the legendary canals, tended to exist only in the eye of the beholder. Indeed, the canals of Mars stand out as an engaging example of man's fertile imagination rushing in to fill a void on the map, a nineteenth-century variation on the medieval Prester John.

In 1877, conditions were particularly favorable for observing Mars. When the planet came unusually close to Earth that summer, Giovanni V. Schiaparelli of Milan took full advantage. He recorded with considerable accuracy the latitude and longitude of sixty-two characteristic features of the surface, which he later incorporated in some of the better contemporary maps. But, peering through his small refractor telescope, Schiaparelli spotted something else, something amazing. He saw what appeared to be fine, dark straight lines on the surface of Mars. Some of the lines appeared to be more than a thousand kilometers long. And, to be visible from Earth, the lines had to be several kilometers wide.

Schiaparelli was not sure what he had discovered. He called the lines *canali,* which in Italian means "channels" and could refer to either natural or artificial features. But the word was often translated into English as "canals." Could it be that artificial waterways laced the Martian landscape? If so, who built them?

An astronomer of considerable repute, Schiaparelli did not jump to any conclusions, and for nine years no one else could see what Schiaparelli claimed to have seen. But, when some other astronomers finally did, the stories of Martian super-civilizations proliferated. Even the most learned scientists had for years spoken of inhabitants of Mars as a matter of accepted fact. Now it seemed that these Martians had engineered an elaborate system of waterways over their entire planet; by comparison, the recently completed Suez Canal was a mere ditch. These Martians, it was assumed, must be superior in many ways to people on Earth.

In 1889, reflecting the climate of opinion, the French Academy of Sciences announced that its newest award, the Guzman Medal, would be presented to the first person "to find the means of communicating with a heavenly body—Mars excluded." The exception of Mars was specified by the award's founder, Anna Émile Guzman, "because that planet appears to be sufficiently well known."*

Foremost among those who nourished ideas of Martians was Percival Lowell. A brother of Amy Lowell, the poet, and A. Lawrence Lowell, the Harvard president, Percival Lowell graduated from Harvard in 1876 and spent many years in the Far East as a businessman, diplomat, and author. Upon learning of Schiaparelli's canals, however, Lowell vowed to devote his life to a thorough study of the planets, particularly Mars. He returned to the United States and, after a careful search, built his observatory in the mountains near Flagstaff, where the clear, dry air favored telescopic viewing.

Between 1894 and 1916, using a 60-centimeter telescope, Lowell made several valuable contributions to the knowledge of Mars. He determined that the Martian atmosphere was considerably thinner than Earth's and that the planet "is very badly off for water." But the canals—he classified more than 700 of them—became his obsession. Accordingly, Lowell's maps of Mars looked like spreading spider webs, the network of thin canal lines embracing the globe. In books and lectures he advanced the theory that the "canals" and "oases" were the work of "intelligent creatures" on a dying planet who were struggling to keep alive by a planet-wide system of irrigation from the water of the melting polar caps. So persuasive was Lowell that even Schiaparelli was won over. He conceded in 1897 that the pattern of canals "presents an indescribable simplicity and symmetry which cannot be the work of chance."

Other scientists remained skeptical. One famous astronomer of the day is said to have received a message from William Randolph Hearst: "Is there life on Mars? Please cable one thousand words." The astronomer's reply to the publisher was "Nobody knows"—repeated five hundred times.

But, by the time of Lowell's death in 1916, his Martians had assumed a certain immortality in the human mind. Edgar Rice Burroughs, the creator of Tarzan, was in the midst of writing a series of popular tales of life on Mars, which he called Barsoom. In 1938, Orson Welles would frighten a nation with a chilling, all-too-realistic radio account of Martians invading New Jersey. Indeed, man's abiding fascination with Mars

*The Guzman Medal went unclaimed until 1969, when it was awarded to the Apollo 11 astronauts for the first landing on the Moon.

and the general paucity of knowledge gave priority to the exploration and mapping of that planet as an early goal of the space age.

Then, through one map, the Mariner 9 map, Mars emerged from the shadows of distance and abstraction and could be seen for the first time as a distinctive and varied world, a place of mystery still, but beyond myth.

Nowhere on the map was there a canal or oasis or any other sign that Lowell's "intelligent creatures" had ever thrived on Mars. The human eye tends to align scattered spots into a line, astronomers say, and this could explain why Schiaparelli and Lowell saw their canals. Or perhaps Mark Twain, writing in another context, had the explanation. "Partialities," he wrote, "often make people see more than really exists."

M apping a place where men have yet to set foot represented a considerable step beyond the plane table and steel tape of ordinary surveying or even the aerial photogrammetry of modern Earth mapping.

For the task of mapping Mars the Jet Propulsion Laboratory designed and built two identical spacecraft—Mariner 8 and Mariner 9. The laboratory had constructed the first successful American satellite, Explorer 1, and three earlier Mariners that scouted the environs of Mars. In 1965, Mariner 4 flew within 10,000 kilometers of Mars and sent back a score of pictures showing a bleak vista of Moonlike craters. In 1969, Mariners 6 and 7 passed within 3,200 kilometers of Mars and returned pictures of about 10 percent of the planet's surface; more craters, more disappointment to those who hoped to find Mars an abode for some forms of life. But these earlier Mariners had made only spot checks of Mars.

Originally, the mission to map Mars was assigned to Mariner 8. But, when its launching rocket failed shortly after liftoff and the spacecraft plunged into the ocean off Cape Canaveral, the burden of the entire mission fell to Mariner 9. Mariner 9's tasks had to be modified to include mapping as well as a survey of variable features on Mars, such as the polar caps, clouds, hazes, and changing surface colorations. Mariner 9 was launched successfully on May 30, 1971, and performed without flaw.

But, as sometimes happens to mundane surveyors, Mariner 9 had to wait out the elements before it could begin its mapping work. At the time of the spacecraft's arrival at Mars, in November, the greatest known Martian storm was raging. Winds of 320 kilometers an hour

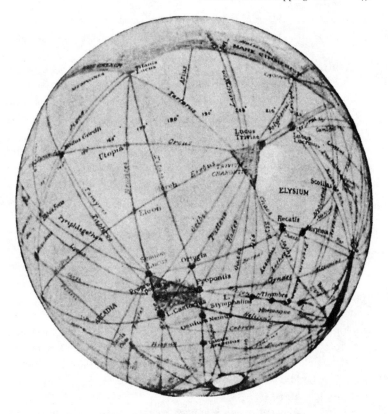

Percival Lowell's depiction of Mars canals

swept the planet, enveloping it almost entirely in a dense cloud of swirling dust. The first pictures were a disappointing blur, except for glimpses of the south polar cap and some volcano peaks, and of no use to mappers. By Christmas, however, the winds had died down, the dust had settled, and Mariner 9 got on with its work.

The spacecraft made a complete orbit of the planet twice a day. Its orbit crossed the equator at a 65° angle, an inclination making it possible to cover the equatorial regions and also come within photographic range of both polar caps. Each orbit took almost exactly 12 hours, or half the time of a single rotation of Earth. This facilitated tracking of the spacecraft by antennas on Earth. And, because the Martian rotational period is slightly longer than Earth's, 24 hours and 37 minutes, Mariner 9's orbital track was 9° farther east each day. This enabled the spacecraft to photograph a different swath on each orbit without changing course. In this way, the spacecraft passed over the same point on the planet several times, each time 19½ days apart.

Mariner 9's altitude was a compromise between the need to get close enough to the surface for high-resolution photography and the requirement not to get so close that the spacecraft might crash into the planet in the next fifty years, a precaution against contaminating Mars with Earth organisms before it could be systematically explored with life-detection probes. Mariner could have flown lower than it did. But, being limited in the number of pictures it could take, it was flown at a moderate altitude so that each photograph would cover more ground and so that the entire planet could be photographed at least once.

On each orbit Mariner 9 was capable of taking up to thirty television pictures. Most of the mapping pictures were taken with the wide-angle camera, which could encompass more territory in a single picture, but at a sacrifice in resolution. The camera's rectangular field of view spanned an area 11° by 14°, but it could detect only features larger than a kilometer in diameter. The narrow-angle camera, with ten times better resolution, could detect features as small as a football field, and was therefore used primarily to photograph objects of special scientific interest, such as canyons and volcanoes. It would not have been possible, in the limited time, to obtain coverage of the entire planet with the narrow-angle camera.

In three 19-day mapping sequences, working from the southern latitudes upward, Mariner 9 transmitted pictures of nearly all the planet. The pictures were tape-recorded on board and radioed to Earth during the approximately ten hours each day that the spacecraft was in contact with the Jet Propulsion Laboratory's deep-space antenna at Goldstone in the Mojave Desert of California. As the durable spacecraft operated far beyond its three-month life expectancy, there was time left to fill in the few gaps in the photographic coverage and to rephotograph particularly interesting features. There was even time to photograph the north polar cap, where the winter hazes did not dissipate until near the end of the mission. Thus, Mariner 9, surpassing all its objectives, provided the mapmakers with 100 percent coverage of Mars.

Mars is extremely difficult to photograph. The light contrasts are

Shaded relief map of Mars, made
from Mariner 9 photography

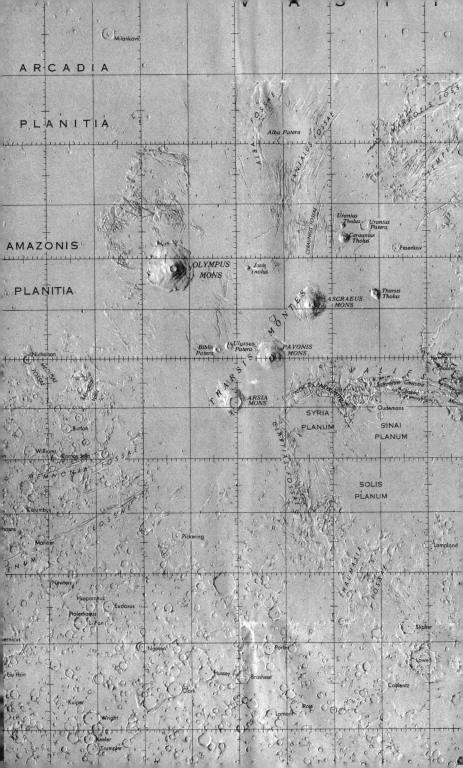

very low. The surface is subdued and, unlike the airless Moon, sometimes is masked by light hazes and dust. To make matters worse, the
pictures lacked uniform perspective. Not all of them were taken at the
same altitude or from the same angle. Often, as a result, round craters
appeared to be oval.

Other distortions were introduced by the Vidicon camera system
itself. The camera optics focused an image on the photosensitive surface of the Vidicon tube in much the same way that a television signal is
projected on the home set. The image was then read, using a scanning
electron beam. But the scan lines were not always true, because of
slight variations in the magnetically deflected beam as it traveled from
the center to the edges of a picture. Devices of this type, although the
most advanced available for the demands of the mission, were thus
unable to reproduce exactly the geometry—the precise relative position
and shape—of a scene that covers a broad area. For example, if Mariner
9 had photographed a grid of streets on Earth from the same altitude,
the streets would appear curved, a barrel-shaped distortion not unlike
that produced by fish-eye lenses used by professional photographers. In
addition, distortions occurred because the erasure of the Vidicon
between exposures was imperfect; each image contained some faint
ghosts of the previous image.

Masursky and his television examination team had anticipated such
problems and were undismayed by what they saw coming from Mariner
9. They counted on a complex computerized process of "decalibration"
to remove most of the distortions and render the pictures usable by
mapmakers. This was done as soon as the pictures reached the Jet
Propulsion Laboratory.

The Mars pictures arrived at the laboratory, by way of the Goldstone
antenna, as so much digital data radioed from the spacecraft and
recorded on magnetic tape. The conversion of the picture to digital form
took place when the spacecraft's scanning electron beam read each
image on the Vidicon tube. Each black-and-white picture was composed of 700 scan lines with 832 points (pixels, or picture elements) on
each line. And each pixel was electronically coded on a scale of gray
from 0 to 511, the darker the photographed feature the lower the number. When the data reached the laboratory, the decoding of the many
numbers produced a reconstruction of the picture that Mariner 9 had
taken.

This technique of reconstructing an image dot by dot is reminiscent
of the art of Georges Seurat, the nineteenth-century French impression-

ist. He laid down rows upon rows of tiny dots of paint to re-create a scene, as in perhaps his most famous work, *Sunday Afternoon on the Island of La Grande Jatte*. His technique came to be known as "pointillism."

To improve the clarity of the Mars pictures produced by space-age pointillism, the laboratory's computers were called upon to "play games with these numbers," as James A. Cutts, a young physicist on the Mariner photo-interpretation team, put it. This involved six major steps.

First, the computers were programmed to correct random data errors, such as missing scan lines and data "spikes" caused by radio noise. Missing lines were replaced by interpolation. An intensity value— a gray scale number—for each pixel in the missing line was assigned by the computer on the basis of the value of pixels in the lines on either side of the gap. A data "spike" was smoothed out by the computer in much the same way. Any pixel deviating by more than plus or minus 50 from the average of the two neighboring values was isolated by the computer and replaced by a pixel based on an average of the neighboring values.

Second, to eliminate the barrel-shaped geometric distortions, the image-processing computer was programmed with algebraic formulas to remove the known distorting factor in the Vidicon system, which was determined in pre-mission tests. Through a series of rapid calculations, the computer could predict where a particular pixel should really be in the pixture. All the pixels in a picture were thus "stretched" to their appropriate positions. Reseau marks, a series of black dots superimposed on each picture, served as additional guides to the computer. The dots represented a pattern of tiny metallic squares fixed at known points on the surface of the Vidicon tube. The reseau marks gave the machine a frame of reference in stretching and squeezing the many pixels into the positions that were truer to reality.

In a third processing step, the computer sought to reduce any distortions from residual images. The computer's mathematical instructions for this were based on ground tests that characterized the Vidicon tube's behavior. Each picture was compared with the immediately preceding exposure. A certain fraction of the picture's light intensity was then subtracted to remove what was assumed to be unerased data from the preceding picture. Although it was not a perfect solution to the problem, the intent was to improve all frames without further degrading any of them.

As a fourth step, the computer dealt with the generally poor contrast of the raw pictures. If, for example, all the pixels in a picture ranged in grays between 220 and 270, there would be only subtle differences in the shadings. Surface features would be almost undistinguishable, which was not uncommon. To overcome this problem, the computer used a mathematical formula by which 220—or whatever the lowest gray reading happened to be—became 0, or total black, and 270 became 511, or total white. By changing the gray scale in this way, it was possible to get more discernible shades of gray in between. Thus a crater or canyon or polar terrace might emerge from the vague shadows and dark streaks of an otherwise unrevealing picture.

Without computer enhancement, Cutts remarked, "We would not have been able to convince people that Lowell's canals were gone forever."

The fifth step in decalibration involved an adjustment of the pictures so that, though taken at different angles, they would have a common perspective. The computer's calculations took into account the position of the reseau marks and Mariner 9's tracking data, which gave the angle at which each picture was taken. With each picture, the spacecraft radioed associated information—the time and date of exposure, altitude, angles, longitude and latitude readings of the center and corners, and the approximate dimensions of the photographed area.

The ideal would have been to have each picture taken from directly overhead. But the computer was able to reconstruct the perspective of pictures taken from angles as much as 70 degrees off the vertical. Craters that had appeared oval were transformed to their truer round shapes.

Finally, since the Mars map had to be a flat representation of a rounded world, the pictures were put through another processing step to stretch the features to a certain projection. This was essential in transforming a picture into a map segment. For this, the computer was programmed with formulas describing the projection and what had to be done to each picture.

Before the mission the mapmakers had decided to use the standard Mercator projection for the planet's broad equatorial band, running from 65° south latitude up to 65° north. A stereographic projection was selected to show the polar regions as they would look from directly overhead. In some cases, Lambert conformal projections were also used for the north and south latitude bands between 30° and 65°.

Every few days a package of magnetic tapes wrapped in aluminum foil would leave the Jet Propulsion Laboratory in Pasadena and go by air freight to the Center of Astrogeology in Flagstaff. These were the decalibrated and enhanced Mariner 9 photographs, reduced to digital data and recorded on tapes. These were the image processors' finished product and the cartographers' raw material for the map of Mars.

Once again, the digital data had to be converted—this time to a photographic negative, not to a television image. This was done by a machine called the Optronics Photowrite. As in playing back a music tape, only with extremely high fidelity, the machine's magnetic head picked up the data and, instead of producing sound, converted the data to electrical impulses. These impulses directed a fine light to expose photographic film on a spinning drum. More than 300 tapes, each containing two to nine pictures, were thus processed during 1972. The negatives were developed and enlarged for use as "map fragments" by Masursky's photomosaic team.

To piece together the 1,500 photographs as precisely as possible, the mosaic team was guided by Mariner tracking data, a new latitude and longitude grid, and a network of 1,300 "control points"—sort of Martian benchmarks.

The coordinate system was established from newly refined knowledge of the planet's spin axis and oblate shape and from measurements placing the equatorial radius at 3393.4 kilometers. A Martian prime meridian, corresponding to the o-longitude line that runs through Greenwich, England, was described as passing through a small but well-defined crater, Airy-o, near the equator. The rest of the grid then fell into place.

But this was only the framework for the map. How could the mosaic team know where each picture, each surface feature, fitted into the framework?

The general coordinates for the broad area covered by each picture were derived from Mariner 9 tracking data. This gave the mosaic team a "footprint plot" for placing the pictures, and some preliminary photomosaics of Mars were so prepared. They provided a rough view of coverage, showed approximate latitudes and longitudes of features, and showed gaps in coverage. But the early photomosaics fell far short of being true maps.

For a map it was necessary to prepare a control net covering the entire planet. Small craters served as the individual control points, those well-defined reference points whose latitude and longitude were carefully plotted. These were identified by Merton E. Davies of the Rand Corporation and David Arthur of the Geological Survey, using a special series of geodesy pictures as well as the mapping pictures returned by the spacecraft. Each crater, to serve as a point, had to be found on at least two pictures and, if possible, on more. Each picture in the net had to contain more than two such craters. The control point, the exact center of a crater, was found through a slow and tedious process, not by computer but by the human eye.

Davies and his associates counted and measured pixels with magnifying glasses until they agreed on where the center of the crater lay and what its exact coordinates were. The pixel coordinates of each point were measured three or four times by two or three different persons.

The agreed-upon coordinates were included in tables that were turned over to the mosaic team. When a member of the team got one of the enhanced mapping pictures, he made sure that reference craters in the picture were placed exactly over where the control point for that crater had been marked. Masursky believes that nearly all visible features on Mars were thus located within 2 or 3 kilometers of their true longitude and latitude. On earlier Mars maps, the few visible surface features were often out of place by 250 to 650 kilometers.

After the entire mosaic was assembled and photographed, artists had some touching up to do, because, as Raymond M. Batson, technical director of the map project, said, the original could "look like a set of fish scales." Artists used erasers and an airbrush, techniques developed by Patricia Bridges* for lunar mapping, to emphasize major features before sending the map to the publishers. On the basis of experience with lunar maps, the cartographers found the airbrush-shaded relief map to have several advantages. It could be prepared quickly by skilled cartographers. It could emphasize only the "high-frequency information." Cartographers were cautioned to resist the temptation to overinterpret any feature, lest the final drawing become an artist's "rendition" rather than a map.

The finished map, published in November 1972, showed the Martian surface at a scale of 1:25,000,000. Other more detailed maps of Mars soon followed. Some showed the "ups and downs" of the Martian topography, which had proved to be much larger than expected. Masursky's cartographers derived the elevation contours from radar surveys and Mariner 9's ultraviolet and infrared measurements of Martian surface atmosphere, which is thickest and densest in places where the elevations are lowest. These maps, in particular, were used as guides in

*Who now worked for the Geological Survey.

the selection of sites for landing the unmanned Viking spacecraft on Mars in the summer of 1976.

Even before the first map was off the drawing boards at Flagstaff, scientists were using it for its intended purpose—the study of Mars. The photographs upon which the map was based and the preliminary photomosaics provided a clear and often surprising portrait of Mars as a distinctive, dynamic planet. It was not as Lowell had imagined, but neither was it the dead world suggested by earlier reconnaissance.

Masursky and other scientists were discovering Mars to be a varied world of sharp relief and many contrasts, one that has been shaped in part by water and therefore may be or may have been suitable for some forms of life. They were seeing, as never before, a world of towering volcanic peaks, chasms longer and deeper than the Grand Canyon, glacial terraces, craters resembling those on the Moon, great expanses of wind-blown plains and sand dunes, deep faults and cracks, and meandering channels that probably were cut by the flow of water. No single mapping expedition ever revealed more of an entire planet.

Casting their eyes over the new map, the scientists identified at least four major geological provinces on Mars—the volcanic regions, the canyonlands, the battered expanses of craters, and the polar terraces.

One of the outstanding features of Mars, and on the map (18° N, 134° W), is Nix Olympica (Snows of Olympus), which was discovered by Mariner 9 to be a volcanic mountain larger than any on Earth. At its base, the volcano is 500 kilometers wide, about twice as wide as the island of Hawaii—Earth's largest volcanic mountain. Nix Olympica (subsequently rechristened Olympus Mons) rises 24,000 meters above the surrounding plain and is capped with a 65-kilometer-wide crater, many times wider than Oregon's Crater Lake. Southeast of Olympus Mons are three somewhat smaller craters, all of which suggest to scientists that Mars had and may still have a hot, active interior.

To the east of the volcanoes, just south of the equator, stretches a canyon unlike any on Earth. The canyon, named Valles Marineris, after Mariner 9, runs some 4,000 kilometers, nearly the distance between Los Angeles and New York and one-sixth the circumference of the Martian globe. It is 120 kilometers wide and 6.5 kilometers deep, which is six times as wide and more than three times as deep as the Grand Canyon. The great canyon and other cracks and faults are strong evidence of the slipping and cracking of the crust during the relatively recent history of Mars.

Although the great Martian canyon does not appear to be associated with erosion, and there are no present signs of liquid water on the planet, other deep cuts in the surface may have once been fast-flowing streams. The most provocative example is a 400-kilometer-long meandering channel (centered at 29° S, 40° W). Unlike Lowell's straight channels, this one twists and turns like the Mississippi and has a number of tributaries.

The largest basin on Mars is Hellas, circular flatlands south of the equator that extend from 310° E to 260° E. Surrounding Hellas is a densely cratered region believed to be the oldest surface feature, the result of ancient impacts by meteorites. Great sand dunes are seen in this area and near the north polar region. The three Mariners that preceded Mariner 9 flew over this vast area, which accounts for their rather unrevealing photography.

The fourth province is a spectacular expanse of stairstep terraces in the two polar regions. The layers in the terraces are estimated to be 25 to 90 meters thick and are probably accumulations of water ice, dry ice (frozen carbon dioxide), dust, and volcanic ash. The map shows the south polar ice cap at its minimum breadth and shows the northern ice cap close to or at its minimum.

The Mariner 9 maps showed the way for future missions. In the summer of 1976, two Viking spacecraft swung into orbit around Mars and dropped landers to the broad plains below. Although the landers found no unambiguous evidence of living organisms in the soil, which was the big question and primary objective, the two Viking orbiters spent several years photographing the varied landscape. Together they transmitted more than 51,000 pictures, mapping 97 percent of the surface in much greater detail than had Mariner 9.

The maps supplied some answers to questions raised by Mariner 9. They traced revealing textures of the cold, arid planet, inspiring scientific speculation about its early history. They showed deep channels resembling dry riverbeds cutting across flood plains. Chaotic terrain suggested an underlying permafrost. These features, and volcanoes, attested to a time when Mars was a much different place, more active geologically, wrapped in a thicker atmosphere, warmer, and presumably well watered by episodic flooding. In a wetter past, life might have emerged, only to vanish in the dry wind or retreat to moist subterranean refuges.

The issue of water on Mars gave scientists a thirst for more exploration. But that was not to occur for another two decades. Political support for such ventures faded with the failure of the Vikings to find life. Beginning in 1997, however, American spacecraft returned to orbit and land on the planet.

A small mapping spacecraft, Mars Global Surveyor, arrived in an orbit of the planet in September 1997. Its cameras brought Mars into much better focus, producing pictures of the landscape that were 20 to 40 times more detailed than anything obtained by Mariner 9 or the Vikings. Looking down at the steep walls of Valles Marineris, Surveyor's camera revealed for the first time distinct layers, each up to 50 meters thick, probably left by periodic lava flows in the distant past. Many of the dry channels, as expected, appeared to be eroded by short-duration floods of water. But at least one narrow meandering valley seemed to be carved by water steadily flowing over a long period of time, evidence of an epoch when Mars had a denser atmosphere and wetter environment, which might possibly have supported primitive life.

As more data arrived, mapmakers produced the first three-dimensional global chart of Mars—a revelation to scientists. Since the time of Mariner 9, they had puzzled over the differences between the relatively flat topography of the northern hemisphere and the cratered highlands of the south. But they were unprepared for what they saw on the map, which was drawn from 27 million precise elevation measurements by Surveyor's laser altimeter. The entire planet slopes from its south to its north pole. The average elevation in the south is 5 kilometers higher than in the north. Here was evidence for the global-scale flow of water that appears to have occurred early in Martian history. Water, running downhill from the south, could have pooled in a huge northern depression, which for a time could have contained a large ocean. On Mars the full range of the highs and lows of topography is 30 kilometers, one and a half times the range of elevations on Earth.

Scientists also became aware that they had vastly underestimated the dimensions of the imposing Hellas basin in the south. The map showed the basin to be 10 kilometers deep and 2,000 wide, big enough to swallow Mount Everest and possibly the largest impact crater in the solar system. The asteroid impact that gouged out Hellas four billion years ago scattered rocky debris over 4,000 kilometers, significantly contributing to the southern hemisphere's higher elevations.

Another Surveyor instrument, its magnetometer, picked up traces of the planet's magnetic history imprinted in the ancient highland crust. Plotted on a map, the data showed alternating bands of magnetic polarity in the rock, which reminded scientists of the alternating patterns of crustal magnetization that flank the mid-ocean ridges on Earth and provided the strongest evidence of Earth's spreading sea floor and continental drift. Studying the magnetic map, scientists recognized telling clues that the phenomenon of plate tectonics, powered by heat from a planet's molten metal core, has existed on another planet besides Earth. If the observations are confirmed by further research, this would be one more indication that Mars, though cold, arid, and placid now, was a more active world in an early epoch. Mars and Earth seem to have started out much alike, then evolved into very different worlds.

All this was accomplished as the Mars Global Surveyor spacecraft was beginning its primary two-year mission of mapping the entire planet with greater precision, the prelude to more spacecraft visits in years to come. Mapmakers were responding once again to the challenge of recording discovery in a distant place and encouraging visions of greater discovery.

Cosmic Cartographers

E arth and the entire solar system are too small to contain the mapping compulsion, the universe too vast to be comprehended without the mapmaker's talents for clarifying spatial relationships.

In antiquity, astronomers brought an extraterrestrial dimension to the map idea by drawing charts of the visible constellations. In form and purpose, these were progenitors of the sky charts familiar to amateur and professional astronomers today. Mercator himself seemed to anticipate the day when maps would play a role in making sense of the universe and cosmology. The purpose of his atlas, Mercator wrote, was to gather "cosmological meditations upon the creation of the universe, and the universe as created." His conception of the universe was parochial, to be sure. For all Mercator's cosmic talk, his maps stuck to terrestrial space, which in his day was challenge enough. Only in the 1920s, with Edwin P. Hubble's discoveries of other galaxies beyond the Milky Way and of the expanding universe, did astronomers begin to consider the possibilities of true cosmic mapmaking.

More than five decades passed before a cosmic map played a critical role in discovery, much as we had come to expect of terrestrial and planetary maps. This cosmic map gave enlarged meaning to the cartographic premise that "anything that can be spatially conceived can be mapped." Never had "anything" encompassed so much space.

Guided by a two-dimensional map of the universe made in the 1950s, Margaret J. Geller and John P. Huchra of the Harvard-Smithsonian Center for Astrophysics in Cambridge, Massachusetts, observed more than 11,000 galaxies in a wedge of the northern celestial hemisphere. Technological advances, including electronic sensors (charge-coupled devices) that improved the light-collecting efficiency of telescopes, enabled them to see farther and more definitively than ever before. Geller and Huchra measured the red shifts of the brighter galaxies—the degree to which each galaxy's light has "stretched" into the red part of the spectrum during its journey to Earth. Because the universe is expanding, as Hubble determined, galaxies with higher red shifts lie farther out in space. These measurements by Geller and Huchra added an essential third dimension to maps of deep space—longitude, latitude, and now estimated distance.

One day in the summer of 1985, Valerie de Lapparent, a Harvard graduate student working with Geller and Huchra, fed the distances and positions of 1,000 galaxies into a computer to produce a graphic representation of their distribution through this slice of sky. The results were totally unexpected. Here was evidence of structure to the universe beyond previous conjecture and imagination. De Lapparent saw the galaxies arranged in a broad pattern resembling a child's drawing of a stickman. Not knowing what to make of the computer printout, she rushed to share her perplexity with Geller and Huchra.

The map of galaxies and clusters of galaxies stunned Huchra, the team's virtuoso of the telescope, who had made most of the observations. "I looked at that thing, and I panicked," he recalled. "My only reaction was, 'Oh, my Lord, I must have made some terrible mistake.' I spent the next month checking my data to find the error."

Geller, more of an interpretive astronomer than an observer, never doubted Huchra's work. She soon recognized that the map was telling them something real and surprising about the universe.

Further observations and analysis led Geller and Huchra to the discovery of the largest structure yet observed in the visible universe. They called it the Great Wall. On the completed map, issued in 1989, the Great Wall of interconnected galaxies stretched at least 500 million light-years long, 200 million light-years wide, and 15 million light-years thick. In order to incorporate so much space, of course, the map and subsequent ones were drawn on a scale rendering Mars and Earth invisible; at this scale, Earth is smaller than an elementary particle and whole galaxies are like tiny islands of dim light, enormous clumps of

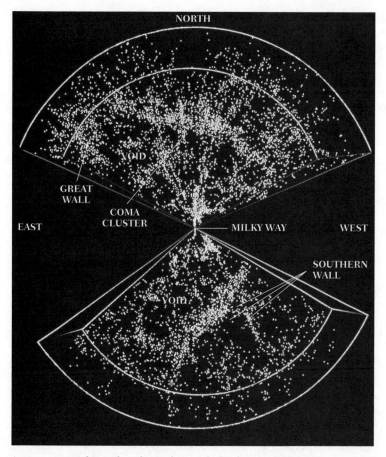

An extension to the southern hemisphere of the landmark astronomical survey that revealed the "Great Wall" suggests such large-scale structures persist over the entire sky.

them surrounded by veritable seas of empty space. Once again, a map had put us in our place.

The emerging pattern of structure, the Great Wall, contradicted astronomy's common wisdom. Previously, astronomers had thought of galaxies being scattered more or less uniformly throughout space. But their earlier observations and maps, being inadequate, had failed them. The new maps proved uniformity on such scales to be illusory. Producing a better map, Geller, Huchra, and de Lapparent opened eyes wider to an understanding of the complex nature of the universe.

After this experience, Geller liked to describe herself as a cosmic cartographer. She identified with her terrestrial counterparts in ancient Babylonia, with all cartographers who have had to make do with insufficient surveys. And she realized that her new maps were only a beginning.

"Cosmic mapping today," Geller said in 1998, "is like trying to picture the whole world from a survey of Rhode Island. But a century from now the entire visible universe may be mapped."

Geller recalled Ptolemy's admonition "to survey the whole in its just proportions" and to represent "in pictures the whole known world together with the phenomena contained therein." With a figurative interpretation of "world," she said, "Ptolemy's statement is a remarkably apt description of the aspirations of those of us who are now mapping the universe."

In a sense, astronomy and cartography had come full circle. Ptolemy set the early standards in both fields. In the second century, he catalogued roughly one-sixth of the stars visible to the unaided eye, which was the best anyone could do then. He and other ancient astronomers looked to the sky to get their bearings on Earth and lay the groundwork for scientific cartography. Eratosthenes had used the light from a star, the Sun, to measure Earth's circumference. Now astronomers had learned from the accumulated experience of terrestrial cartography and could appreciate more than ever the clarifying influence of a good map.

Cosmic maps are different from most terrestrial maps in at least one important respect: they can chart time. Peering ever deeper into space, astronomers see and map galactic structure as it existed billions of years ago. (The light they see originated that long ago and shows galaxies as they existed then, not now.) Astronomers are thus able to plot from direct observations structure in the early universe. The best cartographers can do in mapping time on Earth itself is to reconstruct the positions of land, continents, and bodies of water in past epochs from the indirect evidence of subsurface mapping, patterns of sea-floor spreading, and other data. Unlike astronomers, cartographers of time on Earth are seeing not the past as it actually existed, but as a deduction drawn from present conditions.

By the end of the twentieth century, other astronomers were becoming cosmic cartographers. They charted the universe in all its electromagnetic manifestations—infrared, ultraviolet, radio, X-ray, gamma ray, as well as optical. More slices of the sky were being mapped in three dimensions, producing more illuminating outlines of major structure.

And expanded volumes of space are covered in the new surveys to find out if there are even larger structures of galaxies out there.

An American group, observing at telescopes on Cerro Las Campañas in Chile, surveyed deeper (out to two billion light-years) and wider (85° across the sky) to record the positions of 26,000 galaxies. The maps illustrated even greater changes in galactic structure over time. Some of the structures proved to be several times the size of the Geller-Huchra Great Wall. An Anglo-Australian consortium embarked on a project to map more than a quarter million galaxies. The most ambitious effort was the Sloan Digital Sky Survey, begun in 1998 by the United States, Japan, and Germany. It erected a new wide-field telescope in New Mexico to measure distances to 100 million galaxies and fix their positions, with a much tighter control of systematic errors than in any previous cosmic mapping.

Cosmic cartographers were busy compiling the first really comprehensive atlas of the galaxies. By measuring the distribution of distant galaxies, astronomers were mapping the growth of cosmic architecture over time, the very history of the universe.

The horizons of mapmakers now seem limitless. Remote sensors coax data points from nature. Computers crunch the data and automatically create a picture of some world—geographic, planetary, galactic, or thematic—with a blinding speed no human cartographer could ever match. Armed with new technologies, cartographers can make and revise a map in minutes; even noncartographers sitting at computer terminals are making their own maps. And instantaneous communication brings mapmakers new information and measurements at the speed of light; no more waiting long periods for Captain Cook to return from the Pacific, for the pundits to make their way back from across the Himalayas, or for photographic film to be shipped and developed and slowly converted to maps.

The idea of the map as a means of converting measurements into meaningful patterns now extends to every territory of scientific enterprise, from the cosmic down to the atomic. In *Mapping the Next Millennium: The Discovery of New Geographies,* Stephen Hall describes how biologists map the cells in tissues, the proteins in cells, and the atoms in proteins and how neurobiologists map areas of the brain that light up when we dream. "In short, we find ourselves amid perhaps the greatest reconsideration of the concept of space since an anonymous Babylonian first attempted to organize human knowledge of the physical world by drawing a world map on a clay tablet 2,600 years ago," Hall wrote. "Almost arrogant in the sheer extent of the territory they presume to represent, the new maps are charting the world all over again, creating anew the idea of the frontier."

The range, technology, and character of mapmaking may have changed recently, as demonstrated by spacecraft data about Mars, the new maps of the cosmic Great Wall, and the mapping of the brain, but the purpose of maps has not. Maps embody a perspective of that which is known and a perception of that which may be worth knowing.

Epilogue at Bright Angel

There are still chores in mapping that cannot be done with helicopters, laser, aerial photography, or any other of the modern technologies, not if you want to be faithful to every nuance of the land. "You have to steep yourself in the place, its trails and rocks and vegetation," Brad Washburn was saying, "if you are to produce a map with more authenticity, one that has the right *feel* as well as the right mathematics."

So, at dawn one day, we set out on foot from the south rim. We followed the South Kaibab Trail, the steeper but shorter of two much-traveled routes to the bottom of the Grand Canyon. We hoped by getting an early start to cover as much ground as possible before the Sun got high and too hot.

At each twist in the trail we examined an enlarged aerial photograph, comparing it with what we were seeing. We wanted to be sure that the detail had indeed been recorded by the camera. Sometimes a cliff could be so steep that, in the photographs, many zigs and zags of the trail were lost under overhangs. Or shadows could mask some features, while poor contrasts obscured others. This had been brought home to us a few days before down on the Tonto Trail. No existing maps or high-altitude photographs made clear how the trail winds up a particular side canyon and where exactly it crosses Monument Creek; the route under the vertical cliffs and through the dark shadows could only be traced on foot.

On the Kaibab, when we got to a missing link, we would note some landmark visible in the photograph, a boulder or tree, and measure the distance and twists in the trail from it to a point where the trail reappeared on the photograph. Washburn made careful notes and sketches.

Time does not permit all mappers to be so fastidious, and not all users of maps demand such attention to detail. The points of cliffs and the wiggles of a creek or trail are sometimes rounded off. "Who cares if a creek has nine wiggles in it? you may ask," Washburn remarked. "But if you have this kind of accuracy in all detail, the overall *feeling* of the country will be correct. Great precision in detail always yields extraordinary accuracy in the sense of the whole."

In the beginning, the cliff was particularly steep. It would present a formidable problem to those who had to prepare the finished map.

"If you have a seventy-degree slope, it's awfully steep, but you can still depict it," Washburn said. "But if it's ninety degrees, perpendicular like much of this, there's no room on the map to draw any symbols, because the bottom of the cliff is at precisely the same point as the top, only a thousand feet below. Contour lines would simply be stacked one right on top of the other. We will have to come up with some method of shaded relief drawing. The Swiss are struggling with us to try to find a way to convey the feeling of these cliffs that can't be depicted in the contours. The Alps have a very few really vertical cliffs—the Grand Canyon abounds with them."

Each of us in the party took turns pushing a surveyor's measuring wheel, called a perambulator. We wanted a precise measure of the trail's entire length and also of distances between important points along the way. The wheel's circumference is 5.28 feet, and so each complete turn of the wheel clicks off one-thousandth of a mile on the odometer. Though old and slow compared to the laser, the wheel is incomparable when it comes to measuring a twisting trail.

When a mule train with tourists passed us, a woman was amused by the sight of a presumably rational man pushing a wheel down into the Grand Canyon. "What *are* you doing," she inquired with a broad smile, "riding a unicycle to the bottom?"

Descending deeper into the Canyon, we walked out of a temperate alpine climate of pine, juniper, and cedar and into a desert of mesquite,

Section from Grand Canyon map
by Bradford Washburn

cactus, Mormon tea, and stark skeletons of agave. At the foot of the Redwall, one of the most distinctive layers of the Canyon, where limestone formed out of the ooze of a great sea floor 335 million years ago, we rested in the shade of a boulder.

We ate cold chicken and drank plenty of grapefruit juice. Dehydration is one of the greatest hazards for mappers or anyone else in the Canyon. The place can be an oven of dry air. A hiker can lose pounds of vital moisture and not realize it, because perspiration evaporates immediately into the thirsty air.

By early afternoon, hot and tired, we extended our measurements down into the Inner Gorge, Powell's "grand, gloomy depths." We crossed the footbridge over the Colorado and reached Phantom Ranch, an oasis of cottonwoods and rustic cabins by Bright Angel Creek.

We had stood in wonder at Dana Butte, we had sought to reach beyond wonder to comprehension at Yaki Point and at Bright Angel Point. We had used the amazing tools of the new cartography and, along the Kaibab, the dusty ways of the old. Now we could rest and reflect on our accomplishments.

While we drank cold lemonade and soaked ourselves in Bright Angel Creek, we spoke of the improved map of the Grand Canyon that would come of our efforts, in 1978. Hard work still has its reward. We discussed our small part in man's quest to understand where he is, and possibly why, and this made us feel happy—and a little important. The sum of knowledge was greater by a fraction, *our* fraction. Here, in the magnificent depths of Earth, we talked of Mariner 9, which at that moment was discovering and surveying the grandeur of Mars. We felt a return of appropriate humility.

Washburn spoke of the day, not too many years away, when the instruments of remote sensing, photography, navigation, and computation would be so refined that it should be possible to produce many good, accurate maps without any fieldwork. Gone out of mapping would be much of the hard work and tedium, and cartography's embrace would be wider than ever.

So be it. We would not bolt the door against the search for knowledge. But the prospect brought a twinge of regret apparently as old as man, a regret over some paradise lost while pursuing the fruit of knowledge. We lay back in the gurgling water and, holding on to a boulder in midstream, basked in the soothing eddies. Future mappers, alas, would never know the matchless pleasure of the Bright Angel's massaging waters after a day's work in the blistering heat of the Canyon. They would never experience the euphoria of standing on a place like Dana Butte. They would have to seek their own benchmarks of satisfaction and of wonder, the wonder that moves the human spirit to explore, encompass, and strive to understand.

General

A few books range across large sections of the history of cartography, geodesy, and geographical exploration. Among those consulted in the writing of this book are:

Bagrow, Leo. *History of Cartography.* Revised and enlarged by R. A. Skelton. Cambridge, Mass., 1964. Comprehensive and well illustrated.

Baker, John L. N. *A History of Geographical Discovery and Exploration.* Boston, 1938. Excellent one-volume summary.

Berthon, Simon, and Robinson, Andrew. *The Shape of the World: The Mapping and Discovery of the Earth.* Chicago, 1991. Companion book to a six-part television series by the same title.

Bricker, Charles, and Tooley, R. V. *A History of Cartography: 2500 Years of Maps and Mapmakers.* London, 1969. Concise history with many color illustrations.

Brown, Lloyd A. *The Story of Maps.* Boston, 1949. Scholarly and comprehensive, up to the nineteenth century. Extensive bibliography. Reissued by Dover, New York, 1980.

Crone, Gerald R. *Maps and Their Makers.* 4th edition. London, 1968. Review of "the main stages of cartographic development."

Greenhood, David. *Mapping.* Chicago, 1964. Introduction to mapping and surveying techniques, with considerable how-to-do-it detail for nonprofessionals.

Harley, J. B., and Woodward, David (editors). *The History of Cartography: Cartography in Prehistoric, Ancient, and Medieval Europe and the Mediterranean.* Volume I. Chicago, 1987. The complete work is still in progress, but promises to be the definitive history of maps and mapmaking. At this writing, the published volumes composing the work include: I, Books 1 and 2, *Cartography in the Traditional Islamic and South Asian Societies*, 1992 and 1994; II, Book 3, *Cartography in the Traditional African, American, Arctic, Australian, and Pacific Societies*, 1998 (edited by Woodward and G. Malcolm Lewis). Subsequent volumes are to cover cartography in the Renaissance and Age of Discovery, the European Enlightenment, and in the nineteenth and the twentieth centuries.

Heiskanon, W. A., and Vening Meinesz, F. A. *The Earth and Its Gravity Field.* New York, 1958. Some of the history of geodesy mixed with a more technical exposition.

Raisz, Erwin J. *Mapping the World.* New York, 1956.

Ristow, Walter W. *Guide to the History of Cartography.* Washington, 1973. A Library of Congress bibliography.

Robinson, Arthur H., and Sale, Randall D. *Elements of Cartography.* 3rd edition. New York, 1969. Introduction to the craft of mapmaking.

Skelton, R. A. *Explorers' Maps.* London, 1958. Illustrated.

———. *Maps: A Historical Survey of Their Study and Collecting.* Chicago, 1972. Collection of lectures.

Thompson, Morris M. *Maps for America.* Washington, 1979. A publication of the U.S. Geological Survey. Excellent illustrations of the many kinds of maps, with emphasis on how they are prepared and how they are read and interpreted. A concise glossary.

Thrower, Norman J. W. *Maps and Man.* Englewood Cliffs, N.J., 1972. Summary of the many types and uses of maps.

———. "The Art and Science of Navigation in Relation to Geographical Exploration Before 1900." In Herman R. Friis (editor), *The Pacific Basin: A History of Its Geographical Exploration.* New York, 1963.

Chapter by chapter, the materials and people consulted are the following:

1. The Map Idea

Bagrow, Leo. *History of Cartography.*

Berthon, Simon, and Robinson, Andrew. *The Shape of the World.* Chapter 2.

Black, Jeremy. *Maps and Politics.* Chicago, 1997.

Blaut, J. M. *The Colonizers' Model of the World: Geographical Diffusionism and Eurocentric History.* New York, 1993.

Bulling, A. Gutkind. "Ancient Chinese Maps." In *Expedition* 20, Winter 1978.

Bunbury, Edward. *A History of Ancient Geography.* 2 volumes. London, 1883.

Hall, Stephen S. *Mapping the Millennium.* New York, 1992.

Harley, Brian. "The Map and the Development of the History of Cartography." In Harley and Woodward (editors), *The History of Cartography.* Volume I.

Huang, Yan, and Yang, Xinhe. "China's Oldest Map." In *Archaeology* March–April 1998.

Lyons, Henry. "Sailing Charts of the Marshall Islanders." In *Geographical Journal* 72, October 1928.

Markham, Beryl. *West with the Night.* San Francisco, 1983.

Monmonier, Mark. *How to Lie with Maps.* Chicago, 1991.

Muhly, James D. "Ancient Cartography." In *Expedition* 20, Winter 1978.

Robinson, Arthur H., and Petchenik, Barbara B. *The Nature of Maps: Essays Toward Understanding Maps and Mapping.* Chicago, 1976.

Thrower, Norman J. *Maps and Man.*

Yee, Cordell D. K. "Reinterpreting Traditional Chinese Geographic Maps." In Harley and Woodward (editors), *History of Cartography.* Volume 2, Book 2. 1994.

2. *The Librarian Who Measured Earth*

Brown, Lloyd A. *The Story of Maps.*

Bunbury, Edward A. *A History of Ancient Geography.* Volume 1.

Burton, Harry E. *The Discovery of the Ancient World.* Cambridge, Mass., 1932.

Daly, Charles P. *The Early History of Cartography.* New York, 1879.

Heidel, William A. *The Frame of the Ancient Greek Maps.* American Geographical Society Research Series 20. New York, 1937.

Jones, David E. H. "The Great Museum at Alexandria: Its Ascent to Glory." In *Smithsonian* 2, December 1971.

Sarton, George. *A History of Science.* Cambridge, Mass., 1959. Chapter 2 for the Museum at Alexandria and chapter 6 for Eratosthenes.

Thomson, J. O. *Everyman's Classical Atlas.* London, 1961.

Thrower, Norman J. *Maps and Man.* Chapter 2.

3. *First Principles by Ptolemy*

Bagrow, Leo. *History of Cartography.*

Berthon, Simon, and Robinson, Andrew. *The Shape of the World.* Chapter 2.

Boorstin, Daniel J. *The Discoverers.* New York, 1983. Chapter 14.

Bunbury, Edward. *A History of Ancient Geography.* Volume 2.

Crone, Gerald R. *Maps and Their Makers.* Chapters 1 and 5.

Ptolemy, Claudius. *Geography of Claudius Ptolemy.* Translated and edited by Edward C. Stevenson. New York, 1932.

Thrower, Norman J. *Maps and Man.* Chapter 4.

Warntz, William, and Wolff, Peter. *Breakthroughs in Geography.* New York, 1971. Chapter 2.

4. *The Topography of Myth and Dogma*

Beazley, C. R. *The Dawn of Modern Geography.* 3 volumes. London, 1897. Volume 1 for Solinus and Cosmas. Volume 3 for the portolan charts.

Berthon, Simon, and Robinson, Andrew. *The Shape of the World.* Chapter 4.

Brown, Lloyd A. *The Story of Maps.* Chapter 4.

Crone, Gerald R. *Maps and Their Makers.*

———. "New Light on the Hereford Maps." In *Geographical Journal* 131, December 1965.

Crosby, Alfred W. *The Measure of Reality: Quantification and Western Society, 1250–1600.* Cambridge, Eng., 1997.

Daly, Charles P. *The Early History of Cartography.*

Humphreys, Arthur L. *Old Decorative Maps and Charts.* London, 1926. Good reproductions of medieval maps.

Kimble, George H. T. *Geography in the Middle Ages.* New York, 1968. Reprint of 1938 edition.

Larner, John. *Marco Polo and the Discovery of the World.* New Haven, 1999.

Mandeville, Sir John. *Mandeville's Travels.* Edited by Malcolm Letts. 2 volumes. London, 1953.

Penrose, Boies. *Travel and Discovery in the Renaissance.* Cambridge, Mass., 1955.

Polo, Marco. *The Travels of Marco Polo.* Translated by Ronald Latham. London, 1958.

Silverberg, Robert. *The Realm of Prester John.* Garden City, N.Y., 1972.

Stevenson, Edward L. *Portolan Charts, Their Origin and Characteristics.* New York, 1911.

Woodward, David, "Medieval Mappaemundi." In Harley and Woodward (editors), *The History of Cartography.* Volume 1.

Wright, John K. *Geographical Lore at the Time of the Crusades.* New York, 1965. Reprint of 1925 book.

5. *1492*

Beazley, C. R. *Prince Henry the Navigator.* London, 1895.

Berthon, Simon, and Robinson, Andrew. *The Shape of the World.* Chapter 2.

Buisseret, David (editor). *Monarchs, Ministers, and Maps: The Emergence of Cartography as a Tool of Government in Early Modern Europe.* Chicago, 1992. Introduction and chapter 1.

Crosby, Alfred W. *The Measure of Reality.*

Davies, Arthur. "The Date of Juan de la Cosa's World Map and Its Implications for American Discovery." In *Geographical Journal* 142, March 1976.

Goldstein, Thomas. "Geography in Fifteenth Century Florence." In John Parker (editor), *Merchants and Scholars.* Minneapolis, 1965.

Hale, J. R. *Renaissance Exploration.* New York, 1968.

Lewis, Bernard. *The Muslim Discovery of Europe.* New York, 1982.

Morison, Samuel Eliot. *Admiral of the Ocean Sea: A Life of Christopher Columbus.* 2 volumes. Boston, 1942.

———. *The European Discovery of America: The Northern Voyages, 500–1600.* New York, 1971.

———. *The European Discovery of America: The Southern Voyages, 1492–1616.* New York, 1974.

Parry, J. H. *The Age of Reconnaissance.* Cleveland, 1963.

Penrose, Boies. *Travel and Discovery in the Renaissance.*

Prestage, Edgar. *The Portuguese Pioneers.* London, 1933.

Quinn, David Beers. *England and the Discovery of America, 1481–1620.* New York, 1973.

Ravenstein, E. G. *Martin Behaim: His Life and His Globe.* London, 1908.

Skelton, R. A. *Explorers' Maps.*

———, Marston, Thomas E., and Painter, George D. *The Vinland Map and the Tartar Relation.* New Haven, 1965. A new, expanded edition issued in 1996.

Stevenson, Edward L. *Terrestrial and Celestial Globes.* 2 volumes. New Haven, 1921. Volume 1 for Behaim.

6. *Mercator Squares the Circle*

Alpha, Tau Rho, and Gerin, Marybeth. *A Survey of the Properties and Uses of Selected Map Projections.* United States Geological Survey, Miscellaneous Investigations Series Map I-1096. Washington, 1978.

Black, Jeremy. *Maps and Politics.*

Deetz, Charles H., and Adams, Oscar S. *Elements of Map Projections.* United States Coast and Geodetic Survey Special Publication 68. Washington, 1934.

Gardner, Martin. "On Map Projections (With Special Reference to Some Inspired Ones)." In *Scientific American* 233, November 1975.

Henrikson, Alan K. "All the World's a Map." In *Wilson Quarterly,* Spring 1979.

Humphreys, Arthur L. *Old Decorative Maps and Charts.*

Osley, Arthur S. *Mercator.* London, 1969. Includes material from the memoirs of Walter Ghim, Mercator's neighbor at Duisburg.

Robinson, Arthur H. "Arno Peters and His New Cartography." In *The American Cartographer* 12, 1985.

Snyder, John P. *Flattening the Earth: Two Thousand Years of Map Projections.* Chicago, 1993.

Stevenson, Edward L.. *Terrestrial and Celestial Globes.* Volume 1 for Mercator and his times.

Warntz, William, and Wolff, Peter. *Breakthroughs in Geography.* Includes excerpts from "Text and Translation of the Legends of the Original Chart of the World by Gerhard Mercator Issued in 1569." From *Hydrographic Review* 9, November 1932.

Wilford, J. N. "The Impossible Quest for the Perfect Map." In the *New York Times,* October 25, 1988.

Interviews with Arthur Robinson of the University of Wisconsin; Mark Monmonier of Syracuse University; John B. Garver, Jr., chief cartographer of the National Geographic Society.

7. *The Matter of a Degree*

Barthalot, Raymonde. "The Story of Paris Observatory." In *Sky and Telescope* 59, February 1980.

Brown, Lloyd A. *The Story of Maps.* Chapters 7 and 8.

Fernel, Jean. *Dialogue.* Paris, 1530. Quoted in J. D. Bernal, *The Extension of Man.* Cambridge, Mass., 1972. Page 155.

Fordham, George H. *Some Notable Surveyors and Map-Makers of the Sixteenth, Seventeenth, and Eighteenth Centuries.* Cambridge, Eng., 1929.

Furneaux, Robin. *The Amazon: The Story of a Great River.* London, 1969. Chapter 5 for account of La Condamine and the expedition.

Glass, H. Bentley. "Maupertuis, A Forgotten Genius." In *Scientific American* 193, October 1955.

La Condamine, Charles Marie de. *A Succinct Abridgement of a Voyage Made within the Inland Parts of South-America as it was read to the Academy of Sciences, Paris, April 28, 1745.* London, 1747.

Mathieson, John. "Geodesy: A Brief Historical Sketch." In *Scottish Geographical Magazine* 42, November 1926.

Maupertuis, Pierre L. M. de. *The Figure of the Earth Determined from Observations Made by Order of the French King, at the Polar Circle.* Translation. London, 1738.

Merdinger, Charles J. "Surveying Through the Ages." In *Military Engineer* 46, March–April 1954.

Olmsted, John W. "The Scientific Expedition of Jean Richer to Cayenne (1672–1673)." In *Isis* 34, part 2, 1942.

Taylor, Eva G. R. "The Earliest Account of Triangulation." In *Scottish Geographical Magazine* 43, November 1927.

———. "The Measure of the Degree—300 B.C.–A.D. 1700." In *Geography* 34, September 1949.

8. *The Family That Mapped France*

Barthalot, Raymonde. "The Story of Paris Observatory."

Brown, Lloyd A. *The Story of Maps.* Chapter 9.

———. *Jean Dominique Cassini and His World Map of 1696.* Ann Arbor, 1941.

Crone, Gerald R. *Maps and Their Makers.*

Fordham, George H. *Some Notable Surveyors and Map-Makers of the Sixteenth, Seventeenth, and Eighteenth Centuries.*

Hall, A. R. *The Scientific Revolution, 1500–1800.* London, 1956.

Skelton, R. A. "The Origin of the Ordnance Survey of Great Britain." In *Geographical Journal* 128, December 1962.

9. *John Harrison's Timepiece*

Brown, Lloyd A. *The Story of Maps.* Chapters 7 and 8.

Gould, Rupert T. *The Marine Chronometer: Its History and Development.* London, 1923.

Sobel, Dava. *Longitude.* New York, 1995. Though not used as a source for this chapter, this is a delightful book and highly recommended.

Taylor, Eva G. R. *The Haven Finding Art.* London, 1956.

Thrower, Norman J. W. "The Discovery of Longitude." In *Navigation* 5, no. 8, 1957–8.

Personal correspondence with Neil A. Armstrong.

10. *Surveyors of Sea and Shore*

Anderson, Bern. *The Life and Voyages of Captain George Vancouver.* Seattle, 1960.

Beaglehole, J. C. *The Life of Captain James Cook.* Stanford, 1974. The definitive and admiring biography by the editor of Cook's journals.

Blewitt, Mary. *Surveys of the Seas: A Brief History of British Hydrography.* London, 1957.

Friendly, Alfred. *Beaufort of the Admiralty.* New York, 1977.

Friis, Herman (editor). *The Pacific Basin: A History of Its Geographical Exploration.* New York, 1967. Of particular value are Norman J. W. Thrower, "The Art and Science of Navigation in Relation to Geographical Exploration Before 1900"; Wilcomb E. Washburn, "The Intellectual Assumptions and Consequences of Geographical Exploration in the Pacific"; Richard J. Ruggles, "Geographical Exploration by the English."

Moorehead, Alan. *The Fatal Impact.* New York, 1966.

Skelton, R. A. "James Cook, Surveyor of Newfoundland." Introductory essay in the Grabhorn Press facsimile edition of *A Collection of Charts of the Coasts of Newfoundland and Labradore, first published 1769–70.* London, 1965.

Vancouver, George. *A Voyage of Discovery to the North Pacific Ocean and Round the World.* 3 volumes. London, 1798.

11. *Soldiers, Pundits, and the India Survey*

Edney, Matthew H. *Mapping an Empire: The Geographical Construction of British India, 1765–1843.* Chicago, 1997.

Hayford, J. F. *The Figure of the Earth and Isostasy.* Washington, 1909.

Heaney, G. F. "Rennell and the Surveyors of India." In *Geographical Journal* 134, September 1968.

Hopkirk, Peter. *Trespassers on the Roof of the World.* New York, 1995.

Phillimore, Reginald H. *Historical Records of the Survey of India.* 3 volumes. Dehra Dun, 1945–54.

Records of the Survey of India, Explorations in Tibet and Neighboring Regions, 1865–1879. Dehra Dun, 1915. Volume 8, in two parts, was an important source.

Styles, Showell. *The Forbidden Frontiers: The Survey of India from 1765 to 1949.* London, 1970.

12. *Mapping America: The Boundary Makers*

Bedini, Silvio A. *Thinkers and Tinkers: Early American Men of Science.* New York, 1975. Chapter 6 for Mason and Dixon and chapter 13 for the post-Revolution surveys.

Freeman, Douglas Southall. *George Washington.* New York, 1948. Volume 1, chapters 5, 6, and 7.

Marschner, Francis T. *Boundaries and Records in the Territory of Early Settlement from Canada to Florida.* Washington, 1960.

Mason, A. Hughlett, and Swindler, William F. "Mason and Dixon: Their Line and Its Legend." In *American Heritage* 15, February 1964.

Mason, Charles, and Dixon, Jeremiah. *The Journal of Charles Mason and Jeremiah Dixon, Transcribed from the Original in the U.S. National Archives.* Introduction by A. Hughlett Mason. Philadelphia, 1969.

Meade, Buford K. "Delaware-Maryland Boundary Resurvey in 1961–62." In *Surveying and Mapping*, 24, March 1964.

Pattison, William D. *Beginnings of the American Rectangular Land Survey System, 1784–1800.* University of Chicago, Department of Geography Research Paper Number 50. December 1957.

Van Zandt, Franklin K. *Boundaries of the United States and the Several States.* U.S. Geological Survey Bulletin 1212. Washington, 1966.

13. *Mapping America: Westward the Topographers*

Allen, John Logan. *Passage Through the Garden: Lewis and Clark and the Image of the American Northwest.* Urbana, 1975.

Alter, J. Cecil, and Gregory, Herbert E. (editors). "Diary of Almon Harris Thompson." In *Utah Historical Quarterly* 7, numbers 1, 2, and 3, 1939. Includes biographical sketch of Thompson.

Bartlett, Richard A. *Great Surveys of the American West.* Norman, Okla., 1962.

Darrah, William Culp. "Biographical Sketches and Original Documents of the First Powell Expedition of 1869." In *Utah Historical Quarterly* 15, 1947.

———. *Powell of the Colorado.* Princeton, 1951.

Dellenbaugh, Frederick. *A Canyon Voyage.* New York, 1908.

DeVoto, Bernard. *The Course of Empire.* Boston, 1952.

Egan, Ferol. *Frémont: Explorer for a Restless Nation.* Garden City, N.Y., 1977. Most recent biography.

Frémont, John Charles. *Memoirs of My Life.* Chicago, 1887.

Goetzmann, William H. *Army Exploration in the American West, 1803–1863.* New Haven, 1959. Appendix C is G. K. Warren's description of the methods of compiling his map of 1857 from Warren's "Memoir" in *Pacific Railroad Reports.* 33rd Congress, 2nd Session. Senate Document 78. Volume 11.

———. *Exploration and Empire.* New York, 1966. The most comprehensive treatment of the scientific and cartographic exploration of the American West.

Nevins, Allen. *Frémont: The West's Greatest Adventurer.* 2 volumes. New York, 1923. The standard biography.

Powell, J. W. *The Exploration of the Colorado River and Its Canyons.* New York, 1961. Unabridged reprint of 1895 book entitled *Canyons of the Colorado.*

Preuss, Charles. *Exploring with Frémont.* Translated and edited by Erwin G. and Elisabeth K. Gudde. Norman, Okla., 1958. Includes biographical sketch of Preuss.

Smith, Henry Nash. *Virgin Land: The American West as Symbol and Myth.* New York, 1950.

Stegner, Wallace. *Beyond the Hundredth Meridian.* Boston, 1954.

United States Geological Survey. *History of the Topographic Branch.* Preliminary draft. Undated, but about 1976.

Wallace, Edward S. *The Great Reconnaissance.* Boston, 1955.

Wheat, Carl I. *Mapping the Transmississippi West.* 6 volumes. San Francisco, 1958–63. A definitive work, and well illustrated.

14. *Meters, Meridians, and a New World Map*

Brown, Lloyd A. *The Story of Maps.* Chapter 10.

De Boer, J. "From the Earlier Systems of Measures to the International System of Units." In *The International Bureau of Weights and Measures, 1875–1975.* United States Department of Commerce, National Bureau of Standards Special Publication 420. Washington, 1975.

Klein, H. Arthur. *The World of Measurements.* New York, 1974.

Penck, A. "The Construction of a Map of the World on a Scale of 1:1,000,000." In *Geographical Journal* 1, March 1893.

Perry, John. *The Story of Standards.* New York, 1955.

Thrower, Norman J. W. *Maps and Man.* Chapter 8.

United Nations. "First Progress Report on the International Map of the World on the Millionth Scale (1954)." In *World Cartography* 4, 1954.

Wright, John K. *Geography in the Making.* New York, 1952. A history of the American Geographical Society. The international map is discussed in chapter 9.

15. *The Winged Mappers*

Bates, Marston, and Abbott, Donald P. *Coral Island: Portrait of an Atoll.* New York, 1958.

Brandenberger, A. J. "What Can Photos Tell Us." In *International Science and Technology,* September 1967.

Cahill, Edward H.: "Brock Process of Aerial Mapping." In *Journal of the Optical Society of America* 22, March 1932.

Eder, Josef Maria. *History of Photography.* Translated by Edward Epsteen. New York, 1945.

Heiman, Grover. *Aerial Photography: The Story of Aerial Mapping and Reconnaissance.* New York, 1972.

Landen, David. "History of Photogrammetry in the United States." In *Photogrammetric Engineering,* December 1952.

Thompson, Morris M. "Development of Photogrammetry in the U.S. Geological Survey." Geological Survey Circular 218. 1952.

———. *Maps for America.*

Whitmore, George D.; Thompson, Morris M.; and Speert, Julius L. "Modern Instruments for Surveying and Mapping." In *Science* 130, October 23, 1959.

Interviews with Morris M. Thompson, United States Geological Survey in Reston, Va., and Virgil Kauffman, founder of Aero Service Company, Philadelphia, Pa. Kauffman made available some of the early records of Aero Service and other materials on early aerial mapping.

16. *Radar Over the Amazon*

Bassett, Thomas J. "Cartography and Europe in Nineteenth-Century Africa." In *Geographical Review* 84, 3, July 1994.

Brown, W. M., and Porcello, L. J. "An Introduction to Synthetic-Aperture Radar." In the Institute of Electrical and Electronic Engineers *Spectrum,* September 1969.

Clapp, Nicholas. *The Road to Ubar.* Boston, 1998.

Evans, Diane L.; Stofan, Ellen R.; Jones, Thomas D.; and Godwin, Linda M. "Earth from the Sky." In *Scientific American* 271, December 1994.

Furneaux, Robin. *The Amazon: The Story of a Great River.*

Grafton, Anthony. *New Worlds, Ancient Texts.* Cambridge, Mass., 1992. Goro Dati verse quoted here.

Grinspoon, David Harry. *Venus Revealed.* Reading, Mass., 1997.

Hall, Stephen. *Mapping the New Millennium.*

Hecht, Susanna, and Coburn, Alexander. *The Fate of the Forest: Developers and Defenders of the Amazon.* London, 1989.

Holz, Robert K. (editor). *The Surveillance Science: Remote Sensing of the Environment.* Boston, 1973. Excellent source, though somewhat technical, on remote sensing in general. Part 8 concerns airborne radar imagery.

Jensen, H.; Graham, L. C.; Porcello, L. J.; and Leith, E. "Side-Looking Airborne Radar." In *Scientific American* 237, October 1977.

Jet Propulsion Laboratory. *The Magellan Venus Explorer's Guide.* JPL Publications 90-24. 1990.

——. News release 936-5-23-80. June 1, 1980. Announcement of the airborne radar survey of Guatemala that revealed the existence of Maya canals.

Interviews with Gerard Dietrich, John M. Schmunk, and Homer Jensen, all of whom were involved in the development of radar mapping for Aero Service in the 1970s. Other interviews with Ronald Blom and Diane Evans of the Jet Propulsion Laboratory, Nicholas Clapp of Los Angeles, James W. Head of Brown University.

17. *Deep Horizons*

Adams, F. D. *The Birth and Development of the Geological Sciences.* New York, 1954. Reprint of a 1938 book.

Bishop, Margaret. *Subsurface Mapping.* New York, 1960.

Dohr, Gerhard. *Applied Geophysics: Introduction to Geophysical Prospecting.* New York, 1974.

Fenton, Carroll L. and Mildred A. *Giants of Geology.* Garden City, N.Y., 1952. Chapter 7 on William Smith.

Franklin, Ben A. "Shell Ship, Without Symbol, Seeks Oil off New Jersey." In *The New York Times,* July 31, 1975.

Hammond, Allen L. "Bright Spot: Better Seismological Indicators of Gas and Oil." In *Science* 185, August 9, 1974.

Harrison, J. M. "Nature and Significance of Geological Maps." In C. C. Albritton, Jr. (editor), *The Future of Geology*. New York, 1963.

Jensen, Homer. "The Airborne Magnetometer." In *Scientific American*, 204, June 1961.

Leveson, David. *A Sense of the Earth*. Garden City, N.Y., 1972. Includes an inspired chapter on geologic mapping.

Oliver, Jack. "Exploration of the Continental Basement by Seismic Reflection Profiling." In *Nature* 275, October 12, 1978.

Robinson, G. D., and Spieker, Andrew M. (editors). *Nature to Be Commanded: Earth-Science Maps Applied to Land and Water Management*. United States Geological Survey Professional Paper 950. Reston, Va., 1978.

Segesman, F.; Soloway, S.; and Watson, M. "Well Logging—The Exploration of Subsurface Geology." In *Proceedings* of the Institute of Radio Engineers, 50, November 1962.

Interviews with Norman L. Hatch, Jr., the United States Geological Survey; Ronald K. Cormick, Charles Payson, O. E. Gregson, and Jerry Ware, Conoco, Inc., Ponca City, Okla., and Oklahoma City; Boice Nelson and Robert Peacock, Mobil Oil Company, Dallas, Texas. Correspondence with Carl H. Savit, of Western Geophysical Company, Houston, and with the Consortium for Continental Reflection Profiling (COCORP), supported by the National Science Foundation and based at Cornell University, Ithaca, New York.

18. *A Continent Beneath the Ice*

Bentley, C. R.; Crary, A. P.; Ostenso, N. A.; and Thiel, E. C. "Structure of West Antarctica." In *Science* 131, January 15, 1960.

Campbell, K. J., and Orange, A. S. "A Continuous Profile of Sea Ice and Freshwater Ice Thickness by Impulse Radar." In *Polar Record* 17, January 1974.

Drewry, David J. "Radio Echo Sounding Map of Antarctica." In *Polar Record* 17, January 1975.

Evans, S.; Gudmandsen, P.; Swithinbank, C.; Hattersley-Smith, G.; and Robin, G. de Q. "Glacier Sounding in the Polar Regions: A Symposium." In *Geographical Journal* 135, part 4, December 1969.

Kirwin, Laurence P. *A History of Polar Exploration*. New York, 1960.

Lewis, Richard S. *A Continent for Science*. New York, 1965. Chapter 4 contains an excellent summary of the one- or two-continent riddle.

Oswald, G. K. A., and Robin, G. de Q. "Lakes Beneath the Antarctic Ice Sheet." In *Nature* 245, October 5, 1973.

Press, Frank, and Dewart, Gilbert. "Extent of the Antarctic Continent." In *Science* 129, February 20, 1959.

Robin, Gordon de Q. "The Ice of the Antarctic." In *Scientific American* 207, September 1962.

Sullivan, Walter. *Quest for a Continent*. New York, 1957.

Thiel, Edward C. "Antarctica, One Continent or Two?" In *Polar Record* 10, January 1961.

Whitmore, George D.; Thompson, Morris M.; and Speert, Julius L. "Modern Instruments for Surveying and Mapping."

Woolard, G. P. "The Land of the Antarctica." In *Scientific American* 207, September 1962.

Interview with William R. MacDonald, the United States Geological Survey,
Reston, Va.

19. *Mountains of the Sea*

Ballard, Robert D. "Dive into the Great Rift." In *National Geographic* 147, May
1975.

Dietz, Robert S. "The Underwater Landscape." In C. P. Idyll (editor), *Exploring
the Ocean World*. New York, 1969. Good summary of sounding history.

————, and Knebel, Harley J. "Survey of Ross's Original Deep Sea Sounding
Site." In *Nature* 220, November 23, 1968.

Ericson, David B., and Wollin, Goesta. *The Ever-Changing Sea*. New York, 1967.
Especially chapter 5.

Ewing, Maurice. "Exploring the Mid-Atlantic Ridge." In *National Geographic* 94,
September 1948.

Heezen, Bruce C., and Hollister, Charles D. *The Face of the Deep*. New York,
1971.

————, and Menard, H. W. "Topography of the Deep-Sea Floor." In M. H. Hill
(editor), *The Sea*. New York, 1963.

————; Tharp, Marie; and Ewing, Maurice. "The Floors of the Ocean. I. North
Atlantic." Geological Society of America Special Paper 65. 1959.

Laughton, A. S.; Whitmarsh, R. B.; Rusby, J. S. M.; Somers, M. L.; Revie, J.; and
McCartney, B. S. "A Continuous East-West Fault on the Azores-Gibraltar
Ridge." In *Nature* 237, May 26, 1972.

Lawler, Andrew. "Sea-Floor Data Flow for Postwar Era." In *Science* 270, November 3, 1995.

Menard, H. W. *Anatomy of an Expedition*. New York, 1969.

Rusby, Stuart. "A Long Range Side-Scan Sonar for Use in the Deep Sea." In *International Hydrographic Review* 47, July 1970.

Sandwell, David, and Smith, W. *Measured and Estimated Seafloor Topography
(version 4.2)*. World Data Center for Marine Geology and Geophysics. Research publication RP-1, 1997.

Schlee, Susan. *The Edge of an Unfamiliar World: A History of Oceanography*. New
York, 1973.

Simpson, Sarah. "Life's First Scalding Steps." In *Science News* 155, January 9,
1999.

Stride, A. H. "Mapping the Ocean Floor." In *Science Journal* 6, December
1970.

Sullivan, Walter. *Continents in Motion*. New York, 1974.

Van Andel, Tjeerd. *Tales of an Old Ocean*. New York, 1978.

Vine, F. J., and Matthews, D. H. "Magnetic Anomalies over Oceanic Ridges." In *Nature* 199, September 7, 1963.

Wegener, Alfred. *The Origin of Continents and Oceans.* New York, 1966. Translation of 4th edition, 1929.

Wertenbaker, William. *The Floor of the Sea: Maurice Ewing and the Search to Understand the Earth.* Boston, 1974.

Interviews with Maurice Ewing, University of Texas at Galveston; Bruce C. Heezen and Marie Tharp, Columbia University's Lamont-Doherty Observatory. (Various sizes of the sea-floor map reproduced in this book, as well as other such maps, may be purchased from Marie Tharp, 1 Washington Avenue, South Nyack, N.Y. 10960.) Correspondence with A. S. Laughton, the National Institute of Oceanography, England.

20. *Base Lines Across a Continent*

Dracup, Joseph F. "Geodetic Surveying, 1940–1990." In the National Oceanic and Atmospheric Administration website, http://www.history.noaa.gov/tales/cgs/geodetic2.html. Highlights and people involved in recent geodesy by a retired officer of the Coast and Geodetic Survey.

———. "Geodetic Surveys in the United States: The Beginning and the Next One Hundred Years." In the NOAA website, http://www.history.noaa.gov/tales/cgs/geodetic.html. Early history of geodetic surveys.

Gossett, F. R. *Manual of Geodetic Triangulation.* United States Coast and Geodetic Survey Special Publication 247. Washington, 1959.

Meade, Buford K. "High-Precision Geodimeter Traverse Surveys in the United States." Paper to the XIV General Assembly of the International Union of Geodesy and Geophysics, Lucerne, Switzerland. September 1967.

National Research Council. *Geodesy: Trends and Prospects.* Washington, 1978.

Simmons, Lansing. "A Singular Geodetic Survey." United States Coast and Geodetic Survey Technical Bulletin 13. September 1960.

Stanley, Albert A. "Hassler's Legacy." In *NOAA Magazine* 6, January 1976.

Wilcove, Raymond. "Jasper Bilby's Brainchild." In *ESSA Magazine* 5, July 1970. A magazine of the Department of Commerce that has been superseded by *NOAA Magazine.*

Interviews with Leonard S. Baker, J. D. D'Onofrio, Joseph F. Dracup, and Buford K. Meade, the National Geodetic Survey, Rockville, Md.

21. *Geodesy from Space*

Arnold, K. "The Use of Satellites for Geodetic Studies." In *Space Science Reviews* 7, August 1967.

D'Onofrio, J. D. "The Role of Satellite Doppler Geodesy in NGS." Paper presented at the American Congress on Surveying and Mapping, Fall Convention. Washington, September 10–13, 1974.

Dracup, Joseph F. "Geodetic Surveying, 1940–1990."

Kaula, W. M. *Theory of Satellite Geodesy.* Waltham, Mass., 1966.

King-Hele, Desmond. "The Shape of the Earth." In *Science* 192, June 25, 1976.

———. "The Shape of the Earth." In *Scientific American* 217, October 1967.

Kleppner, Daniel, and Taubes, Gary. "The Global Positioning System." In *Beyond Discovery: The Path from Research to Human Benefit.* National Academy of Sciences. Washington, 1996.

National Aeronautics and Space Administration. *Significant Achievements in Satellite Geodesy 1958–1964.* NASA SP-94. Washington, 1966.

O'Keefe, J. A.; Eckels, Ann; and Squires, R. K. "Vanguard Measurements Give Pear-Shaped Component of Earth's Figure." In *Science* 129, February 27, 1959.

Schmid, Hellmut H. "Satellite vs. Classic Geodetic Triangulation." In *Surveying and Mapping* 28, March 1968.

Swanson, Lawrence W., and Yeager, J. Austin. "A New World Geodetic Network." In *Polar Record* 15, March 1971.

Interviews with J. D. D'Onofrio, Hellmut H. Schmid, and Lawrence W. Swanson, National Geodetic Survey, Rockville, Md.; Irwin I. Shapiro, Massachusetts Institute of Technology; Brian Skinner of the United States Army Corps of Engineers; Greg Desmond, United States Geological Survey.

22. *Mapping from Space*

Bird, J. B., and Morrison, A. "Space Photography and Its Geographical Applications." In *Geographical Review* 54, October 1964.

Bylinsky, Gene. "From a High-Flying Technology—A Fresh View of Earth." In *Fortune* 78, June 1, 1978.

Colvocoresses, Alden P. "Evaluation of the Cartographic Applications of ERTS-1 Imagery." In *The American Cartographer* 2, April 1975.

———. "Surveying the Earth from 20,000 Miles." In *Proceedings* of the Annual Convention of Photographic Scientists and Engineers—Image Technology. Washington, 1969.

MacDonald, William R. "New Space Technology Advances Knowledge of the Remote Polar Regions." In *Third Earth Resources Technology Satellite Symposium.* Volume 3, NASA SP-357. Goddard Space Flight Center, May 1974.

Snyder, John P. *Flattening the Earth.*

———. "The Space Oblique Mercator Projection." In *Photogrammetric Engineering and Remote Sensing* 44, May 1978.

Swithinbank, Charles. "A New Map of Alexander Island, Antarctica." In *Polar Record* 17, May 1974.

Williams, Richard S., Jr., and Carter, William D. (editors). *ERTS-1: A New Window on Our Planet.* United States Geological Survey Professional Paper 929. Washington, 1976.

Interviews with Paul D. Lowman, Jr., William Nordberg, John A. O'Keefe, and Nicholas M. Short, NASA's Goddard Space Flight Center, Greenbelt, Md.; John P. Snyder, Madison, N.J.; Alden P. Colvocoresses and Stephen Guptill of the United States Geological Survey, Reston, Va.; Joel L. Morrison of the United States Census Bureau in Washington.

23. *Dynamic Maps: A New Geography*

Cohen, Patricia. "Geography Redux: Where You Live Is What You Are." In the *New York Times,* March 21, 1998. Reviews the revival of academic and intellectual interest in geography.

Davis, David E. *GIS for Everyone.* Redlands, Calif., 1999. An introduction to digital mapping and the potential of geographic information systems, published by a leader in the field, Environmental Systems Research Institute, Inc.

Dobson, Jerome E. "Automated Geography." In *The Professional Geographer* 35, May 1983.

———. "The Geographic Revolution: A Retrospective on the Age of Automated Geography." In *The Professional Geographer* 45, November 1993.

———. "Advice on Going Digital." In *GISWorld,* August 1991.

———. "GIS Advances Geography's Field Tradition." In *GISWorld,* January 1994.

——— et al. "A Global Population Database for Estimating Populations at Risk." Draft manuscript, July 1999, for publication in *Photogrammetric Engineering and Remote Sensing.*

Gould, Peter. "Geography 1957–1977: The Augean Period." In *Annals* of the Association of American Geographers, 69, March 1979.

———. "The New Geography." In *Harper's Magazine* 238, March 1969.

Morrison, Joel L. "The Paradigm Shift in Cartography: The Use of Electronic Technology, Digital Spatial Data, and Future Needs." Keynote speech at Spatial Data Handling Conference, Edinburgh, September 1994.

Muehrcke, Phillip. *Thematic Cartography.* Commission on College Geography Resource Paper No. 19. Association of American Geographers. Washington, 1972.

Peucker, Thomas K. *Computer Cartography.* Commission on College Geography Resource Paper No. 17. Association of American Geographers. Washington, 1972.

Tobler, Waldo. *Spatial Interaction Patterns.* International Institute for Applied Systems Analysis, RR-75-19. Laxenberg, Austria, 1975.

Warntz, William, and Wolff, Peter. *Breakthroughs in Geography.* Chapter on Halley's thematic mapping.

Wilford, J. N. "Revolutions in Mapping." In *National Geographic* 193, February 1998.

Interviews with John A. Kelmelis and Stephen Guptill, United States Geological Survey; Joel L. Morrison, United States Census Bureau; David Woodward, University of Wisconsin; Peter Gould, Pennsylvania State University; Eric Teicholz and Allan H. Schmidt, Harvard University Laboratory for Computer Graphic and Spatial Analysis; Jerome E. Dobson, Oak Ridge National Laboratory.

24. *Extraterrestrial Mapping: The Moon*

Alpert, Mark. "The Little Spacecraft That Could." *Scientific American* 280, June 1999.

Apollo II Mission Commentary. Tape 232, 12:55 CDT, July 19, 1969. Tape 236, 13:56 CDT, July 19, 1969.

Doyle, Frederick J. "Photogrammetric Mapping of the Moon." Unpublished monograph.

Kopal, Zdenek. "Topography of the Moon." In Zdenek Kopal (editor), *Physics and Astronomy of the Moon.* New York, 1962.

————, and Carder, Robert W. *Mapping of the Moon: Past and Present.* Astrophysics and Space Sciences Library, 50. Dordrecht, Holland, and Boston, 1974.

Levin, Ellis; Viele, Donald D.; and Eldrenkamp, Lowell B. "The Lunar Orbiter Missions to the Moon." In *Scientific American* 218, May 1968.

Lewis, H. A. G. (editor). *The Times Atlas of the Moon.* London, 1969. Includes chapter on history of lunar mapping.

Norman, Paul E. "Out-of-This-World Photogrammetry." In *Photogrammetric Engineering* 35, July 1969.

Schimerman, Lawrence A. "The Expanding Apollo Control System." Paper to XVI General Assembly of the International Astronomic Union, Grenoble. August 1976.

Zink, Judith Ann. *Lunar Cartography: 1610–1962.* Unpublished M.A. thesis. University of California, Los Angeles. 1963.

Interviews with Paul E. Norman, NASA's Johnson Space Center, Houston, Texas; Raymond M. Batson, United States Geological Survey, Flagstaff, Ariz.; Robert W. Carder, formerly of the Air Force Aeronautical Chart and Information Center, St. Louis, Mo. Correspondence with Ewen A. Whitaker, University of Arizona, Tucson, and Lawrence A. Schimerman, Defense Mapping Agency, St. Louis.

25. *Extraterrestrial Mapping: Mars*

Batson, Raymond M. "Cartographic Products for the Mariner 9 Mission." In *Journal of Geophysical Research* 78, July 10, 1973. Complete issue devoted to Mariner 9 results.

Davies, Merton, and Arthur, David W. G. *Martian Surface Coordinates, April 1973.* Rand Corporation Publication R-1252-JPL.

Icarus 26, November 1975. Complete issue devoted to Mars.

Levi, Barbara Goss. "New Data from Mars Hint at a Dynamic Past." In *Physics Today,* June 1999.

Levinthal, E. C. "Mariner 9—Image Processing and Products." In *Icarus* 18, January 1973.

Lowell, Percival. *Mars and Its Canals.* New York, 1906.

Raeburn, Paul. *Mars: Uncovering the Secrets of the Red Planet.* Washington, 1998.

Weaver, Kenneth F. "Journey to Mars." In *National Geographic* 143, February 1973.

Wuethrich, Bernice. "A New Look at the Martian Landscape." In *Science* 284, May 28, 1999.

Interviews with Arden Albee of the California Institute of Technology, Pasadena; Michael H. Carr, David W. G. Arthur, Raymond M. Batson, and Harold Masursky, United States Geological Survey, Flagstaff, Ariz.; Merton E. Davies, Rand Corporation, Santa Monica, Calif.; James A. Cutts, Richard M. Goldstein, and William B. Green, Jet Propulsion Laboratory, Pasadena, Calif.

26. *Cosmic Cartographers*

Geller, Margaret J., and Huchra, John P. "Mapping the Universe." In *Science* 246, November 17, 1989.

Greenstein, George. "Our Address in the Universe." In *Harvard Magazine,* January–February 1994.

Hall, Stephen. *Mapping the Millennium.*

———. "Uncommon Landscapes: Maps in a New Age of Scientific Discovery." In *The Sciences,* September–October 1991.

Wilford, J. N. "Revolutions in Mapping."

Epilogue at Bright Angel

Bradford Washburn and others spent many weeks on the field survey at the Grand Canyon, primarily in 1972. The resulting map, *The Heart of the Grand Canyon,* was published as a supplement to *National Geographic* 154, July 1978.

ACKNOWLEDGMENTS

I have, in the Preface, acknowledged my debt to Bradford A. Washburn, whose work and companionship were the inspiration that got me started on this book back in the early 1970s. I am also grateful to the many people I consulted through personal and telephone interviews and by letter, many of whom are identified in the Bibliographical Notes. To the following I am indebted for their efforts in arranging interviews and tracking down information and illustrations: Graham Berry, Frank Bristow, Frank Colella, Frank Forrester, John E. Heaney, Alice Hudson, Bob Johns, Nancy Kandoian, James Lynch, John G. Stringer, Raymond E. Wilcove, and Ann Wise.

Those who were kind enough to read and comment on chapters or parts of chapters were Robert W. Carder, Alden P. Colvocoresses, James A. Cutts, Jerome Dobson, Joseph F. Dracup, Peter Gould, Norman L. Hatch, Jr., Homer Jensen, Paul D. Lowman, Jr., Harold Masursky, Lawrence A. Schimerman, John P. Synder, Charles Swithinbank, Morris M. Thompson, and Ewen A. Whitaker.

The following organizations were especially helpful: U.S. Geological Survey, National Geodetic Survey, Goddard Space Flight Center, Jet Propulsion Laboratory, Johnson Space Center, Lamont-Doherty Earth Observatory of Columbia University, Harvard University Laboratory of Computer Graphics, Aero Services Corp., Conoco, Inc., New York Public Library (especially its excellent Map Division), Library of Congress, and the libraries of the California Institute of Technology and the *New York Times*.

In addition, I wish to express my appreciation to the editors of *The New York Times* for the opportunity to accompany the Grand Canyon mapping expedition and to report on many of the space-age cartographic developments covered in the final chapters.

An assignment by *National Geographic* magazine encouraged and enabled me to revisit the subject in 1997. It was a pleasure to work with the magazine's editors. For the article, published in February 1998, I received considerable help from interviews with Greg Desmond, Stephen C. Guptill, and John A. Kelmelis, U.S. Geological Society, Reston, Va.; David Woodward, University of Wisconsin, Madison; Joel L. Morrison, U.S. Census Bureau, Washington; Ralph Ehrenberg, Library of Congress, Washington; Arnold Nesselrath, Vatican Museums, Rome; David Sandwell, University of California, San Diego; and Margaret Geller, Harvard-Smithsonian Center for Astrophysics, Cambridge, Mass.

Several fellow councilors of the American Geographical Society and my longtime editor at Knopf, Ashbel Green, nudged me to revise and expand the book to include the many cartographic developments in the years since the first edition in 1981.

Finally, without my wife, Nancy, always patient and encouraging, I doubt that I could have kept my bearings through the long course from Syene to Mars and the cosmos.

Page numbers in *italics* denote maps or illustrations.